Lecture Notes in Chemistry

Volume 97

The Lecture Notes in Chemistry

The series Lecture Notes in Chemistry (LNC) reports new developments in chemistry and molecular science-quickly and informally, but with a high quality and the explicit aim to summarize and communicate current knowledge for teaching and training purposes. Books published in this series are conceived as bridging material between advanced graduate textbooks and the forefront of research. They will serve the following purposes:

- provide an accessible introduction to the field to postgraduate students and nonspecialist researchers from related areas,
- provide a source of advanced teaching material for specialized seminars, courses and schools, and
- be readily accessible in print and online

The series covers all established fields of chemistry such as analytical chemistry, organic chemistry, inorganic chemistry, physical chemistry including electrochemistry, theoretical and computational chemistry, industrial chemistry, and catalysis. It is also a particularly suitable forum for volumes addressing the interfaces of chemistry with other disciplines, such as biology, medicine, physics, engineering, materials science including polymer and nanoscience, or earth and environmental science.

Both authored and edited volumes will be considered for publication. Edited volumes should however consist of a very limited number of contributions only. Proceedings will not be considered for LNC.

The year 2010 marks the relaunch of LNC.

More information about this series at http://www.springer.com/series/632

Akihiro Morita

Theory of Sum Frequency Generation Spectroscopy

Springer

Akihiro Morita
Tohoku University
Sendai, Miyagi, Japan

ISSN 0342-4901 ISSN 2192-6603 (electronic)
Lecture Notes in Chemistry
ISBN 978-981-13-1606-7 ISBN 978-981-13-1607-4 (eBook)
https://doi.org/10.1007/978-981-13-1607-4

Library of Congress Control Number: 2018949922

Printed on acid-free paper

This Springer imprint is published by the registered company Springer Nature Singapore Pte Ltd.
The registered company address is: 152 Beach Road, #21-01/04 Gateway East, Singapore 189721, Singapore

To

the late Professor
Shigeki Kato

and

my parents,
Keiichi and Haruko Morita

Preface

Surface-specific nonlinear spectroscopy, such as second harmonic generation (SHG) or sum frequency generation (SFG), has been growing to be a popular tool of interface characterization. These techniques have microscopic sensitivity to the interfaces at a monolayer scale and are applicable to a variety of interfaces as long as the interfaces are accessible by optical probe lights and signal. These techniques are particularly useful to wet and/or soft interfaces, or even buried interfaces, which are hard to be probed by most conventional surface science techniques. Therefore, these techniques have large potential to expand our applicability of interface characterization in wide areas of science and engineering, such as electrochemistry, polymer science, colloid chemistry, heterogeneous atmospheric chemistry, etc. The last decade witnessed great advances in technical aspects, and the surface nonlinear spectroscopy is presently not just for limited experts of spectroscopy but available to wide researchers who want to use these techniques in their fields using commercial apparatus.

Currently I think that a major bottleneck to achieve further advancement in the nonlinear spectroscopy lies in their difficulties to interpret the observed spectra. Typical conventional analysis tries to interpret a spectrum by decomposing it into some bands and by assigning these bands to various species at the interface. However, such empirical analysis has apparently of limited utility, and spectral decomposition is often quite ambiguous. To overcome such difficulties, therefore, reliable support of theoretical analysis is strongly desirable in the field of surface nonlinear spectroscopy. Recent advances in theory have made us possible to directly "calculate" the spectra using molecular modeling and molecular dynamics (MD) simulation, which allows for simultaneous understanding of the observed spectra and interface structure in unprecedented details. I believe that close collaboration of spectroscopic measurement and MD simulation will be a main avenue in the further nonlinear spectroscopy of interfaces.

The present book aims at explaining the basic principles of theory and computation of surface nonlinear spectroscopy, mainly developed by the author's group. This book was originally intended to newcomers in our laboratory and in summer schools to have them familiarize with fundamental understanding of nonlinear spectroscopy and computation of it. However, the present book has been significantly expanded for wide readers who are interested in theoretical aspects of surface nonlinear spectroscopy. The topics covered by this book include the traditional (electromagnetic) theory of surface nonlinear spectroscopy, quantum description and modeling of nonlinear susceptibility, method of molecular modeling, and computational scheme of SFG spectroscopy by MD simulation. The topics also include recent efforts to deepen our understanding of the nonlinear susceptibility, including quadrupole contribution, $\chi^{(3)}$ effect, and chiral application. A few late chapters are devoted to describe recent applications of the MD analysis of SFG spectra for aqueous and organic interfaces. These examples demonstrate that the combination of SFG spectroscopy and MD simulation is in fact quite powerful to obtain clear insights into the interfaces. This book provides some problems and solutions to help the reader to fully understand the theoretical aspects. The problems mostly treat derivations of key formulas involved in the theory. I have shown detailed solutions to these problems in the end of each chapter, which I think are an integral part of this book.

I would like to acknowledge my current and former group members who helped with improving the manuscript. In particular, Prof. Tatsuya Ishiyama extended the applications of SFG analysis to various aqueous and organic interfaces. Dr. Kazuya Shiratori made a key contribution to develop the theory of quadrupole in Chap. 7. Drs. Lin Wang and Vladimir Sokolov developed the molecular modeling of various organic species. Dr. Tatsuya Joutsuka performed the theoretical analysis of $\chi^{(3)}$ effect in Chap. 8. I also thank former students of my laboratory who carried out the MD simulation of SFG spectroscopy, including Takako Imamura, Yuji Sato, Tatsuya Kawaguchi, Yusuke Takei, Yuri Mizukoshi, Hiromi Sawai, Takashi Ishihara, Shogo Tanaka, Kengo Saito, Tomonori Hirano, and Wataru Mori. Mr. Yamato Sato read this manuscript and provided useful feedbacks from a viewpoint of student. The computational studies of SFG spectroscopy inevitably become collaborative works with my colleagues including experimentalists. I am grateful to my collaborators, such as Drs. Tahei Tahara, Shoichi Yamaguchi, Satoshi Nihonyanagi, Takayuki Miyamae, Yukio Ouchi, Takaaki Ishibashi, Toshiki Sugimoto, Yoshiyasu Matsumoto, Mischa Bonn, Francesco Paesani, Heather Allen, Franz Geiger, Michiel Sprik, Kenichi Inoue, and Shen Ye. I also thank Dr. Toshiki Sugimoto for permitting me to use original figures of ice and Mr. Ko Hosokawa and Ms. Miyuki Tonosaki for helping with typesetting. I greatly appreciate the comments of Profs. Tony Heinz and Ron Shen for clarifying the discussion in Chap. 2. I am also grateful to Prof. Casey Hynes for getting me interested in this field in relation to heterogeneous atmospheric chemistry and his continued encouragement.

Sendai, Japan
April 2018

Akihiro Morita

Contents

Chapter 1
Introduction

Abstract A brief introduction of the sum frequency generation (SFG) spectroscopy, the main theme of this book, is provided. An overview of second-order nonlinear optical processes, including SFG and second-harmonic generation (SHG), is presented with emphasizing their spatial and temporal characteristics and symmetries. Fundamental features of visible-infrared SFG vibrational spectroscopy as a tool of interface characterization are summarized in comparison with infrared and Raman vibrational spectroscopies. Some examples of visible-infrared SFG vibrational spectra are also illustrated to show typical spectroscopic information drawn from the SFG spectroscopy.

Keywords SFG · SHG · Surface sensitivity · Vibrational spectroscopy

This chapter gives a brief introduction of the sum frequency generation (SFG) spectroscopy, the main theme of this book, including the concept of second-order nonlinear optical process, visible-infrared SFG spectroscopy, and some typical experimental examples. The SFG spectroscopy has been reviewed in previous literature from various aspects [4, 10, 18–21]. The present book deals with the theory and computation of SFG spectroscopy, though the collaboration with experiments will be a vital part of this book. The fundamental theory of SFG is common to second harmonic generation (SHG). The two spectroscopies are based on the second-order nonlinear optical processes where two photons with frequencies ω_1 and ω_2 generates a light with the sum frequency $\Omega = \omega_1 + \omega_2$. SHG is considered as a special case of $\omega_1 = \omega_2$. The following theory mostly deals with SFG, while the general discussion also holds for SHG.

1.1 Sum Frequency Generation

A material system under external electric field $\boldsymbol{E}$ induces dipole moment, or polarization, in general. The polarization $\boldsymbol{P}$ is defined as the dipole moment per

A. Morita, *Theory of Sum Frequency Generation Spectroscopy*,
Lecture Notes in Chemistry 97, https://doi.org/10.1007/978-981-13-1607-4_1

a unit volume of a bulk material. The definition of $\boldsymbol{P}$ depends on the system in question. When the system indicates a surface, $\boldsymbol{P}$ is defined as the dipole moment per a unit area. For a single molecule, $\boldsymbol{P}$ denotes the dipole moment of the molecule. In any case, the induced polarization $\boldsymbol{P}$ is represented as a power series of the electric field [3, 14, 17],

$$P_p = \sum_q \chi^{(1)}_{pq} E_q + \sum_{q,r} \chi^{(2)}_{pqr} E_q E_r + \sum_{q,r,s} \chi^{(3)}_{pqrs} E_q E_r E_s + \cdots \quad (p, q, r, s = x \sim z) \tag{1.1}$$

where the suffixes p, q, r, s denote the Cartesian components $x \sim z$. Note that $\boldsymbol{P}$ and $\boldsymbol{E}$ are vector quantities with one spatial suffix. The first term describes the linear response of polarization with respect to the field, where $\boldsymbol{\chi}^{(1)}$ is a second-rank tensor called linear susceptibility. (When the system is a molecule, it is called polarizability.) The higher-order terms, describing nonlinear response of the polarization, become substantial when the electric field is sufficiently intense. The second term involving $\boldsymbol{\chi}^{(2)}$ is responsible to the second-order nonlinear optical processes, such as SHG or SFG. $\boldsymbol{\chi}^{(2)}$ is a third-rank tensor called second-order nonlinear susceptibility, or hyperpolarizability for a molecule. The following discussion will mainly focus on this second term including $\boldsymbol{\chi}^{(2)}$ in relation to the surface nonlinear spectroscopy.

[Problem 1.1] If we suppose that material properties such as $\boldsymbol{\chi}^{(2)}$ are invariant by inversion, show $\boldsymbol{\chi}^{(2)} = 0$. This indicates that the second-order optical processes in Eq. (1.1) are forbidden for a centrosymmetric material.

Next we consider time-dependent electric field $\boldsymbol{E}(t)$, and accordingly generalize Eq. (1.1) to treat time-dependent polarization $\boldsymbol{P}(t)$. The second-order term in Eq. (1.1) is generalized to the time-dependent form $P_p^{(2)}(t)$ as follows,

$$P_p^{(2)} = \sum_{q,r}^{x\sim z} \chi^{(2)}_{pqr} E_q E_r$$

$$\longrightarrow \; P_p^{(2)}(t) = \int_{-\infty}^{t} dt' \int_{-\infty}^{t} dt'' \sum_{q,r}^{x\sim z} \chi^{(2)}_{pqr}(t, t', t'') E_q(t') E_r(t''), \tag{1.2}$$

where the modified form allows for non-local response in the time domain. The range of integral is restricted to $t' \leq t$ and $t'' \leq t$ t by the causality, since the electric fields at t' and t'' can influence on the polarization at a later time t. The time-dependent fields and polarization can be described with the Fourier series,

$$P_p(t) = \sum_k P_p(\omega_k) \exp(-i\omega_k t), \quad E_q(t) = \sum_k E_q(\omega_k) \exp(-i\omega_k t) \tag{1.3}$$

where $\boldsymbol{P}(\omega_k)$ and $\boldsymbol{E}(\omega_k)$ denote the amplitudes of frequency ω_k. The Fourier form of Eq. (1.3) is convenient to deal with oscillating electric fields of light in the following. Note that $\boldsymbol{P}(t)$ and $\boldsymbol{E}(t)$ are real quantities. Accordingly, the summation with k in Eq. (1.3) must include a pair of the ω_k term and its complex conjugate ω_{-k} (i.e. $\omega_{-k} = -\omega_k$ and $\boldsymbol{P}(\omega_{-k}) = \boldsymbol{P}(\omega_k)^*$) except for the static component $\omega = 0$.

By inserting Eq. (1.3) into (1.2), the second-order polarization coefficient $\boldsymbol{P}^{(2)}$ of sum frequency Ω is expressed as follows,

$$P_p^{(2)}(\Omega = \omega_1 + \omega_2) = \sum_{q,r}^{x \sim z} \chi_{pqr}^{(2)}(\Omega, \omega_1, \omega_2) E_q(\omega_1) E_r(\omega_2). \tag{1.4}$$

[Problem 1.2] Derive Eq. (1.4) from Eqs. (1.2) and (1.3).

During the derivation, make use of the fact that the coefficient $\boldsymbol{\chi}^{(2)}(t, t', t'')$ in Eq. (1.2) does not depend on the origin of time, i.e. $\boldsymbol{\chi}^{(2)}(t, t', t'') = \boldsymbol{\chi}^{(2)}(t + \Delta, t' + \Delta, t'' + \Delta)$ with an arbitrary time shift by Δ. (In other words, $\boldsymbol{\chi}^{(2)}(t, t', t'')$ is a function of the time intervals, $\tau' \equiv t - t'$ and $\tau'' \equiv t - t''$.)

Explain that only the component of sum frequency, $\Omega = \omega_1 + \omega_2$, appears in the left-hand side of Eq. (1.4), when the right-hand side is composed of $E(\omega_1)$ and $E(\omega_2)$.

Equation (1.4) means that two oscillating electric fields with ω_1 and ω_2 generate the oscillating polarization with the sum frequency $\Omega = \omega_1 + \omega_2$. $\boldsymbol{\chi}^{(2)}(\Omega, \omega_1, \omega_2)$ is the second-order nonlinear susceptibility which depends on the frequencies Ω, ω_1 and ω_2. This property, including its frequency dependence, is characteristic of materials. The oscillating polarization $\boldsymbol{P}^{(2)}(\Omega)$ emits the electromagnetic wave of the frequency Ω, according to the theory of electrodynamics [8]. The SFG spectroscopy detects this electromagnetic light as the signal.

The above mechanism of SFG emission is valid for the case of $\boldsymbol{\chi}^{(2)} \neq 0$. This condition means that the material should not have the inversion symmetry to be SFG active. Most bulk materials of gas or liquid are isotropic and thus invariant to inversion, indicating that these materials generate no SFG signal. Bulk crystals having the inversion symmetry are not SFG active either for the same reason. However, if two isotropic bulk phases form an interface, the inversion symmetry necessarily breaks down in the vicinity of the interface, which generally results in $\boldsymbol{\chi}^{(2)} \neq 0$. For such a system involving the interface, the SFG signal selectively stems from the interface. The interface sensitivity of the SFG spectroscopy is attributed to the symmetry reason that interfaces generally lose the inversion symmetry.

Another important feature of the SFG spectroscopy is its coherent nature, since the oscillating polarization with Ω is a consequence of coherent superposition of two electric fields oscillating with ω_1 and ω_2. The coherent nature is manifested in the directionality of the emitted SFG signal. When two monochromatic laser lights with frequencies ω_1 and ω_2 are incident to the interface system, the SFG signal is observed to a certain direction (see Fig. 2.1). The strong directionality of the SFG emission is convenient to experimental detection of its weak signal. This is a consequence of interference among oscillating polarizations in spatially different regions, which is analogous to the Bragg's law of diffraction [11].

The coherent nature also implicates that the SFG signals emitted from different oscillators (vibrating atomic groups) can interfere each other, which may complicate the assignment of experimental SFG spectra. In the case that an observed SFG spectrum originates from different oscillators in the same frequency region, the interference of the overlapping source signals may enhance or suppress the intensity of the SFG signal owing to the phase relation. Therefore, the observed SFG intensity is not amenable to straightforward decomposition into the source oscillators. (In other words, weak SFG signal does not necessarily mean that there is no or weak SFG source. The weak intensity could be a consequence of cancelling interference.) The interpretation of the SFG spectroscopy has to take account of the interference effects, as discussed later.

1.2 Visible-Infrared SFG Vibrational Spectroscopy

The most common application of SFG to surface nonlinear spectroscopy employs the combination of incident visible and infrared lights, and presents the SFG signal as a function of infrared frequency ω_2 with the visible frequency ω_1 fixed. Such measurement is considered as a vibrational spectroscopy of the interface. Some features of the vibrational SFG spectroscopy are briefly summarized below.

1. It is possible to observe molecular species at the interface selectively, even though overwhelming amount of the same species exists in the isotropic bulk region. The sensitivity of detection is fairly high, and a submonolayer amount of the surface species can be easily detected.
2. The vibrational spectra provide highly specific information to the molecular species at the interface. The frequency shift also offers useful information on local environment of the interface. Molecular orientation at the interface can be measured, by choosing proper combinations of light polarizations of visible, infrared and SFG.
3. Since the interface selectivity is entirely attributed to the symmetry reasons, no vacuum condition is necessary in principle. The optical measurement is suitable to in-situ detection of a variety of interfaces. This technique is also applicable to buried interfaces, such as liquid-liquid or liquid-solid, as long as the interfaces are accessible by light.
4. This technique is suitable to study ultrafast dynamics at the interfaces, by making use of high temporal resolution of pulse lasers.

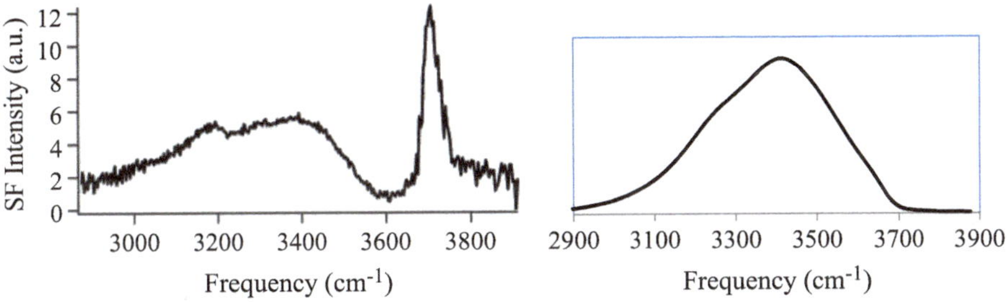

Fig. 1.1 (Left panel) SFG spectrum of water surface in SSP configuration [16]. (Right panel) Infrared absorption spectrum of water [2]. (Reprinted with permission from Ref. [16]. Copyright 2003 American Chemical Society.)

These features render the SFG spectroscopy a versatile tool of surface characterization, particularly to wet or soft interfaces [12]. Microscopic measurement of wet interfaces has been limited due to scarcity of available experimental means, compared to that of solid surfaces. The SFG spectroscopy is considered as a unique, powerful technique to explore the wet or soft interfaces.

To exemplify the SFG spectroscopy, two typical visible-infrared SFG spectra are illustrated, water and alkyl chains. Left panel of Fig. 1.1 displays the SFG spectrum of liquid-vapor interface of water in the O-H stretching frequency region, 2900 $\sim$ 3900 cm^{-1}. In the right panel, the conventional infrared (IR) absorption spectrum of bulk water is shown for comparison. Comparing the two spectra of O-H stretching vibrations in Fig. 1.1, qualitative difference in spectral structure is apparent. These spectra essentially reflect the differences in structure of surface and bulk water. The SFG spectrum is composed of a sharp band at about 3700 cm^{-1} and a broad, red-shifted band at 3000 $\sim$ 3600 cm^{-1}. The former band of SFG finds no apparent counterpart in the IR spectrum, and is attributed to the O-H moieties at the topmost surface layer which are free from hydrogen bonds, called free O-H or dangling O-H. The latter, broad band is assigned to hydrogen-bonding O-H moieties in a few monolayer regions at the surface. The clear difference of SFG and IR spectra in Fig. 1.1 demonstrates the surface sensitivity of the SFG spectroscopy to detect structural features characteristic of surface water.

[Problem 1.3] Answer two frequencies of O-H stretching vibration of an isolated water molecule (symmetric and anti-symmetric stretching). Where are these frequencies located in the spectra of Fig. 1.1?

Another example of SFG and IR is given in Fig. 1.2, C-H stretching vibrational spectra of self-assembled monolayer of alkane thiol on gold substrate. Panel (a) is the observed SFG spectrum which involves the strong background from gold substrate, and panel (b) is the calibrated SFG spectrum of C-H stretching with the gold background eliminated. Comparing these panels (a) or (b) with the

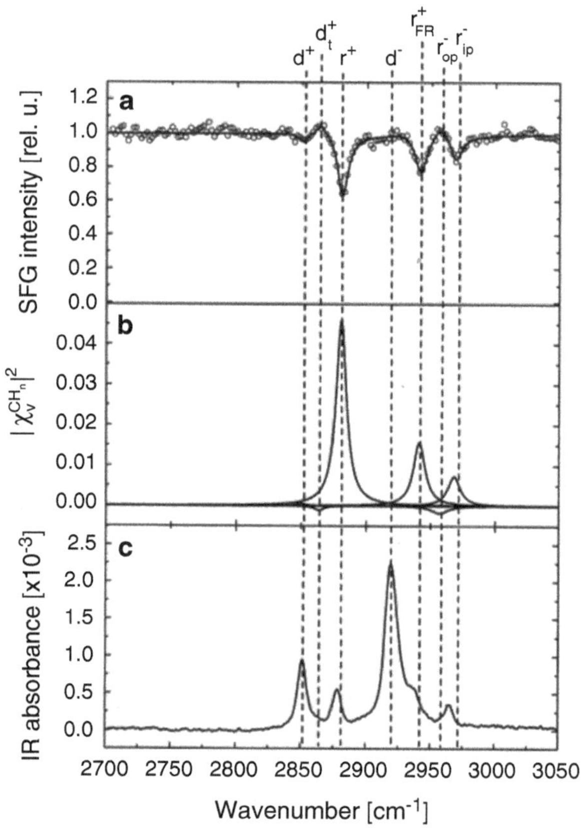

Fig. 1.2 Vibrational spectra of self-assembled monolayer of docosane thiol on gold substrate. (**a**) SFG spectrum, (**b**) SFG spectrum of C-H vibration after calibrating the background of gold, (**c**) IR absorption spectrum [4]. (Reprinted with permission from Ref. [4]. Copyright 2001, American Institute of Physics.)

conventional IR absorption spectrum in panel (c), one may notice a remarkable feature that the location of vibrational bands of SFG does not correspond to that of IR, implying that the two spectroscopies detect different vibrational bands. The IR spectroscopy is dominated by methylene groups since the number of methylene groups is much larger than that of methyl in the alkane thiol monolayer, whereas the SFG spectroscopy selectively observes the outermost methyl groups. It is known that the SFG yields little signal of methylene for a well-ordered all-trans monolayer, though the vibrational bands of methylene emerges for a disordered monolayer with gauche defects [1, 18]. This sensitive change in the spectra provides a useful clue to characterize the order of self-assembled monolayers. The orientational angle of alkyl chains is also widely analyzed by the SFG measurements with different combinations of light polarizations [6, 7, 22, 23].

1.3 Solutions to Problems

1.3.1 *Inversion Symmetry of* $\chi^{(2)}$

[Problem 1.1] If we suppose that material properties such as $\chi^{(2)}$ are invariant by inversion, show $\chi^{(2)} = 0$. This indicates that the second-order optical processes in Eq. (1.1) are forbidden for a centrosymmetric material.

We impose electric field on the material and consider the induced polarization. The second term in Eq. (1.1) is denoted by $P_p^{(2)}$,

$$P_p^{(2)} = \sum_{q,r} \chi_{pqr}^{(2)} E_q E_r, \tag{1.5}$$

which is the source of the second-order optical process. If we operate inversion upon the coordinate system, vector quantities such as $P_p^{(2)}$, E_q or E_r change their signs. However, material properties such as $\chi^{(2)}$ are assumed to be invariant by the inversion. Therefore, the inversion operation transforms Eq. (1.5) to

$$-P_p^{(2)} = \sum_{q,r} \chi_{pqr}^{(2)} (-E_q)(-E_r). \tag{1.6}$$

Both Eqs. (1.5) and (1.6) should hold simultaneously, which necessarily leads to $\chi^{(2)} = 0$. This means that $\boldsymbol{P}^{(2)} = 0$ in Eq. (1.5) for a centrosymmetric material.

1.3.2 *Time and Frequency Domains*

[Problem 1.2] Derive Eq. (1.4) from Eqs. (1.2) and (1.3).

During the derivation, make use of the fact that the coefficient $\chi^{(2)}(t, t', t'')$ in Eq. (1.2) does not depend on the origin of time, i.e. $\chi^{(2)}(t, t', t'') = \chi^{(2)}(t+\Delta, t'+\Delta, t''+\Delta)$ with an arbitrary time shift by Δ. (In other words, $\chi^{(2)}(t, t', t'')$ is a function of the time intervals, $\tau' \equiv t - t'$ and $\tau'' \equiv t - t''$.)

Explain that only the component of sum frequency, $\Omega = \omega_1 + \omega_2$, appears in the left-hand side of Eq. (1.4), when the right-hand side is composed of $E(\omega_1)$ and $E(\omega_2)$.

The time-dependent polarization of Eq. (1.2) is represented using the Fourier series of the electric fields as

$$P_p^{(2)}(t) = \int_{-\infty}^{t} dt' \int_{-\infty}^{t} dt'' \sum_{q,r}^{x-z} \chi_{pqr}^{(2)}(t, t', t'') E_q(t') E_r(t'') \tag{1.2}$$

$$= \sum_k \sum_l \int_{-\infty}^{t} dt' \int_{-\infty}^{t} dt'' \sum_{q,r}^{x-z} \chi_{pqr}^{(2)}(t, t', t'') E_q(\omega_k) E_r(\omega_l)$$
$$\exp(-i\omega_k t') \exp(-i\omega_l t'').$$

We note that $\boldsymbol{\chi}^{(2)}(t, t', t'')$ is invariant with respect to the origin of time, and thus it is a function of time intervals, $\tau' \equiv t - t'$ and $\tau'' \equiv t - t''$. Accordingly, the variables of integration are transformed from (t', t'') to (τ', τ''), which leads to

$$P_p^{(2)}(t) = \sum_k \sum_l \int_0^{\infty} d\tau' \int_0^{\infty} d\tau''$$
$$\sum_{q,r}^{x-z} \chi_{pqr}^{(2)}(t, t - \tau', t - \tau'') E_q(\omega_k) E_r(\omega_l) \exp(-i\omega_k(t - \tau')) \exp(-i\omega_l(t - \tau''))$$
$$= \sum_k \sum_l \sum_{q,r}^{x-z} \left\{ \int_0^{\infty} d\tau' \int_0^{\infty} d\tau'' \chi_{pqr}^{(2)}(t, t - \tau', t - \tau'') \exp(i\omega_k \tau') \exp(i\omega_l \tau'') \right\}$$
$$\cdot E_q(\omega_k) E_r(\omega_l) \exp(-i(\omega_k + \omega_l)t). \tag{1.7}$$

In Eq. (1.7) we notice that the quantity in the curly parentheses is independent of t, but depends on ω_k and ω_l. The quantity in the curly parentheses is denoted by $\boldsymbol{\chi}^{(2)}(\omega_k + \omega_l, \omega_k, \omega_l)$, i.e.

$$\chi_{pqr}^{(2)}(\omega_k + \omega_l, \omega_k, \omega_l)$$
$$= \int_0^{\infty} d\tau' \int_0^{\infty} d\tau'' \chi_{pqr}^{(2)}(t, t - \tau', t - \tau'') \exp(i\omega_k \tau') \exp(i\omega_l \tau''). \tag{1.8}$$

Equation (1.8) defines the relation of second-order susceptibility in the time domain, $\boldsymbol{\chi}^{(2)}(t, t', t'')$, to that in the frequency domain, $\boldsymbol{\chi}^{(2)}(\omega_k + \omega_l, \omega_k, \omega_l)$. Using this notation in the frequency domain, the second-order polarization $\boldsymbol{P}^{(2)}(t)$ in Eq. (1.7) is represented by

$$P_p^{(2)}(t) = \sum_k \sum_l \sum_{q,r}^{x-z} \chi_{pqr}^{(2)}(\omega_k + \omega_l, \omega_k, \omega_l) E_q(\omega_k) E_r(\omega_l) \exp(-i(\omega_k + \omega_l)t)$$
$$= \sum_k \sum_l P_p^{(2)}(\omega_k + \omega_l) \exp(-i(\omega_k + \omega_l)t). \tag{1.9}$$

The term that includes $\exp(-i(\omega_k + \omega_l)t)$ in Eq. (1.9) indicates the oscillating component at the frequency $\omega_k + \omega_l$. Therefore, this term corresponds to the polarization of sum frequency,

$$P_p^{(2)}(\omega_k + \omega_l) = \sum_{q,r}^{x-z} \chi_{pqr}^{(2)}(\omega_k + \omega_l, \omega_k, \omega_l) E_q(\omega_k) E_r(\omega_l). \tag{1.10}$$

This equation coincides with Eq. (1.4), by replacing $\omega_k, \omega_l, \omega_k + \omega_l$ with $\omega_1, \omega_2, \Omega$, respectively. Using monochromatic lights of ω_1 and ω_2, the sum frequency signal at $\Omega = \omega_1 + \omega_2$ occurs by the second-order process.

In relation to the above derivation of the sum frequency polarization, we can also derive the difference frequency generation (DFG), a related second-order nonlinear process to SFG. We have learned in Sect. 1.1 that the light field of frequency ω, $\boldsymbol{E}(\omega)\exp(-i\omega t)$ in Eq. (1.3), is accompanied with its complex conjugate, $\boldsymbol{E}(\omega)^* \exp(i\omega t)$. Therefore, the combinations of light fields at ω_1 ($\neq 0$) and ω_2 ($\neq 0$) actually give rise to four possible phase factors, $\exp(-i(\pm\omega_1 \pm \omega_2)t)$. All the possible combinations of ω_1 and ω_2 are presented in analogous forms to Eq. (1.4):

$$P_p^{(2)}(\omega_1 + \omega_2) = \sum_{q,r}^{x-z} \chi_{pqr}^{(2)}(\omega_1 + \omega_2, \omega_1, \omega_2) E_q(\omega_1) E_r(\omega_2), \tag{1.11}$$

$$P_p^{(2)}(\omega_1 - \omega_2) = \sum_{q,r}^{x-z} \chi_{pqr}^{(2)}(\omega_1 - \omega_2, \omega_1, -\omega_2) E_q(\omega_1) E_r(\omega_2)^*, \tag{1.12}$$

$$P_p^{(2)}(-\omega_1 + \omega_2) = \sum_{q,r}^{x-z} \chi_{pqr}^{(2)}(-\omega_1 + \omega_2, -\omega_1, \omega_2) E_q(\omega_1)^* E_r(\omega_2), \tag{1.13}$$

$$P_p^{(2)}(-\omega_1 - \omega_2) = \sum_{q,r}^{x-z} \chi_{pqr}^{(2)}(-\omega_1 - \omega_2, -\omega_1, -\omega_2) E_q(\omega_1)^* E_r(\omega_2)^*. \tag{1.14}$$

Among the above four terms, Eqs. (1.11) and (1.14) including $\omega_1 + \omega_2$ and $-\omega_1 - \omega_2$ correspond to SFG, whereas Eqs. (1.12) and (1.13) including $\omega_1 - \omega_2$, and $-\omega_1 + \omega_2$ to DFG.

1.3.3 *Red Shift of O-H Frequency*

[Problem 1.3] Answer two frequencies of O-H stretching vibration of an isolated water molecule (symmetric and anti-symmetric stretching). Where are these frequencies located in the spectra of Fig. 1.1?

Two frequencies of the O-H stretching vibrations of an isolated water molecule are $3657\,\mathrm{cm}^{-1}$ (symmetric stretching mode) and $3756\,\mathrm{cm}^{-1}$ (anti-symmetric stretching mode) [5]. In the SFG spectrum (left panel) of Fig. 1.1, these frequencies are located in the sharp band at about $3700\,\mathrm{cm}^{-1}$, which is assigned to the free O-H moieties. In the infrared spectrum (right panel) of Fig. 1.1, they are located at the high-frequency edge of the broad band in $3000 \sim 3700\,\mathrm{cm}^{-1}$.

It is well known that the hydrogen bond formation gives rise to substantial red shift of O-H frequency, and that the amount of the red shift reflects the strength of the hydrogen bonds [9, 13, 15]. In the SFG spectrum (left panel), the frequency of the free O-H vibration at the water surface retains the original frequency of isolated water molecules since it is free from the hydrogen bond.

Bibliography

1. Bain CD (1999) Non-linear optical techniques in modern characterization methods of surfactant systems. Marcel Dekker, New York
2. Bertie JE, Lan Z (1996) Infrared intensities of liquids XX: the intensity of the OH stretching band of liquid water revisited. Appl Spectrosc 50:1047–1057
3. Boyd RW (2003) Nonlinear optics. Academic, Amsterdam
4. Buck M, Himmelhaus M (2001) Vibrational spectroscopy of interfaces by infrared-visible sum frequency generation. J Vac Sci Technol A 19:2717–2736
5. Haynes WM, editor (2012) CRC Handbook of chemistry and physics, 93rd edn. CRC Press, Boca Raton
6. Hirose C, Akamatsu A, Domen K (1992) Formulas for the analysis of surface sum-frequency generation spectrum by CH stretching modes of methyl and methylene groups. J Chem Phys 96:997–1004
7. Hirose C, Hiroyoshi Y, Akamatsu N, Domen K (1993) Orientation analysis by simulation of vibrational sum frequency generation spectrum: CH stretching bands of the methyl group. J Phys Chem 97:10064–10069
8. Jackson JD (1998) Classical electrodynamics. Wiley, New York
9. Jeffrey GA (1997) An introduction to hydrogen bonding. Oxford University Press, Oxford
10. Lambert AG, Davies PB, Neivandt DJ (2005) Implementing the theory of sum frequency generation vibrational spectroscopy: a tutorial review. Appl Spec Rev 40:103–145
11. McQuarrie DA, Simon JD (1997) Physical chemistry – a molecular approach. University of Science Books, Sausalito
12. Miranda PB, Shen YR (1999) Liquid interfaces: a study by sum-frequency vibrational spectroscopy. J Phys Chem B 103:3292–3307
13. Morita A, Hynes JT (2000) A theoretical analysis of the sum frequency generation spectrum of the water surface. Chem Phys 258:371–390
14. Mukamel S (1995) Principles of nonlinear optical spectroscopy. Oxford University Press, New York
15. Pimentel GC, McClellan AL (1960) The hydrogen bond. W. H. Freemann, San Francisco
16. Raymond EA, Tarbuck TL, Brown MG, Richmond GL (2003) Hydrogen-bonding interactions at the vapor/water interface investigated by vibrational sum-frequency spectroscopy of HOD/H_2O/D_2O mixtures and molecular dynamics simulations. J Phys Chem B 107:546–556
17. Shen YR (1984) The principle of nonlinear optics. Wiley, New York
18. Shen YR (1994) Surface spectroscopy by nonlinear optics. In: Hänsch T, Inguscio M (eds) Frontiers in laser spectroscopy. Proceedings of international school of physics, vol CXX. North Holland, Amsterdam, pp 139–165

19. Shen YR (2016) Fundamentals of sum-frequency spectroscopy. Cambridge University Press, Cambridge
20. Tian CS, Shen YR (2014) Recent progress on sum-frequency spectroscopy. Surf Sci Rep 69:105–131
21. Vidal F, Tadjeddine A (2005) Sum-frequency generation spectroscopy of interfaces. Rep Prog Phys 68:1095–1127
22. Wei X, Hong SC, Goto T, Shen YR (2000) Nonlinear optical studies of liquid crystal alignment on a rubbed polyvinyl alcohol surface. Phys Rev E 62:5160–5172
23. Zhuang X, Miranda PB, Kim D, Shen YR (1999) Mapping molecular orientation and conformation at interfaces by surface nonlinear optics. Phys Rev B 59:12632–12640

Chapter 2
Electrodynamics at Interface

Abstract This chapter deals with electrodynamic aspects of the interfacial SFG spectroscopy. In the SFG processes, two incident light fields generate nonlinear polarization at the interface, which in turn emits the sum-frequency signal. We delineate the whole SFG processes from macroscopic viewpoint of electromagnetics. This theory forms a basis of the SFG spectroscopy in order to establish quantitative relation between the nonlinear polarization at interface and the observed light fields. The SFG spectroscopy of interface probe essentially investigates the former through observing the latter. The present overview of the SFG processes helps us with comprehending the essential factors in observable SFG spectra.

Keywords Maxwell equations · Boundary conditions · Fresnel factor · Source polarization

The fundamental theory of the SFG spectroscopy consists of two aspects. One is the macroscopic description of electrodynamic SFG process, where the incident lights induce the nonlinear polarization at the interface which in turn generates the sum frequency signal. Since the SFG spectroscopy investigates the nonlinear polarization of interfaces through measuring the incident and emitted lights outside the interface, understanding of the relation between observed lights and the source polarization at the interface is of crucial significance in the analysis of SFG spectroscopy. The other aspect is the microscopic theory of material properties, particularly the frequency-dependent nonlinear susceptibility $\boldsymbol{\chi}^{(2)}(\Omega, \omega_1, \omega_2)$. This theory is necessary to connect the observed nonlinear polarization to the microscopic structure of interfaces. This chapter deals with the former, macroscopic theory of SFG mechanism, while the theory of material properties will be treated in the following chapters.

A. Morita, *Theory of Sum Frequency Generation Spectroscopy*,
Lecture Notes in Chemistry 97, https://doi.org/10.1007/978-981-13-1607-4_2

2.1 Electromagnetic Fields at Interface

In the macroscopic description of electromagnetic fields, the interface region between two bulk media is often modeled with three layers, consisting of two bulk media and the interface, illustrated in Fig. 2.1. The two bulk media are assumed to be centrosymmetric, and their optical properties are represented with different dielectric constants, $\varepsilon^{\alpha}(\omega)$ and $\varepsilon^{\beta}(\omega)$. (The dispersion with the frequency ω is taken into account.) The interface layer represents a transient region between the two bulk media, and accordingly its dielectric constant is given with a phenomenological parameter $\varepsilon'(\omega)$. The thickness of the transient region is usually much smaller than the typical scale of light wavelength. We discuss spatial configuration of electromagnetic fields near the interface on the basis of the three-layer model in Fig. 2.1.

In this chapter, we assume that the nonlinear source polarization $\boldsymbol{P}^{(2)}$ is generated only in the interface layer. The oscillating source polarization $\boldsymbol{P}^{(2)}$ emits the sum frequency signal of electromagnetic field. This sum frequency field is related to the source polarization by solving the Maxwell equations under proper boundary conditions in Sect. 2.1 [2]. In Sect. 2.2, the nonlinear source polarization $\boldsymbol{P}^{(2)}$ is derived from the incident visible and infrared electromagnetic fields with nonlinear susceptibility of the interface $\boldsymbol{\chi}^{(2)}$. The combined discussion of the two subsections describes the whole SFG process that the incident lights induce the nonlinear polarization (in Sect. 2.2), which in turn generates the sum frequency signal (in Sect. 2.1). We also note that the following discussion will be expanded in Chap. 7 to incorporate quadrupole contributions in the bulk region.

2.1.1 Maxwell Equations

First we present the Maxwell equations to describe the electromagnetic fields in Fig. 2.1. The electromagnetic fields are generated from the nonlinear source

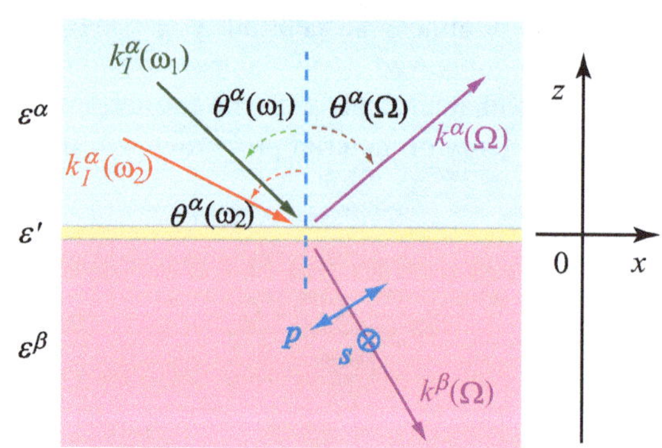

Fig. 2.1 Spatial configuration of lights near the interface. The two incident fields of visible (green) and infrared (red) frequencies and the sum frequency signals (purple) of reflected and transmitted directions are illustrated

polarization $\boldsymbol{P}^{(2)}(\boldsymbol{r}, t)$. This source polarization is equivalent to the following charge density $\rho(\boldsymbol{r}, t)$ and current density $\boldsymbol{j}(\boldsymbol{r}, t)$ in the Maxwell equations,

$$\rho(\boldsymbol{r}, t) = -\nabla \cdot \boldsymbol{P}^{(2)}(\boldsymbol{r}, t), \qquad \boldsymbol{j}(\boldsymbol{r}, t) = \frac{\partial \boldsymbol{P}^{(2)}(\boldsymbol{r}, t)}{\partial t} \tag{2.1}$$

We assume that no other source of charge or current is present in Fig. 2.1. Then the Maxwell equations are given in the cgs Gauss unit system [1, 3],[1]

$$\text{(a)} \qquad \nabla \cdot \boldsymbol{D} = 4\pi\rho = -4\pi\nabla \cdot \boldsymbol{P}^{(2)}, \tag{2.2}$$

$$\text{(b)} \qquad \nabla \times \boldsymbol{E} + \frac{1}{c}\frac{\partial \boldsymbol{B}}{\partial t} = 0, \tag{2.3}$$

$$\text{(c)} \qquad \nabla \cdot \boldsymbol{B} = 0, \tag{2.4}$$

$$\text{(d)} \qquad \nabla \times \boldsymbol{H} - \frac{1}{c}\frac{\partial \boldsymbol{D}}{\partial t} = \frac{4\pi}{c}\boldsymbol{j} = \frac{4\pi}{c}\frac{\partial \boldsymbol{P}^{(2)}}{\partial t}, \tag{2.5}$$

where c is the light velocity in vacuo.

The nonlinear source polarization $\boldsymbol{P}^{(2)}$ is distributed in the interface layer at around $z = 0$ in Fig. 2.1, where the z axis is normal to the interface. In the macroscopic description of electromagnetic fields by Eqs. (2.2), (2.3), (2.4), and (2.5), the thickness of the interface layer is considered to be much thinner than the order of light wavelength (~100 nm), and consequently the spatial distribution of $\boldsymbol{P}^{(2)}$ can be represented with a delta function,

$$\boldsymbol{P}^{(2)}(\boldsymbol{r}, t) = \boldsymbol{P}^{S}(x, y, t)\,\delta(z). \tag{2.6}$$

Equations (2.2), (2.3), (2.4), (2.5) and (2.6) with proper boundary conditions (at $z = 0$ and $z \to \pm\infty$) determine the electromagnetic fields, on condition that the surface polarization $\boldsymbol{P}^{S}(x, y, t)$ is given. Note that Eq. (2.6) using the delta function is a macroscopic description of the interface polarization. The microscopic distribution of the induced polarization along the z axis will be treated in Chap. 7.

2.1.2 Boundary Conditions at Interface

The boundary conditions for electromagnetic fields between two different dielectric media are derived with the help of the Gauss and Stokes theorems [1, 3]. We derive the boundary conditions for $\boldsymbol{B}$, $\boldsymbol{D}$, $\boldsymbol{E}$, and $\boldsymbol{H}$ at the interface $z = 0$ in Fig. 2.1 by taking account of the source polarization $\boldsymbol{P}^{s}$ at $z = 0$. The present boundary conditions at $z = 0$ are somewhat different from the conventional ones due to the source polarization [2].

[1] The cgs Gauss units are used throughout for the electrodynamics in this chapter, except for otherwise noted.

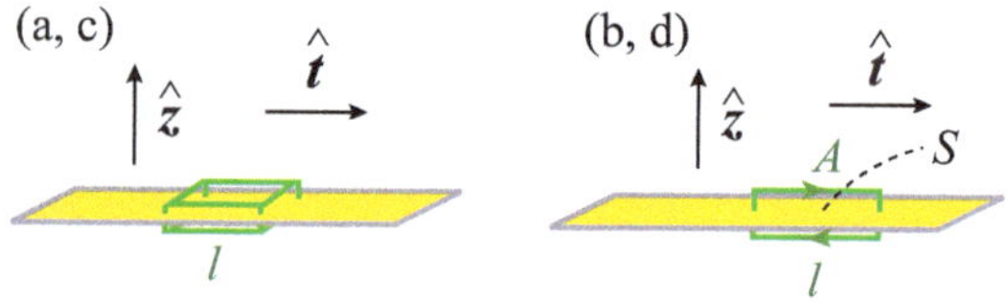

Fig. 2.2 Schematic configurations of volume integral (left) and line integral (right) to derive the boundary conditions at the interface. The left panel is used to the Maxwell equations of (a) and (c), and the right panel to (b) and (d)

The matching condition for $\boldsymbol{B}$ or $\boldsymbol{D}$ at $z = 0$ is derived from the Maxwell equations of (c) (Eq. (2.4)) and (a) (Eq. (2.2)), respectively. These equations are integrated in the volume region V, as depicted in the left panel of Fig. 2.2. The volume region V is an infinitely thin, square plate with the side length l, and it contains the interface. The surface of this region is denoted by σ, and $\hat{\boldsymbol{n}}$ is the unit normal vector at the surface element $d\sigma$. Then the Gauss divergence theorem leads to the following conditions of Eqs. (2.7) and (2.8) from (c) and (a), respectively.

(c)

$$\int_V (\nabla \cdot \boldsymbol{B}) d\boldsymbol{r} = 0 = \int_\sigma \boldsymbol{B} \cdot \hat{\boldsymbol{n}} d\sigma$$

Therefore,

$$\Delta B_z = B_z(z = +0) - B_z(z = -0) = 0 \tag{2.7}$$

(a)

$$\int_V (\nabla \cdot \boldsymbol{D}) d\boldsymbol{r} = -4\pi \int_V \left[\nabla \cdot \left(\boldsymbol{P}^S(x, y)\delta(z) \right) \right] d\boldsymbol{r} = \int_\sigma \boldsymbol{D} \cdot \hat{\boldsymbol{n}} d\sigma$$

Therefore,

$$l^2 \Delta D_z = -4\pi \int_\sigma \left(\boldsymbol{P}^S(x, y)\delta(z) \right) \cdot \hat{\boldsymbol{n}} d\sigma = -4\pi \left(\frac{\partial P_x^S}{\partial x} l + \frac{\partial P_y^S}{\partial y} l \right) l$$

$$\Delta D_z = -4\pi \nabla_t \cdot \boldsymbol{P}^S \tag{2.8}$$

where $\nabla_t = \hat{\boldsymbol{x}} \frac{\partial}{\partial x} + \hat{\boldsymbol{y}} \frac{\partial}{\partial y}$ denotes the spatial derivative along the tangential direction t, and $\hat{\boldsymbol{x}}$, $\hat{\boldsymbol{y}}$ are the unit vectors along the x, y directions, respectively. (Hereafter the superscript ^ designates a unit vector, except otherwise noted.) Note that the symbol of time t should be distinguished from the unit vector $\hat{\boldsymbol{t}}$ along the tangential direction.

[Problem 2.1] Explain the derivation of Eqs. (2.7) and (2.8), by taking account of the fact that the interface polarization $\boldsymbol{P}^{(2)}$ is singular at $z = 0$.

We also note that the solution of $\nabla \cdot \boldsymbol{B}(\boldsymbol{r}) = 0$ (Eq. (2.4)) is regular, as the system contains no magnetic source at the interface.

The matching condition for $\boldsymbol{E}$ or $\boldsymbol{H}$ at $z = 0$ is derived from the Maxwell equations of (b) (Eq. (2.3)) and (d) (Eq. (2.5)), respectively. These equations are integrated on the surface of loop A depicted in the right panel of Fig. 2.2. The loop A is of rectangular shape and it crosses the interface. The Stokes theorem and Eq. (2.3) lead to

(b)

$$\oint_A \boldsymbol{E} \cdot d\boldsymbol{l} = \int_S (\nabla \times \boldsymbol{E}) \cdot (\hat{z} \times \hat{t}) ds = -\int_S \left(\frac{1}{c} \frac{\partial \boldsymbol{B}}{\partial t} \right) \cdot (\hat{z} \times \hat{t}) ds = 0,$$

where S is the surface area encircled by the loop A (see the right panel of Fig. 2.2). Then the last expression on the surface integral vanishes in the limit of infinitesimal area of S, as $\boldsymbol{B}$ remains finite at the interface. The line integral is expanded as follows,

$$\begin{aligned}\oint_A \boldsymbol{E} \cdot d\boldsymbol{l} &= \int_{t-l/2}^{t+l/2} \{E_t(z = +0, t') - E_t(z = -0, t')\} dt' \\ &\quad + \int_{-0}^{+0} \left\{ E_z \left(z, t - \frac{l}{2} \right) - E_z \left(z, t + \frac{l}{2} \right) \right\} dz \\ &= \{E_t(z = +0, t) - E_t(z = -0, t)\} l + \int_{-0}^{+0} \left(\frac{\partial}{\partial t} E_z(z, t) \right) (-l) dz + o(l) \\ &= 0\end{aligned}$$

where l is the small tangential length of the loop A, and the loop center is located at $(z = 0, t)$. The line integral from $z = -0$ to $+0$ may have a non-zero value, as the path crosses the singular source polarization $\boldsymbol{P}^{(2)}(x, y, z) = \boldsymbol{P}^S(x, y)\delta(z)$, i.e.

$$\int_{-0}^{+0} E_z dz = \int_{-0}^{+0} \frac{D_z}{\varepsilon' dz} = -\int_{-0}^{+0} \frac{4\pi}{\varepsilon'} P_z^{(2)} dz = -\frac{4\pi}{\varepsilon'} P_z^S.$$

In the above infinitesimal integral from $z = -0$ to $+0$, only the singular component of integrand could give a non-zero value. The singularity is located at the interface, where the dielectric constant is ε' and $D_z = \varepsilon' E_z$ holds. The third expression is derived from the assumption that $\boldsymbol{D} + 4\pi \boldsymbol{P}^{(2)}$ is regular at any point and hence $\int_{-0}^{+0} (D_z + 4\pi P_z^{(2)}) dz = 0$. Therefore, the line integral of $\boldsymbol{E}$ results in

$$\oint_A \boldsymbol{E} \cdot d\boldsymbol{l} = \{E_t(z = +0) - E_t(z = -0)\}\, l + \frac{4\pi}{\varepsilon'} \left(\hat{\boldsymbol{t}} \cdot \nabla_t P_z^S\right) l + o(l) = 0,$$

and in the limit of $l \to 0$, the matching condition for E_t is derived

$$\Delta E_t = E_t(z = +0) - E_t(z = -0) = -\frac{4\pi}{\varepsilon'} \hat{\boldsymbol{t}} \cdot \nabla_t P_z^S(x, y). \tag{2.9}$$

It is noted that the tangential component of the electric field, E_t, is not continuous at the boundary $z = 0$, unlike the conventional boundary condition of electric field, due to the source polarization $\boldsymbol{P}^S$.

The same argument also holds for the matching condition for $\boldsymbol{H}$. The Stokes theorem and Eq. (2.5) lead to

(d)

$$\oint_A \boldsymbol{H} \cdot d\boldsymbol{l} = \int_S (\nabla \times \boldsymbol{H}) \cdot (\hat{z} \times \hat{t}) ds = \int_S \frac{1}{c} \left(\frac{\partial \boldsymbol{D}}{\partial t} + 4\pi \frac{\partial \boldsymbol{P}^{(2)}}{\partial t} \right) \cdot (\hat{z} \times \hat{t}) ds.$$

Noting that H_z and D_t are regular at the interface, the following condition is obtained for H_t:

$$\Delta H_t = \frac{4\pi}{c} \hat{\boldsymbol{t}} \cdot \left(\frac{\partial \boldsymbol{P}^S(x, y)}{\partial t} \times \hat{z} \right). \tag{2.10}$$

[Problem 2.2] Derive Eq. (2.10), after the discussion associated to Eq. (2.9).

The above equations (2.7), (2.8), (2.9), and (2.10) define the boundary conditions of electromagnetic fields at the interface, $z = 0$, which involves the nonlinear source polarization.

2.1.3 SFG Signal Emitted from Interface

Let us consider the spatial geometry of lights in Fig. 2.1, where all the wave vectors of visible (ω_1), infrared (ω_2) and sum frequency ($\Omega = \omega_1 + \omega_2$) lights are on the xz plane. $\boldsymbol{k}^i(\omega)$ designates the wave vector of light at frequency ω ($= \Omega, \omega_1, \omega_2$) in the bulk region i ($= \alpha$ or β). Suppose that two pump lights of ω_1 and ω_2 are incident from the bulk region $i = \alpha$ onto the interface. These incident electric fields at ω_1 and ω_2 in the region $i = \alpha$ are expressed by

$$\boldsymbol{E}_I^\alpha(\omega_1) \exp(i\boldsymbol{k}_I^\alpha(\omega_1) \cdot \boldsymbol{r} - i\omega_1 t) + c.c.,$$

$$\boldsymbol{E}_I^{\alpha}(\omega_2)\exp(i\boldsymbol{k}_I^{\alpha}(\omega_2)\cdot\boldsymbol{r}-i\omega_2 t)+c.c. \tag{2.11}$$

respectively, where the subscript I indicates the incident field, and $c.c.$ denotes the complex conjugate. The two fields generate the second-order nonlinear polarization of sum frequency Ω at the interface. The phase matching conditions in both spatial and temporal senses in Eq. (1.4) require the following form of the sum frequency polarization $\boldsymbol{P}^{(2)}(\boldsymbol{r},t)$,

$$\boldsymbol{P}^{(2)}(\boldsymbol{r},t)=\boldsymbol{P}^{S}(\Omega)\exp(ik_x(\Omega)x-i\Omega t)\delta(z)+c.c. \tag{2.12}$$

where $k_x(\Omega)=k_{I,x}^{\alpha}(\omega_1)+k_{I,x}^{\alpha}(\omega_2)$. Here the nonlinear susceptibility at the interface is assumed to be uniform along the x direction. $\boldsymbol{P}^{(S)}(\Omega)$ denotes the amplitude of the surface polarization, and will be discussed in Sect. 2.2.

The nonlinear source polarization of Eq. (2.12) emits the electromagnetic wave at the frequency Ω to the regions $i=\alpha$ and $i=\beta$. (The sum frequency signals observed in the regions $i=\alpha$ and $i=\beta$ are called reflection type and transmission type, respectively.) Their wave vectors, $\boldsymbol{k}^{\alpha}(\Omega)$ and $\boldsymbol{k}^{\beta}(\Omega)$, are defined as

$$\boldsymbol{k}^{\alpha}(\Omega)=k_x(\Omega)\hat{\boldsymbol{x}}+q^{\alpha}(\Omega)\hat{\boldsymbol{z}},\quad q^{\alpha}(\Omega)=\sqrt{\varepsilon^{\alpha}(\Omega)K^2-k_x(\Omega)^2},$$
$$\boldsymbol{k}^{\beta}(\Omega)=k_x(\Omega)\hat{\boldsymbol{x}}-q^{\beta}(\Omega)\hat{\boldsymbol{z}},\quad q^{\beta}(\Omega)=\sqrt{\varepsilon^{\beta}(\Omega)K^2-k_x(\Omega)^2}, \tag{2.13}$$

where $K=\Omega/c$ is the wavenumber of sum frequency light in vacuo. $q^{\alpha}(\Omega)$ and $q^{\beta}(\Omega)$ are the absolute z components of $\boldsymbol{k}^{\alpha}(\Omega)$ and $\boldsymbol{k}^{\alpha}(\Omega)$, respectively (i.e. $k_z^{\alpha}(\Omega)=q^{\alpha}(\Omega)$, $k_z^{\beta}(\Omega)=-q^{\beta}(\Omega)$ in Fig. 2.1). Note that the x components of the wavevectors are identical in both media, $k_x(\Omega)=k_x^{\alpha}(\Omega)=k_x^{\beta}(\Omega)$, due to the spatial phase matching condition, while the z components $q^{\alpha}(\Omega)$ and $q^{\beta}(\Omega)$ are determined by Eq. (2.13), depending of the dielectric constants of the two media. With these wavevectors, the emitted electromagnetic fields at the frequency Ω takes the following forms,

$$\text{Electric field:}\quad \boldsymbol{E}^{i}(\Omega)\exp(i\boldsymbol{k}^{i}(\Omega)\cdot\boldsymbol{r}-i\Omega t)+c.c. \qquad (i=\alpha,\beta)$$
$$\text{Magnetic field:}\quad \frac{c}{\Omega}\left(\boldsymbol{k}^{i}(\Omega)\times\boldsymbol{E}^{i}(\Omega)\right)\exp(i\boldsymbol{k}^{i}(\Omega)\cdot\boldsymbol{r}-i\Omega t)+c.c. \tag{2.14}$$

where the amplitudes $\boldsymbol{E}^{i}(\Omega)$ $(i=\alpha,\beta)$ should be determined from the nonlinear source polarization of $\boldsymbol{P}^{S}(\Omega)$. The electromagnetic fields of Eq. (2.14) should satisfy the boundary conditions at $z=0$, Eqs. (2.7), (2.8), (2.9), and (2.10) in Sect. 2.1.2. These conditions derive the relation between $\boldsymbol{E}^{i}(\Omega)$ and $\boldsymbol{P}^{S}(\Omega)$,

$$\boldsymbol{E}^{i}(\Omega)=\frac{2\pi K^2}{iq^{i}}\left\{\boldsymbol{P}^{S}(\Omega)-\left(\hat{\boldsymbol{k}}^{i}(\Omega)\cdot\boldsymbol{P}^{S}(\Omega)\right)\hat{\boldsymbol{k}}^{i}(\Omega)\right\} \tag{2.15}$$

under a simple assumption of $\varepsilon^{\alpha}(\Omega)=\varepsilon^{\beta}(\Omega)=\varepsilon'(\Omega)=1$. (The case of general dielectric constants will be treated in the next subsection.) Equation (2.15) can be

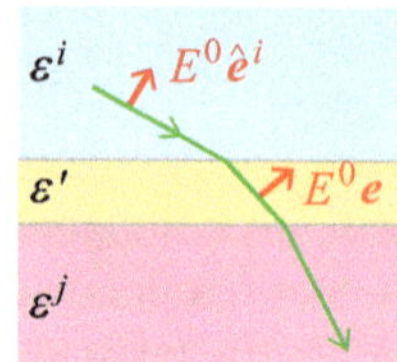

Fig. 2.3 Change in electric field passing through the interface layer

equivalently represented as

$$\hat{e}^i(\Omega) \cdot E^i(\Omega) = \frac{2\pi K^2}{iq^i}\left(\hat{e}^i(\Omega) \cdot P^S(\Omega)\right) \tag{2.16}$$

using the unit vector of the electric field $\hat{e}^i(\Omega) = E^i(\Omega)/\left|E^i(\Omega)\right|$. Notice that $\hat{e}^i(\Omega) \cdot \hat{k}^i(\Omega) = 0$ for the transverse wave of light.

[Problem 2.3] Derive Eq. (2.16) from the boundary conditions (2.7), (2.8), (2.9), and (2.10).

2.1.4 Fresnel Factor

Then we discuss a general case that the dielectric constants of bulk media and interface, $\varepsilon^\alpha, \varepsilon^\beta, \varepsilon'$, are different. When an electromagnetic wave passes a boundary of different dielectric constants, the light is reflected or refracted as illustrated in Fig. 2.3. The change of its wave vector and electromagnetic field at the boundary is described with the Fresnel transformation. To extend Eq. (2.15) or (2.16) in the general case of different dielectric constants, the Fresnel factor has to be considered. The ordinary Fresnel factor between two bulk media is described in Appendix A.2. Here we extend the Fresnel factor in the three-layer model in Fig. 2.3.

Consider the situation of Fig. 2.3 that an incident light wave passes from medium i $(= \alpha$ or $\beta)$ to $j(\neq i)$ through the interface layer. The wavevector varies in these regions, and the amplitude of the incident electric field in the medium i, $E_I^i = E^0\hat{e}^i$, is transformed to $E^0 e$ inside the interface layer. In such case, the vector e is related to $\hat{e}^i$ as

$$e = F^{i\to j} \cdot \hat{e}^i \tag{2.17}$$

with the tensor $F^{i\to j}$ called the Fresnel factor. This Fresnel factor $F^{i\to j}$ in Eq. (2.17) is written in a matrix form by

$$\boldsymbol{F}^{i\to j}=\begin{pmatrix}\dfrac{2\varepsilon^i q^j}{\varepsilon^j q^i+\varepsilon^i q^j} & & \\ & \dfrac{2q^i}{q^i+q^j} & \\ & & \dfrac{2\varepsilon^i\varepsilon^j}{\varepsilon'}\dfrac{q^i}{\varepsilon^j q^i+\varepsilon^i q^j}\end{pmatrix} \tag{2.18}$$

The detailed derivation of Eq. (2.18) is presented in Appendix A.3.

Note 1. The Fresnel factor is a function of frequency ω of the light field, since the wavevectors and dielectric constants depend on the frequency ω.

Note 2. $\boldsymbol{e}$ in the left hand side of Eq. (2.17) is not necessarily a unit vector, while $\hat{\boldsymbol{e}}^i$ is a unit vector by definition. The change in the absolute amplitude is incorporated in the definition of $\boldsymbol{e}$.

Note 3. $\boldsymbol{e}$ describes the electric field *inside the interface layer*, in the present three-layer model [2]. Distinguish from the conventional Fresnel transformation between two bulk media.

If one assumed $\varepsilon' = \varepsilon^j$, the dielectric property of interface layer should be identical to the bulk j. Accordingly the present three-layer model would become the conventional two-layer model with dielectric constants ε^i and ε^j. Thereby the Fresnel factor in Eq. (2.18) coincides with the conventional Fresnel factor between two bulk media.

$$F_{zz}^{i\to j}=\frac{2\varepsilon^i\varepsilon^j}{\varepsilon'}\frac{q^i}{\varepsilon^j q^i+\varepsilon^i q^j}\to\frac{2\varepsilon^i q^i}{\varepsilon^j q^i+\varepsilon^i q^j}\qquad\left(\varepsilon'\to\varepsilon^j\right)$$

By taking account of the Fresnel transformation in Eq. (2.18), the expression of Eq. (2.16) is modified in the case of different dielectric constants to

$$\hat{\boldsymbol{e}}^i(\Omega)\cdot\boldsymbol{E}^i(\Omega)=\frac{2\pi K^2}{iq^i}\left(\boldsymbol{e}(\Omega)\cdot\boldsymbol{P}^S(\Omega)\right). \tag{2.19}$$

Although Eq. (2.19) resembles the previous expression of Eq. (2.16), note that $\hat{\boldsymbol{e}}^i(\Omega)$ in the right hand side of Eq. (2.16) is replaced with $\boldsymbol{e}(\Omega)=\boldsymbol{F}^{i\to j}(\Omega)\cdot\hat{\boldsymbol{e}}^i(\Omega)$ in Eq. (2.19).

[Problem 2.4] Derive Eq. (2.19) in the general dielectric constants ε^α, ε^β, ε' on the basis of the boundary conditions (2.7), (2.8), (2.9), and (2.10).

Equation (2.19) is the general formula to connect the surface nonlinear polarization $\boldsymbol{P}^S(\Omega)$ to the irradiated electric field $\boldsymbol{E}^i(\Omega)$ in the bulk medium i.

2.2 Response to Incident Lights

In the previous subsection we have discussed the electromagnetic process that the nonlinear polarization $\boldsymbol{P}^{(2)}(\boldsymbol{r},t)$ leads to the sum frequency field $\boldsymbol{E}^{i}(\Omega)$ by Eq. (2.19). Next we specify the nonlinear polarization from the incident visible and infrared fields. The surface polarization $\boldsymbol{P}^{S}(\Omega)$ is given using the third-rank tensor of frequency-dependent nonlinear susceptibility $\boldsymbol{\chi}^{(2)}(\Omega,\omega_1,\omega_2)$ by

$$\boldsymbol{P}^{S}(\Omega) = \boldsymbol{\chi}^{(2)}(\Omega,\omega_1,\omega_2) : \boldsymbol{E}(\omega_1)\boldsymbol{E}(\omega_2)$$
$$\left(\text{or } P^{S}_{p}(\Omega) = \sum_{q,r}^{x\sim z} \chi^{(2)}_{pqr}(\Omega,\omega_1,\omega_2)E_q(\omega_1)E_r(\omega_2) \quad (p,q,r = x\sim z)\right) \tag{2.20}$$

where $\boldsymbol{E}(\omega_1)$ and $\boldsymbol{E}(\omega_2)$ denote the amplitudes of the electric fields of visible and infrared lights, respectively, in the interface. This equation is equivalent to Eq. (1.4), though Eq. (2.20) implicitly takes account of the phase matching condition of Eqs. (2.11) and (2.12) along the surface.

Equation (2.20) gives a relation among the interfacial properties, i.e. $\boldsymbol{P}^{S}(\Omega)$, $\boldsymbol{\chi}^{(2)}$, $\boldsymbol{E}(\omega_1)$, $\boldsymbol{E}(\omega_2)$. However, $\boldsymbol{E}(\omega_1)$ and $\boldsymbol{E}(\omega_2)$ in Eq. (2.20) indicate the electric fields *inside the interface*, which are not amenable to direct experimental measurement. For convenience to interpret experimental measurements, Eq. (2.20) is modified using the incident electric fields in the bulk media, instead of using $\boldsymbol{E}(\omega_1)$ and $\boldsymbol{E}(\omega_2)$, the electric fields inside the interface. This modification is carried out by the Fresnel transformation in Eqs. (2.17) and (2.18), and the transformed fields are summarized in Table 2.1.

In this table, the visible and infrared lights are incident from the bulk $i1$ and $i2$ ($= \alpha$ or β), respectively. (Note that Fig. 2.1 illustrates the incident geometry of $i1 = i2 = \alpha$, though other incident geometries are allowed.) And

$$\boldsymbol{e}(\omega_1) = \boldsymbol{F}^{i1\to j1}(\omega_1)\cdot\hat{\boldsymbol{e}}^{i1}(\omega_1), \quad \boldsymbol{e}(\omega_2) = \boldsymbol{F}^{i2\to j2}(\omega_2)\cdot\hat{\boldsymbol{e}}^{i2}(\omega_2)$$

are defined after Eq. (2.17). Based on the above Fresnel transformation, Eq. (2.20) is rewritten using the absolute incident amplitudes in the bulk $i1$ and $i2$, $E^{\mathrm{i1}}_{I}(\omega_1)$ and $E^{\mathrm{i2}}_{I}(\omega_2)$, by

Table 2.1 Relations of electric fields in the bulk and in the interface

	In bulk medium	In interface layer
Visible ω_1 ($i1 \to$ interface)	$\hat{\boldsymbol{e}}^{i1}(\omega_1)E^{i1}_{I}(\omega_1) = \boldsymbol{E}^{i1}_{I}(\omega_1)$	$\boldsymbol{e}(\omega_1)E^{i1}_{I}(\omega_1) = \boldsymbol{E}(\omega_1)$
Infrared ω_2 ($i2 \to$ interface)	$C\hat{\boldsymbol{e}}^{i2}(\omega_2)E^{i2}_{I}(\omega_2) = \boldsymbol{E}^{i2}_{I}(\omega_2)$	$\boldsymbol{e}(\omega_2)E^{i2}_{I}(\omega_2) = \boldsymbol{E}(\omega_2)$

$$\boldsymbol{P}^S(\Omega) = \left\{\boldsymbol{\chi}^{(2)}(\Omega, \omega_1, \omega_2) : \boldsymbol{e}(\omega_1)\boldsymbol{e}(\omega_2)\right\} E_I^{i1}(\omega_1)E_I^{i2}(\omega_2) \tag{2.21}$$

By substituting $\boldsymbol{P}^S(\Omega)$ in Eq. (2.19) with Eq. (2.21), the relation between the incident fields (visible and infrared) and the output field (sum frequency) is given by

$$\begin{aligned}\hat{\boldsymbol{e}}^i(\Omega) \cdot \boldsymbol{E}^i(\Omega) &= \frac{2\pi K^2}{iq^i}\left\{\boldsymbol{e}(\Omega) \cdot \boldsymbol{\chi}^{(2)}(\Omega, \omega_1, \omega_2) : \boldsymbol{e}(\omega_1)\boldsymbol{e}(\omega_2)\right\} E_I^{i1}(\omega_1)E_I^{i2}(\omega_2) \\ &= \frac{2\pi\Omega \sec\theta^i(\Omega)}{ic\sqrt{\varepsilon^i(\Omega)}}\left\{\boldsymbol{e}(\Omega) \cdot \boldsymbol{\chi}^{(2)}(\Omega, \omega_1, \omega_2) : \boldsymbol{e}(\omega_1)\boldsymbol{e}(\omega_2)\right\} E_I^{i1}(\omega_1)E_I^{i2}(\omega_2) \\ &= \frac{2\pi\Omega \sec\theta^i(\Omega)}{ic\sqrt{\varepsilon^i(\Omega)}}\chi_{\text{eff}}^{(2)} E_I^{i1}(\omega_1)E_I^{i2}(\omega_2)\end{aligned} \tag{2.22}$$

where $\theta^i(\Omega)$ denotes the angle of sum frequency emission in Fig. 2.1, and $\sec\theta^i(\Omega)$ is given by Eq. (2.13),

$$\sec\theta^i(\Omega) = \frac{|\boldsymbol{k}^i(\Omega)|}{|k_x^i(\Omega)|} = \frac{\sqrt{\varepsilon^i(\Omega)}K}{q^i(\Omega)} \quad \left(K = \frac{\Omega}{c}\right).$$

In Eq. (2.22) the effective $\chi^{(2)}$ amplitude

$$\chi_{\text{eff}}^{(2)} = \boldsymbol{e}(\Omega) \cdot \boldsymbol{\chi}^{(2)}(\Omega, \omega_1, \omega_2) : \boldsymbol{e}(\omega_1)\boldsymbol{e}(\omega_2) \tag{2.23}$$

is introduced, which will be further discussed in Sect. 3.3 and Chap. 7.

Equation (2.22) indicates the relation between the electric fields of incident and output lights. This can be converted to the relation between the light intensities as follows. The intensity of an electromagnetic wave is represented with its irradiance $I^i(\omega)$, which is the magnitude of the pointing vector $|(c/4\pi)(\boldsymbol{E} \times \boldsymbol{H})|$. Therefore, the irradiance $I^i(\omega)$ at frequency ω in the medium i is given by

$$I^i(\omega) = \frac{c\sqrt{\varepsilon^i(\omega)}}{2\pi}\left|E_I^i(\omega)\right|^2 \tag{2.24}$$

using the formulas of electromagnetic wave in Eq. (2.14). Consequently, the relation between the irradiances of input and output lights becomes

$$\begin{aligned}I^i(\Omega) &= \frac{8\pi^3\Omega^2 \sec^2\theta^i(\Omega)}{c^3\sqrt{\varepsilon^i(\Omega)\varepsilon^{i1}(\omega_1)\varepsilon^{i2}(\omega_2)}}\left|\chi_{\text{eff}}^{(2)}\right|^2 I^{i1}(\omega_1)I^{i2}(\omega_2) \\ &= \frac{8\pi^3\Omega^2 \sec^2\theta^i(\Omega)}{c^3\sqrt{\varepsilon^i(\Omega)\varepsilon^{i1}(\omega_1)\varepsilon^{i2}(\omega_2)}}\left|\boldsymbol{e}(\Omega) \cdot \boldsymbol{\chi}^{(2)}(\Omega, \omega_1, \omega_2):\boldsymbol{e}(\omega_1)\boldsymbol{e}(\omega_2)\right|^2 I^{i1}(\omega_1)I^{i2}(\omega_2).\end{aligned} \tag{2.25}$$

This is the general formula of SFG intensity $I^i(\Omega)$ emitted into the medium i, as a function of the incident visible and infrared intensities in the media $i1$ and $i2$, $I^{i1}(\omega_1)$ and $I^{i2}(\omega_2)$, respectively [2, 5].

2.3 Summary of Factors in SFG Spectra

The general formula of Eq. (2.25) allows us to summarize the following factors to determine the SFG spectra, (A)–(C).

(A) Surface nonlinear susceptibility tensor $\boldsymbol{\chi}^{(2)}(\Omega, \omega_1, \omega_2)$.
This quantity reflects the molecular species as well as their orientation at the interface. This is the most important factor to govern the features of SFG spectra. The $\boldsymbol{\chi}^{(2)}$ depends on the incident frequencies ω_1 and ω_2, and its dependence on the infrared frequency ω_2 is essential for the application of SFG to the vibrational spectroscopy. The next section will focus on the microscopic interpretation of this quantity.

(B) Dielectric constants $\varepsilon^i(\omega)$.
Equation (2.25) involves the dielectric constants of three regions (bulk 1, bulk 2, and interface) at three frequencies (Ω, ω_1, and ω_2);

$$\varepsilon^\alpha(\omega),\ \varepsilon^\beta(\omega),\ \varepsilon'(\omega), \qquad \text{where} \qquad \omega = \Omega,\ \omega_1,\ \omega_2.$$

These values are also incorporated in the Fresnel factors and the vectors, $\boldsymbol{e}(\Omega)$, $\boldsymbol{e}(\omega_1)$, and $\boldsymbol{e}(\omega_2)$. The dielectric constants in the bulk media, $\varepsilon^\alpha(\omega)$ and $\varepsilon^\beta(\omega)$, are measurable quantities, whereas the dielectric constant at the interface, $\varepsilon'(\omega)$, needs some modeling to be determined. This issue will be discussed in Sect. 5.3.

(C) Spatial geometry of lights.
The SFG measurement includes the following three factors, (c1)–(c3), about the light geometry.

(c1) Light Polarization.
Since the light wave is transverse, each light wave may have either S or P polarization. In the geometry illustrated in Fig. 2.1 where the wave vector lies on the xz plane, the S polarization means the case that the electric field oscillates along the y direction (perpendicular to the xz plane), while the P means the case that the electric field is in the xz plane. In the SFG measurements, the complete set of polarization specifies those of three lights, i.e. SFG, visible and infrared, and is thereby designated with three letters, such as SSP, PPP, SPS, etc. The three letters correspond to the polarizations of SFG, visible and infrared, respectively, in order.

(c2) Incident angles $(\theta^{i1}(\omega_1), \theta^{i2}(\omega_2))$ and incident media $(i1,\ i2)$ of the visible and infrared lights.

(c3) Direction of SFG.
The SFG signal is emitted in two directions in $i = \alpha$ and β. The geometry of SFG detection is accordingly twofold, whether one observes the reflected signal or transmitted signal.

Finally, we briefly discuss the heterodyne measurement of SFG, instead of the conventional (homodyne) measurement of SFG signal intensity. The general formula of Eq. (2.25) describes the intensity of SFG signal, $I^i(\Omega)$ in Eq. (2.24). The light intensity is proportional to the square of the electric field, and thus Eq. (2.25) includes the square of the effective $\chi^{(2)}$, $|\chi^{(2)}_{\text{eff}}|^2$. The result of Eq. (2.25) corresponds to the homodyne measurement of SFG intensity.

Recent development of the phase-sensitive or heterodyne SFG measurement allows for detecting the amplitude and phase of the SFG signal [4, 6]. The phase information of $\chi^{(2)}$ is useful to the interpretation of SFG signals, and allows for detailed comparison between experimental and computational results of SFG signals, as we discuss in subsequent chapters. The heterodyne SFG signal can detect the output SFG field itself, that is $\hat{\boldsymbol{e}}^i(\Omega) \cdot \boldsymbol{E}^i(\Omega)$ in Eq. (2.22), which includes $\chi^{(2)}_{\text{eff}}$ (without taking the square). Accordingly, the heterodyne signal is also governed by the same factors (A)–(C) mentioned above. The discussion in this section on the mechanism of SFG emission holds for the heterodyne SFG spectroscopy as well, while the difference from the conventional SFG lies in the detection method of SFG signals.

2.4 Solutions to Problems

2.4.1 *Boundary Condition at Interface (1)*

[Problem 2.1] Explain the derivation of Eqs. (2.7) and (2.8), by taking account of the fact that the interface polarization $\boldsymbol{P}^{(2)}$ is singular at $z = 0$.

We also note that the solution of $\nabla \cdot \boldsymbol{B}(\boldsymbol{r}) = 0$ (Eq. (2.4)) is regular, as the system contains no magnetic source at the interface.

Equation (2.7) We integrate the Maxwell equation of (c) (Eq. (2.4)) in the small volume V (the left panel of Fig. 2.2) and apply the Gauss' divergence theorem as

$$\int_V (\nabla \cdot \boldsymbol{B}) d\boldsymbol{r} = \int_\sigma \boldsymbol{B} \cdot \boldsymbol{n}\, d\sigma = \{B_z(z = +0) - B_z(z = -0)\}\, l^2 + \int_{\text{side}} \boldsymbol{B} \cdot \boldsymbol{n}\, d\sigma = 0.$$

The term for the surface integral on the side, $\int_{\text{side}} \boldsymbol{B} \cdot \boldsymbol{n}\, d\sigma$, can be neglected in the limit of infinitesimal thickness, as $\boldsymbol{B}(\boldsymbol{r})$ is regular. Therefore,

$$\Delta B_z = B_z(z = +0) - B_z(z = -0) = 0. \tag{2.7}$$

Equation (2.8) We integrate the Maxwell equation of (a) (Eq. (2.2)) in the volume V,

$$\int_V (\nabla \cdot \boldsymbol{D}) d\boldsymbol{r} = -4\pi \int_V \left[\nabla \cdot \left(\boldsymbol{P}^S(x, y)\delta(z)\right)\right] d\boldsymbol{r}$$

and apply the Gauss' theorem to both sides of this equation. Thus the following relation for the surface integrals is obtained,

$$\int_\sigma \boldsymbol{D} \cdot \boldsymbol{n} d\sigma = -4\pi \int_\sigma \left(\boldsymbol{P}^S(x, y)\delta(z)\right) \cdot \boldsymbol{n}\, d\sigma.$$

The surface integrals on σ are explicitly represented as follows,

$$\begin{aligned}
&\{D_z(z = +0) - D_z(z = -0)\}\, l^2 + \int_{\text{side}} \boldsymbol{D} \cdot \boldsymbol{n}\, d\sigma \\
&= -4\pi \left[\left\{P_z^S \delta(z = +0) - P_z^S \delta(z = -0)\right\} l^2 + \int_{\text{side}} \boldsymbol{P}^S(x, y)\delta(z) \cdot \boldsymbol{n}\, d\sigma\right].
\end{aligned} \tag{2.26}$$

In the left-hand side of Eq. (2.26), the surface integral on the side planes is neglected in the limit of infinitesimal thickness. This is because the tangential component of $\boldsymbol{D}$, $\boldsymbol{D}_t$, has no singularity in the interface (see the discussion in Appendix A.1). In the right-hand side of Eq. (2.26), the surface integral on the top and bottom planes has no contribution as there is no nonlinear polarization $\boldsymbol{P}^S$ outside the interface, whereas the integral on the side planes could remain finite due to the singular component of $\boldsymbol{P}^{(2)}$ (δ-function in Eq. (2.6)) in the interface. Consequently, Eq. (2.26) becomes

$$\begin{aligned}
&\{D_z(z = +0) - D_z(z = -0)\}\, l^2 = \Delta D_z\, l^2 \\
&= -4\pi \int_{\text{side}} \boldsymbol{P}^S(x, y)\delta(z) \cdot \boldsymbol{n}\, d\sigma \\
&= -4\pi \int_{y-l/2}^{y+l/2} \left\{P_x^S\left(x + \frac{l}{2}, y'\right) - P_x^S\left(x - \frac{l}{2}, y'\right)\right\} dy'
\end{aligned}$$

$$
\begin{aligned}
&- 4\pi \int_{x-l/2}^{x+l/2} \left\{ P_y^S \left(x', y + \frac{l}{2} \right) - P_y^S \left(x', y - \frac{l}{2} \right) \right\} dx' \\
&= -4\pi \left(\frac{\partial P_x^S}{\partial x} l + \frac{\partial P_y^S}{\partial y} l \right) l + o(l).
\end{aligned}
$$

By taking the limit of $l \to 0$, this equation coincides with Eq. (2.8), $\Delta D_z = -4\pi \nabla_t \cdot \boldsymbol{P}^S$.

2.4.2 Boundary Condition at Interface (2)

[Problem 2.2] Derive Eq. (2.10), after the discussion associated to Eq. (2.9).

We carry out surface integral of Eq. (2.5) within the small area S in the right panel of Fig. 2.2, and apply the Stokes' theorem to $\boldsymbol{H}$;

$$
\begin{aligned}
&\int_S (\nabla \times \boldsymbol{H}) \cdot (\hat{z} \times \hat{\boldsymbol{t}}) ds = \int_S \frac{1}{c} \left(\frac{\partial \boldsymbol{D}}{\partial t} + 4\pi \frac{\partial \boldsymbol{P}^{(2)}}{\partial t} \right) \cdot (\hat{z} \times \hat{\boldsymbol{t}}) ds \\
&= \oint_A \boldsymbol{H} \cdot d\boldsymbol{l} = l \Delta H_t + \int_{-0}^{+0} \left\{ H_z \left(z, t - \frac{l}{2} \right) - H_z \left(z, t + \frac{l}{2} \right) \right\} dz. \qquad (2.27)
\end{aligned}
$$

In Eq. (2.27), the surface integral of $\partial \boldsymbol{D}/\partial t$ and the line integral of H_z are neglected in the limit of infinitesimal area of S, since H_z and D_t are not singular there. Consequently, Eq. (2.27) is simplified to be

$$
\begin{aligned}
l \Delta H_t &= \frac{4\pi}{c} \int_S \frac{\partial \boldsymbol{P}^{(2)}}{\partial t} \cdot (\hat{z} \times \hat{\boldsymbol{t}})\, ds \\
&= \frac{4\pi}{c} \int_S \left(\frac{\partial}{\partial t} \boldsymbol{P}^S(x, y) \delta(z) \right) \cdot (\hat{z} \times \hat{\boldsymbol{t}})\, ds = \frac{4\pi}{c} \int_S \left(\frac{\partial \boldsymbol{P}^S}{\partial t} \times \hat{z} \right) \cdot \hat{\boldsymbol{t}} \delta(z) ds \\
&= \frac{4\pi}{c} l \left(\frac{\partial \boldsymbol{P}^S}{\partial t} \times \hat{z} \right) \cdot \hat{\boldsymbol{t}}.
\end{aligned}
$$

Therefore, Eq. (2.10) is obtained.

2.4.3 Boundary Condition at Interface (3)

[Problem 2.3] Derive Eq. (2.16) from the boundary conditions (2.7), (2.8), (2.9), and (2.10).

Here we have assumed $\varepsilon^\alpha(\Omega) = \varepsilon^\beta(\Omega) = \varepsilon'(\Omega) = 1$. The wave vectors of sum frequency Ω of reflection $\boldsymbol{k}^\alpha(\Omega)$ and transmission $\boldsymbol{k}^\beta(\Omega)$ in Eq. (2.13) are

$$\boldsymbol{k}^\alpha(\Omega) = \begin{pmatrix} p \\ 0 \\ q \end{pmatrix} \quad \text{and} \quad \boldsymbol{k}^\beta(\Omega) = \begin{pmatrix} p \\ 0 \\ -q \end{pmatrix},$$

respectively, where $p = k_x^\alpha(\Omega) = k_x^\beta(\Omega)$, and $q = q^\alpha(\Omega) = q^\beta(\Omega) = \sqrt{K^2 - p^2}$. In the following we only treat the sum frequency components of field and polarization, and thus omit "(Ω)" from the notations of $\boldsymbol{k}^\alpha(\Omega)$, $\boldsymbol{E}^\alpha(\Omega)$, $\boldsymbol{P}^S(\Omega)$, etc.

Each emitted light with the wave vector $\boldsymbol{k}^i$ ($i = \alpha, \beta$) may have two polarizations, namely P and S, which are represented with unit vectors of the electric fields, $\hat{\boldsymbol{e}}^i_\mathrm{P}$ and $\hat{\boldsymbol{e}}^i_\mathrm{S}$, respectively. $\hat{\boldsymbol{e}}^i_\mathrm{P}$, $\hat{\boldsymbol{e}}^i_\mathrm{S}$, and $\boldsymbol{k}^i$ are orthogonal each other, and $\hat{\boldsymbol{e}}^i_\mathrm{S}$ is parallel to the y axis in Fig. 2.1. Accordingly, these polarization vectors are

$$\hat{\boldsymbol{e}}^\alpha_\mathrm{P} = \frac{1}{K}\begin{pmatrix} -q \\ 0 \\ p \end{pmatrix}, \quad \hat{\boldsymbol{e}}^\alpha_\mathrm{S} = \begin{pmatrix} 0 \\ 1 \\ 0 \end{pmatrix}, \quad \hat{\boldsymbol{e}}^\beta_\mathrm{P} = \frac{1}{K}\begin{pmatrix} q \\ 0 \\ p \end{pmatrix}, \quad \hat{\boldsymbol{e}}^\beta_\mathrm{S} = \begin{pmatrix} 0 \\ 1 \\ 0 \end{pmatrix}.$$

Then the amplitudes of electric and magnetic fields, $\boldsymbol{E}^i$ and $\boldsymbol{H}^i = (c/\Omega)\, \boldsymbol{k}^i \times \boldsymbol{E}^i$, in Eq. (2.14) are represented by

$$\boldsymbol{E}^i = \begin{pmatrix} E^i_x \\ E^i_y \\ E^i_z \end{pmatrix} \qquad (i = \alpha, \beta),$$

$$\boldsymbol{H}^\alpha = \frac{c}{\Omega}\begin{pmatrix} -qE^\alpha_y \\ qE^\alpha_x - pE^\alpha_z \\ pE^\alpha_y \end{pmatrix}, \quad \boldsymbol{H}^\beta = \frac{c}{\Omega}\begin{pmatrix} qE^\beta_y \\ -qE^\beta_x - pE^\beta_z \\ pE^\beta_y \end{pmatrix}.$$

The boundary conditions for the electromagnetic fields at $z = 0$ are derived from Eqs. (2.7), (2.8), (2.9), and (2.10), and expressed using the electric field amplitudes in the following five equations, (2.28), (2.29), (2.30), (2.31), and (2.32).

- From Eq. (2.7), condition for $B_z = \mu H_z$:

$$B_z: \qquad \mu\frac{c}{\Omega}\left\{pE^\beta_y - pE^\alpha_y\right\} = 0$$

$$\text{Therefore,} \quad E^\beta_y - E^\alpha_y = 0 \tag{2.28}$$

- From Eq. (2.8), condition for $D_z = E_z$:

$$D_z: \qquad E_z^\beta - E_z^\alpha = -4\pi i p P_x^S \tag{2.29}$$

- From Eq. (2.9), conditions for E_x and E_y:

$$E_x: \qquad E_x^\beta - E_x^\alpha = -4\pi i p P_z^S \tag{2.30}$$

$$E_y: \qquad E_y^\beta - E_y^\alpha = 0 \quad \text{(same as Eq. (2.28))}$$

- From Eq. (2.10), conditions for H_x and H_y:

$$H_x: \qquad \frac{c}{\Omega}\left\{\left(qE_y^\beta\right) - \left(-qE_y^\alpha\right)\right\} = \frac{4\pi}{c}(-i\Omega)P_y^S$$

$$\text{Therefore,} \quad q\left(E_y^\beta + E_y^\alpha\right) = -4\pi i K^2 P_y^S \tag{2.31}$$

$$H_y: \qquad \frac{c}{\Omega}\left\{\left(-qE_x^\beta - pE_z^\beta\right) - \left(qE_x^\alpha - pE_z^\alpha\right)\right\} = -\frac{4\pi}{c}(-i\Omega)P_x^S$$

$$\text{Therefore,} \quad q\left(E_x^\beta + E_x^\alpha\right) + p\left(E_z^\beta - E_z^\alpha\right) = -4\pi i K^2 P_x^S \tag{2.32}$$

(Note the right-hand side of Eq. (2.32) includes an extra minus sign due to the property of vector product, $x \times z = -z \times x$.) Using Eqs. (2.28), (2.29), (2.30), (2.31), and (2.32), both P and S components of $\boldsymbol{E}^\alpha$, $\boldsymbol{E}^\beta$ in Eq. (2.14) are obtained as follows.

P components Eqs. (2.29), (2.30), and (2.32) derive

$$E_x^\beta = 2\pi i\left(-qP_x^S - pP_z^S\right). \tag{2.33}$$

This equation, along with the relation $\boldsymbol{k}^\beta \cdot \boldsymbol{E}^\beta = pE_x^\beta - qE_z^\beta = 0$, leads to

$$E_z^\beta = \frac{p}{q}E_x^\beta = 2\pi i\frac{p}{q}\left(-qP_x^S - pP_z^S\right). \tag{2.34}$$

Therefore, E_x^α and E_z^α are also determined from Eqs. (2.29), (2.30).

$$E_x^\alpha = 2\pi i\left(-qP_x^S + pP_z^S\right), \quad E_z^\alpha = -2\pi i\frac{p}{q}\left(-qP_x^S + pP_z^S\right). \tag{2.35}$$

It is readily confirmed that Eqs. (2.33), (2.34), and (2.35) satisfy the condition of Eq. (2.16) for the P polarization;

$$\hat{\boldsymbol{e}}_\mathrm{P}^\alpha \cdot \boldsymbol{E}^\alpha = \frac{1}{K}\left(-qE_x^\alpha + pE_z^\alpha\right) = \frac{2\pi i}{K}\left\{-q(-qP_x^S + pP_z^S) - p\frac{p}{q}(-qP_x^S + pP_z^S)\right\}$$

$$= -\frac{2\pi i}{Kq}(q^2 + p^2)\left(-qP_x^S + pP_z^S\right) = \frac{2\pi K^2}{iq}\left(\hat{\boldsymbol{e}}_\mathrm{P}^\alpha \cdot \boldsymbol{P}^S\right),$$

$$\hat{e}_{\rm P}^{\beta} \cdot \boldsymbol{E}^{\beta} = \frac{1}{K}\left(qE_x^{\beta} + pE_z^{\beta}\right) = \frac{2\pi i}{K}\left\{qE_x^{\beta} + p\left(\frac{p}{q}E_x^{\beta}\right)\right\}$$
$$= \frac{2\pi i}{Kq}(q^2+p^2)\left(-qP_x^S - pP_z^S\right) = \frac{2\pi K^2}{iq}\left(\hat{e}_{\rm P}^{\beta} \cdot \boldsymbol{P}^S\right).$$

S components Eqs. (2.28) and (2.31) derive

$$E_y^{\alpha} = E_y^{\beta} = \frac{2\pi K^2}{iq}P_y^S. \tag{2.36}$$

E_y^{α} and E_y^{β} in Eq. (2.36) satisfy the condition of Eq. (2.16) in the S polarization;

$$\hat{e}_{\rm S}^{\alpha} \cdot \boldsymbol{E}^{\alpha} = E_y^{\alpha} = \frac{2\pi K^2}{iq}P_y^S = \frac{2\pi K^2}{iq}\left(\hat{e}_{\rm S}^{\alpha} \cdot \boldsymbol{P}^S\right),$$
$$\hat{e}_{\rm S}^{\beta} \cdot \boldsymbol{E}^{\beta} = E_y^{\beta} = \frac{2\pi K^2}{iq}P_y^S = \frac{2\pi K^2}{iq}\left(\hat{e}_{\rm S}^{\beta} \cdot \boldsymbol{P}^S\right).$$

2.4.4 Electric Field and Interfacial Polarization

[Problem 2.4] Derive Eq. (2.19) in the general dielectric constants ε^{α}, ε^{β}, ε' on the basis of the boundary conditions (2.7), (2.8), (2.9), and (2.10).

Let us generalize the solution of Problem 2.3 without resort to the previous assumption $\varepsilon^{\alpha}(\Omega) = \varepsilon^{\beta}(\Omega) = \varepsilon'(\Omega) = 1$. The following derivation is rather similar to that in Problem 2.3, and note the differences from the previous derivation.

For arbitrary values of $\varepsilon^{\alpha}(\Omega)$, $\varepsilon^{\beta}(\Omega)$, $\varepsilon'(\Omega)$, the wavevectors $\boldsymbol{k}^{\alpha}(\Omega)$, $\boldsymbol{k}^{\beta}(\Omega)$ for the reflected and transmitted lights, respectively, become

$$\boldsymbol{k}^{\alpha} = \begin{pmatrix} p \\ 0 \\ q^{\alpha} \end{pmatrix}, \quad \boldsymbol{k}^{\beta} = \begin{pmatrix} p \\ 0 \\ -q^{\beta} \end{pmatrix},$$

where

$$p = k_x^{\alpha}(\Omega) = k_x^{\beta}(\Omega) \quad \text{and} \quad \begin{cases} q^{\alpha} = \sqrt{\varepsilon^{\alpha}K^2 - p^2} \\ q^{\beta} = \sqrt{\varepsilon^{\beta}K^2 - p^2} \end{cases} \tag{2.37}$$

The unit vectors for S and P polarizations are expressed by

$$\hat{e}_{\rm P}^{\alpha} = \frac{1}{\sqrt{\varepsilon^{\alpha}}K}\begin{pmatrix} -q^{\alpha} \\ 0 \\ p \end{pmatrix}, \quad \hat{e}_{\rm S}^{\alpha} = \begin{pmatrix} 0 \\ 1 \\ 0 \end{pmatrix}, \quad \hat{e}_{\rm P}^{\beta} = \frac{1}{\sqrt{\varepsilon^{\beta}}K}\begin{pmatrix} q^{\beta} \\ 0 \\ p \end{pmatrix}, \quad \hat{e}_{\rm S}^{\beta} = \begin{pmatrix} 0 \\ 1 \\ 0 \end{pmatrix}$$

Then the amplitudes of electric and magnetic fields, $\boldsymbol{E}^i$ and $\boldsymbol{H}^i = (c/\Omega)\,\boldsymbol{k}^i \times \boldsymbol{E}^i$, in Eq. (2.14) are represented by

$$\boldsymbol{E}^i = \begin{pmatrix} E_x^i \\ E_y^i \\ E_z^i \end{pmatrix} \qquad (i = \alpha, \beta),$$

$$\boldsymbol{H}^\alpha = \frac{c}{\Omega}\begin{pmatrix} -q^\alpha E_y^\alpha \\ q^\alpha E_x^\alpha - pE_z^\alpha \\ pE_y^\alpha \end{pmatrix}, \qquad \boldsymbol{H}^\beta = \frac{c}{\Omega}\begin{pmatrix} q^\beta E_y^\beta \\ -q^\beta E_x^\beta - pE_z^\beta \\ pE_y^\beta \end{pmatrix}.$$

The boundary conditions for the electromagnetic fields at $z = 0$ are derived from Eqs. (2.7), (2.8), (2.9), and (2.10), and summarized in the following five equations, (2.38), (2.39), (2.40), (2.41), and (2.42).

- From Eq. (2.7), boundary condition for $B_z = \mu H_z$:

$$B_z: \qquad \frac{c}{\Omega}\left\{pE_y^\beta - pE_y^\alpha\right\} = 0$$

$$\text{Therefore,} \qquad E_y^\beta - E_y^\alpha = 0 \tag{2.38}$$

- From Eq. (2.8), condition for $D_z = \varepsilon E_z$:

$$D_z: \qquad \varepsilon^\beta E_z^\beta - \varepsilon^\alpha E_z^\alpha = -4\pi i p P_x^S \tag{2.39}$$

- From Eq. (2.9), conditions for E_x and E_y:

$$E_x: \qquad E_x^\beta - E_x^\alpha = -\frac{4\pi i p}{\varepsilon'} P_z^S \tag{2.40}$$

$$E_y: \qquad E_y^\beta - E_y^\alpha = 0 \qquad \text{(same as Eq. (2.38))}$$

- From Eq. (2.10), conditions for H_x and H_y:

$$H_x: \qquad \frac{c}{\Omega}\left\{\left(q^\beta E_y^\beta\right) - \left(-q^\alpha E_y^\alpha\right)\right\} = \frac{4\pi}{c}(-i\Omega)P_y^S$$

$$\text{Therefore,} \qquad q^\beta E_y^\beta + q^\alpha E_y^\alpha = -4\pi i K^2 P_y^S \tag{2.41}$$

$$H_y: \qquad \frac{c}{\Omega}\left\{\left(-q^\beta E_x^\beta - pE_z^\beta\right) - \left(q^\alpha E_x^\alpha - pE_z^\alpha\right)\right\} = -\frac{4\pi}{c}(-i\Omega)P_x^S$$

$$\text{Therefore,} \qquad q^\beta E_x^\beta + q^\alpha E_x^\alpha + p\left(E_z^\beta - E_z^\alpha\right) = -4\pi i K^2 P_x^S \tag{2.42}$$

(Note that the right-hand side of Eq. (2.42) includes an extra minus sign due to the property of vector product $x \times z = -z \times x$.) Using Eqs. (2.38), (2.39), (2.40), (2.41), and (2.42), both P and S components of $\boldsymbol{E}^\alpha$, $\boldsymbol{E}^\beta$ in Eq. (2.14) are obtained as follows.

P components Equation (2.42) is modified with Eqs. (2.39) and (2.40) so as to eliminate E^α,

$$q^\beta E_x^\beta + q^\alpha \left(E_x^\beta + \frac{4\pi i p}{\varepsilon'} P_z^S \right) + p E_z^\beta - p \frac{1}{\varepsilon^\alpha} \left(\varepsilon^\beta E_z^\beta + 4\pi i p P_x^S \right) = -4\pi i K^2 P_x^S.$$

Therefore, using Eq. (2.37),

$$\begin{aligned}(q^\alpha + q^\beta) E_x^\beta + p \left(1 - \frac{\varepsilon^\beta}{\varepsilon^\alpha} \right) E_z^\beta &= -4\pi i \left(K^2 - \frac{p^2}{\varepsilon^\alpha} \right) P_x^S - 4\pi i p \frac{q^\alpha}{\varepsilon'} P_z^S \\ &= -\frac{4\pi i (q^\alpha)^2}{\varepsilon^\alpha} \left(P_x^S + \frac{\varepsilon^\alpha p}{\varepsilon' q^\alpha} P_z^S \right).\end{aligned}$$

Using the relation $\boldsymbol{k}^\beta \cdot \boldsymbol{E}^\beta = p E_x^\beta - q^\beta E_z^\beta = 0$, we insert $E_x^\beta = q^\beta E_z^\beta / p$ into the above equation and obtains

$$\left\{ (q^\alpha + q^\beta) \frac{q^\beta}{p} + p \left(1 - \frac{\varepsilon^\beta}{\varepsilon^\alpha} \right) \right\} E_z^\beta = -\frac{4\pi i (q^\alpha)^2}{\varepsilon^\alpha} \left(P_x^S + \frac{\varepsilon^\alpha p}{\varepsilon' q^\alpha} P_z^S \right).$$

Therefore,

$$E_z^\beta = -\frac{4\pi i p q^\alpha}{\varepsilon^\alpha q^\beta + \varepsilon^\beta q^\alpha} \left(P_x^S + \frac{\varepsilon^\alpha p}{\varepsilon' q^\alpha} P_z^S \right), \tag{2.43}$$

$$E_x^\beta = \frac{q^\beta}{p} E_z^\beta = -\frac{4\pi i q^\alpha q^\beta}{\varepsilon^\alpha q^\beta + \varepsilon^\beta q^\alpha} \left(P_x^S + \frac{\varepsilon^\alpha p}{\varepsilon' q^\alpha} P_z^S \right). \tag{2.44}$$

Consequently, using Eqs. (2.39) and (2.40), we also obtain

$$E_z^\alpha = \frac{4\pi i p q^\beta}{\varepsilon^\alpha q^\beta + \varepsilon^\beta q^\alpha} \left(P_x^S - \frac{\varepsilon^\beta p}{\varepsilon' q^\beta} P_z^S \right), \tag{2.45}$$

$$E_x^\alpha = -\frac{4\pi i q^\alpha q^\beta}{\varepsilon^\alpha q^\beta + \varepsilon^\beta q^\alpha} \left(P_x^S - \frac{\varepsilon^\beta p}{\varepsilon' q^\beta} P_z^S \right). \tag{2.46}$$

It is confirmed that Eqs. (2.43) and (2.44) satisfy the condition of Eq. (2.19) in the P polarization as follows.

$$\begin{aligned}\hat{\boldsymbol{e}}_\mathrm{P}^\beta \cdot \boldsymbol{E}^\beta &= \frac{1}{\sqrt{\varepsilon^\beta} K} \left(q^\beta E_x^\beta + p E_z^\beta \right) = \frac{1}{\sqrt{\varepsilon^\beta} K} \left(q^\beta \frac{q^\beta}{p} E_z^\beta + p E_z^\beta \right) = \frac{\sqrt{\varepsilon^\beta} K}{p} E_z^\beta \\ &= -\sqrt{\varepsilon^\beta} K \frac{4\pi i q^\alpha}{\varepsilon^\alpha q^\beta + \varepsilon^\beta q^\alpha} \left(P_x^S + \frac{\varepsilon^\alpha p}{\varepsilon' q^\alpha} P_z^S \right),\end{aligned}$$

$$
\begin{aligned}
\frac{2\pi K^2}{iq^\beta}\boldsymbol{e}_{\mathrm{P}} \cdot \boldsymbol{P}^S &= \frac{2\pi K^2}{iq^\beta}\frac{1}{\sqrt{\varepsilon^\beta}K}\left(F^{\beta\to\alpha}_{xx} q^\beta P^S_x + F^{\beta\to\alpha}_{zz} p P^S_z\right) \\
&= -\sqrt{\varepsilon^\beta}K\frac{4\pi i q^\alpha}{\varepsilon^\alpha q^\beta + \varepsilon^\beta q^\alpha}\left(P^S_x + \frac{\varepsilon^\alpha p}{\varepsilon' q^\alpha}P^S_z\right),
\end{aligned}
$$

where $\boldsymbol{e}_P = \boldsymbol{F}^{\beta\to\alpha} \cdot \hat{\boldsymbol{e}}^\beta_P$, and $\boldsymbol{F}^{\beta\to\alpha}$ is given in Eq. (2.18). Therefore, Eq. (2.19) is proved for the β phase,

$$
\hat{\boldsymbol{e}}^\beta_{\mathrm{P}} \cdot \boldsymbol{E}^\beta = \frac{2\pi K^2}{iq^\beta}\boldsymbol{e}_{\mathrm{P}} \cdot \boldsymbol{P}^S. \tag{2.47}
$$

The same argument can be applied to $\boldsymbol{E}^\alpha$ in Eqs. (2.45) and (2.46);

$$
\begin{aligned}
\hat{\boldsymbol{e}}^\alpha_{\mathrm{P}} \cdot \boldsymbol{E}^\alpha &= \frac{1}{\sqrt{\varepsilon^\alpha}K}\left(-q^\alpha E^\alpha_x + pE^\alpha_z\right) = \frac{1}{\sqrt{\varepsilon^\alpha}K}\left(q^\alpha \frac{q^\alpha}{p}E^\alpha_z + pE^\alpha_z\right) = \frac{\sqrt{\varepsilon^\alpha}K}{p}E^\alpha_z \\
&= \sqrt{\varepsilon^\alpha}K\frac{4\pi i q^\beta}{\varepsilon^\alpha q^\beta + \varepsilon^\beta q^\alpha}\left(P^S_x - \frac{\varepsilon^\beta p}{\varepsilon' q^\beta}P^S_z\right), \\
\frac{2\pi K^2}{iq^\alpha}\boldsymbol{e}_{\mathrm{P}} \cdot \boldsymbol{P}^S &= \frac{2\pi K^2}{iq^\alpha}\frac{1}{\sqrt{\varepsilon^\alpha}K}\left(-F^{\alpha\to\beta}_{xx} q^\alpha P^S_x + F^{\alpha\to\beta}_{zz} p P^S_z\right) \\
&= \sqrt{\varepsilon^\alpha}K\frac{4\pi i q^\beta}{\varepsilon^\alpha q^\beta + \varepsilon^\beta q^\alpha}\left(P^S_x - \frac{\varepsilon^\beta p}{\varepsilon' q^\beta}P^S_z\right),
\end{aligned}
$$

where $\boldsymbol{e}_P = \boldsymbol{F}^{\alpha\to\beta} \cdot \hat{\boldsymbol{e}}^\alpha_P$. $\boldsymbol{F}^{\alpha\to\beta}$ is also given in Eq. (2.18). Therefore,

$$
\hat{\boldsymbol{e}}^\alpha_{\mathrm{P}} \cdot \boldsymbol{E}^\alpha = \frac{2\pi K^2}{iq^\alpha}\boldsymbol{e}_{\mathrm{P}} \cdot \boldsymbol{P}^S. \tag{2.48}
$$

S components Equations (2.38) and (2.41) derive

$$
E^\alpha_y = E^\beta_y = \frac{4\pi K^2}{i(q^\alpha + q^\beta)}P^S_y \tag{2.49}
$$

E^β_y in Eq. (2.49) satisfy the condition of Eq. (2.19) in the S polarization.

$$
\hat{\boldsymbol{e}}^\beta_{\mathrm{S}} \cdot \boldsymbol{E}^\beta = E^\beta_y = \frac{2\pi K^2}{iq^\beta}\frac{2q^\beta}{q^\alpha + q^\beta}P^S_y = \frac{2\pi K^2}{iq^\beta}F^{\beta\to\alpha}_{yy}P^S_y = \frac{2\pi K^2}{iq^\beta}\boldsymbol{e}_{\mathrm{S}} \cdot \boldsymbol{P}^S. \tag{2.50}
$$

The same argument holds for E^α_y in Eq. (2.49),

$$
\hat{\boldsymbol{e}}^\alpha_{\mathrm{S}} \cdot \boldsymbol{E}^\alpha = E^\alpha_y = \frac{2\pi K^2}{iq^\alpha}\frac{2q^\alpha}{q^\alpha + q^\beta}P^S_y = \frac{2\pi K^2}{iq^\alpha}F^{\alpha\to\beta}_{yy}P^S_y = \frac{2\pi K^2}{iq^\alpha}\boldsymbol{e}_{\mathrm{S}} \cdot \boldsymbol{P}^S. \tag{2.51}
$$

The above equations (2.47), (2.48), (2.50), and (2.51) prove the relation of Eq. (2.19) for both P and S polarizations in either phase of α or β.

Appendix

A.1 Singularity of Source Polarization

The boundary conditions of electromagnetic fields in Sect. 2.1 are affected by the singularity of the nonlinear source polarization at the interface. The singularity is a consequence of taking the limit of $L \to 0$ for the interface thickness L. Here we present some supplementary argument on the singularity issue.

Figure 2.4 show a picture of the three-layer model with a finite thickness L. The thickness should be large enough to define the electric field and polarization in the layer, while it is small compared to the scale of the light wavelengths. The nonlinear source polarization $\boldsymbol{P}^{(2)}$ inside the interface layer is given by

$$\boldsymbol{P}^{(2)}(x, y, z) = \begin{cases} \dfrac{\boldsymbol{P}^S(x, y)}{L} & \left(-\dfrac{L}{2} < z < \dfrac{L}{2}\right) \\ 0 & \text{(otherwise)} \end{cases} \tag{2.52}$$

Note that the limit of $L \to 0$ recovers Eq. (2.6). In deriving Eq. (5.18) in Sect. 2.1 and Problem 2.1, we used the assumption that the tangential component of the electric displacement, D_t, has no singularity at the limit, though the nonlinear source polarization, $P_t^{(2)}$, may have a singular component of $\delta(z)$. In the following we explain these subtle assumptions.[2]

Let us clarify the assumption by deriving the boundary condition of Eq. (2.8). The Gauss divergence theorem is applied to Eq. (2.2) in the green rectangular box in Fig. 2.4, and

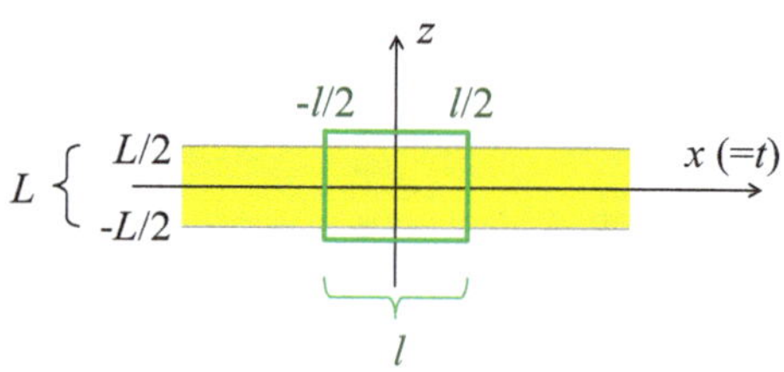

Fig. 2.4 Geometry of the three layers with a finite thickness L of the interface. The cross section of the rectangular box is illustrated in green

[2] This problem is motivated by Ref. [2]. The author is grateful to Profs. Tony Heinz and Ron Shen for clarifying this issue.

$$D_z\left(\frac{L}{2}\right)l^2 - D_z\left(-\frac{L}{2}\right)l^2 + \int_{-L/2}^{L/2} dz \left(\frac{\partial D_x}{\partial x} + \frac{\partial D_y}{\partial y}\right) l^2$$
$$= -4\pi \int_{-L/2}^{L/2} dz \left(\frac{\partial P_x^{(2)}}{\partial x} + \frac{\partial P_y^{(2)}}{\partial y}\right) l^2$$
$$= -4\pi l^2 L \left(\frac{\partial}{\partial x}\frac{P_x^S}{L} + \frac{\partial}{\partial y}\frac{P_y^S}{L}\right) = -4\pi l^2 \left(\frac{\partial P_x^S}{\partial x} + \frac{\partial P_y^S}{\partial y}\right). \quad (2.53)$$

By dividing with l^2 and taking the limit of $L \to 0$, we obtain Eq. (2.8),

$$\Delta D_z = \lim_{L\to 0}\left\{D_z\left(\frac{L}{2}\right) - D_z\left(-\frac{L}{2}\right)\right\} = -4\pi\left(\frac{\partial P_x^S}{\partial x} + \frac{\partial P_y^S}{\partial y}\right). \quad (2.8)$$

One may notice that the above derivation includes the assumption that

$$\lim_{L\to 0}\int_{-L/2}^{L/2} dz \left(\frac{\partial D_x}{\partial x} + \frac{\partial D_y}{\partial y}\right) = 0,$$

which means that the tangential component of $\boldsymbol{D}$ (D_x or D_y) is regular in the interface layer. If $\boldsymbol{D}_t$ had a singular component $D_t^S(x, y)\delta(z)$, then the above derivation should lead to the following boundary condition,

$$\Delta D_z + \nabla_t \cdot \boldsymbol{D}_t^S = -4\pi\nabla_t \cdot \boldsymbol{P}^S \quad \left(\text{or } \Delta D_z = -\nabla_t \cdot \left(4\pi\boldsymbol{P}^S + \boldsymbol{D}_t^S\right)\right),$$

instead of Eq. (2.8). Actually, the term with $\boldsymbol{D}_t^S$ does not appear in Eq. (2.8), indicating that $\boldsymbol{D}_t^S$ has no singularity.

This assumption stems from physical consideration of the polarization. In the present theory, the entire polarization induced in the material is classified into the nonlinear polarization $\boldsymbol{P}^{(2)}$ and the remaining, linear polarization $\boldsymbol{P}$. The former is treated as the source of charge and current in the Maxwell equations (see Eq. (2.1)), while the latter is regarded as conventional linear polarization included in the definition of electric displacement $\boldsymbol{D} = \boldsymbol{E} + 4\pi\boldsymbol{P}$. When we adopt the three-layer model with infinitely thin interface in Fig. 2.1, singular source of $\boldsymbol{P}^{(2)}$ is allowed to be located at the interface in the macroscopic radiation theory. However, the distinction between $\boldsymbol{P}^{(2)}$ and $\boldsymbol{P}$ is somewhat arbitrary in a microscopic sense. If we did not distinguish $\boldsymbol{P}^{(2)}$ and $\boldsymbol{P}$ and treated them just as polarization in the material, the electric displacement would be alternatively defined as

$$\boldsymbol{D}^\circ = \boldsymbol{E} + 4\pi(\boldsymbol{P} + \boldsymbol{P}^{(2)}) = \boldsymbol{D} + 4\pi\boldsymbol{P}^{(2)}. \quad (2.54)$$

If we employ $\boldsymbol{D}^\circ$ defined as such, it should be regular everywhere because the system contains no other source of charge or current than $\boldsymbol{P}^{(2)}$. Figure 2.5 summarizes the sources of electric field and definitions of $\boldsymbol{E}$, $\boldsymbol{D}$ and $\boldsymbol{D}^\circ$ in the present discussion.

Fig. 2.5 Various sources of electric fields and their relation to $\boldsymbol{E}$, $\boldsymbol{D}$, and $\boldsymbol{D}^{\circ}$

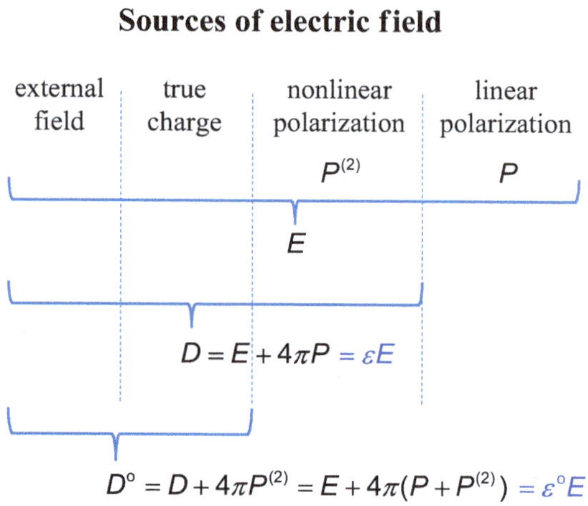

Since the present theory of SFG radiation in Sect. 2.1 assumes that $\boldsymbol{P}^{\mathrm{S}}$ is the only source of nonlinear radiation generated at the interface, we can safely assume that $\boldsymbol{D}_t$ has no singular component without loss of generality. If there were a singular component, it should be incorporated into $\boldsymbol{P}^{(2)}$ so that $4\pi\,\boldsymbol{P}^S + \boldsymbol{D}_t^S$ is regarded as a new $4\pi\,\boldsymbol{P}^S$.

A.2 Fresnel Factors for Two-Layer Model

The ordinary Fresnel factors at an interface of two bulk media are presented as follows. Suppose the two bulk media have dielectric constants of ε_1 and ε_2, or the refractive indices of $n_1 = \sqrt{\varepsilon_1}$, $n_2 = \sqrt{\varepsilon_2}$, respectively. Figure 2.6 illustrates the situation that the incident light propagates from the medium i $(= 1, 2)$ to j $(\neq i)$, or reflects back to the medium i. The incident and transmitted angles are denoted with θ_i and θ_j. The amplitudes of incident, reflected and transmitted electric fields are E^0, E^r and E^t, respectively.

First, the Snell's law holds between the incident angle θ_i and transmitted one θ_j,

$$n_i \sin\theta_i = n_j \sin\theta_j. \tag{2.55}$$

Therefore, $\cos\theta_j$ is expressed by

$$\cos\theta_j = \sqrt{1 - \sin^2\theta_j} = \sqrt{1 - \frac{n_i^2}{n_j^2}\sin^2\theta_i}. \tag{2.56}$$

This value becomes imaginary in the condition of total reflection.

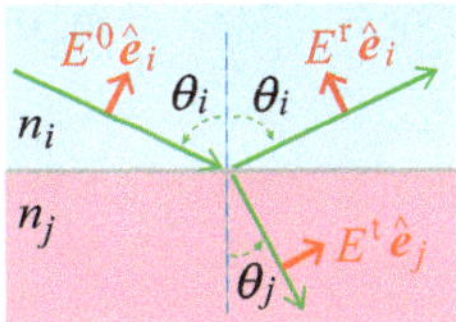

Fig. 2.6 Geometry of lights (of P polarization) in the two-layer model

The ratios of reflected and transmitted field amplitudes to the incident one, $r = E^r/E^0$ and $t = E^t/E^0$, depend on the polarization, P or S. Accordingly, these ratios are denoted with r_P (r_S) and t_P (t_S) at the P (S) polarization. These ratios are derived from the ordinary boundary conditions of electromagnetic fields at a plane interface, and the results are summarized below [3].

- P polarization:

$$r_P = \frac{n_j \cos\theta_i - n_i \cos\theta_j}{n_j \cos\theta_i + n_i \cos\theta_j}, \qquad t_P = \frac{2n_i \cos\theta_i}{n_j \cos\theta_i + n_i \cos\theta_j}, \tag{2.57}$$

- S polarization:

$$r_S = \frac{n_i \cos\theta_i - n_j \cos\theta_j}{n_i \cos\theta_i + n_j \cos\theta_j}, \qquad t_S = \frac{2n_i \cos\theta_i}{n_i \cos\theta_i + n_j \cos\theta_j}. \tag{2.58}$$

where $\cos\theta_j$ is given in Eq. (2.56).

One can readily prove that t_P and t_S in Eqs. (2.57) and (2.58) are equivalent to Eq. (2.18) by putting $\varepsilon' = \varepsilon^j$;

$$\boldsymbol{F}^{i\to j} = \begin{pmatrix} \dfrac{2\varepsilon^i q^j}{\varepsilon^j q^i + \varepsilon^i q^j} & & \\ & \dfrac{2q^i}{q^i + q^j} & \\ & & \dfrac{2\varepsilon^i q^i}{\varepsilon^j q^i + \varepsilon^i q^j} \end{pmatrix}. \tag{2.59}$$

$\boldsymbol{F}^{i\to j}$ relates the electric field of incident and transmitted lights in Fig. 2.6 by $E^t\hat{\boldsymbol{e}}_j = \boldsymbol{F}^{i\to j} \cdot E^0\hat{\boldsymbol{e}}_i$. For P polarization, the electric field amplitudes are

$$E^0\hat{\boldsymbol{e}}_i = E^0 \begin{pmatrix} \cos\theta_i \\ 0 \\ \sin\theta_i \end{pmatrix}, \qquad E^t\hat{\boldsymbol{e}}_j = E^t \begin{pmatrix} \cos\theta_j \\ 0 \\ \sin\theta_j \end{pmatrix} = t_P E^0 \begin{pmatrix} \cos\theta_j \\ 0 \\ \sin\theta_j \end{pmatrix},$$

and accordingly

$$F_{xx}^{i\to j} = t_P \frac{\cos\theta_j}{\cos\theta_i}, \qquad F_{zz}^{i\to j} = t_P \frac{\sin\theta_j}{\sin\theta_i}. \tag{2.60}$$

For S polarization, the electric field amplitudes are

$$E^0\hat{\boldsymbol{e}}_i = E^0\begin{pmatrix}0\\1\\0\end{pmatrix}, \qquad E^t\hat{\boldsymbol{e}}_j = E^t\begin{pmatrix}0\\1\\0\end{pmatrix} = t_S E^0\begin{pmatrix}0\\1\\0\end{pmatrix},$$

and accordingly

$$F_{yy}^{i\to j} = t_S. \tag{2.61}$$

$F_{xx}^{i\to j}$, $F_{yy}^{i\to j}$, and $F_{zz}^{i\to j}$ in Eqs. (2.60) and (2.61) lead to Eq. (2.59), by using the following notations,

$$q^i = \frac{n_i\omega}{c}\cos\theta_i, \quad q^j = \frac{n_j\omega}{c}\cos\theta_j, \quad \varepsilon^i = n_i^2, \quad \varepsilon^j = n_j^2.$$

$q^{i\,(j)}$ denotes the z component of the wavevector in the medium i (j), and $\varepsilon^{i\,(j)}$ is the dielectric constant in the medium i (j).

In the following, we examine the Fresnel coefficients at glass-air interface for example, where we suppose $n_i = 1.49$ for the glass (i) and $n_j = 1$ for the air (j). The critical angle of total reflection is derived from $\sin\theta_c = n_j/n_i = 1/1.49$ and hence $\theta_c = 42.2°$. The Fresnel coefficients r_P, r_S, t_P, t_S in Eqs. (2.57) and (2.58) are plotted as a function of θ_i in Fig. 2.7. We notice that these coefficients become

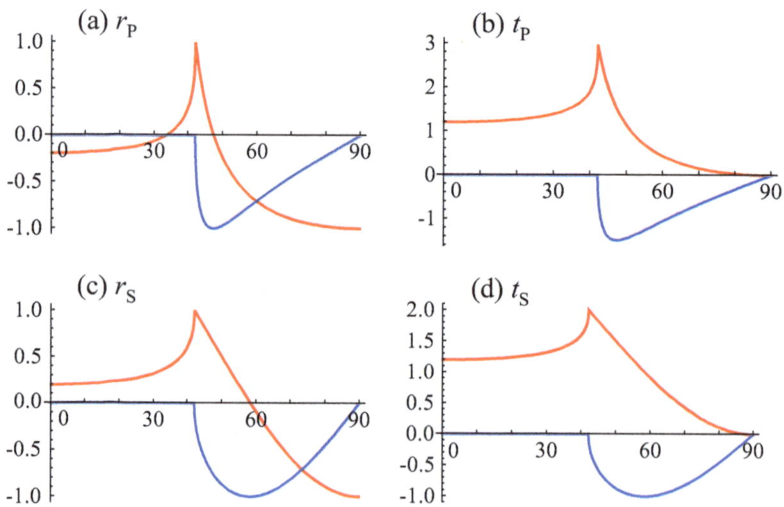

Fig. 2.7 The Fresnel coefficients (**a**) r_P, (**b**) t_P, (**c**) r_S, (**d**) t_S for the glass ($n_i = 1.49$) – air ($n_j = 1$) interface in Eqs. (2.57) and (2.58). The abscissas denote the incident angle θ_i in degrees. Red lines stand for the real part while the blue lines for the imaginary part

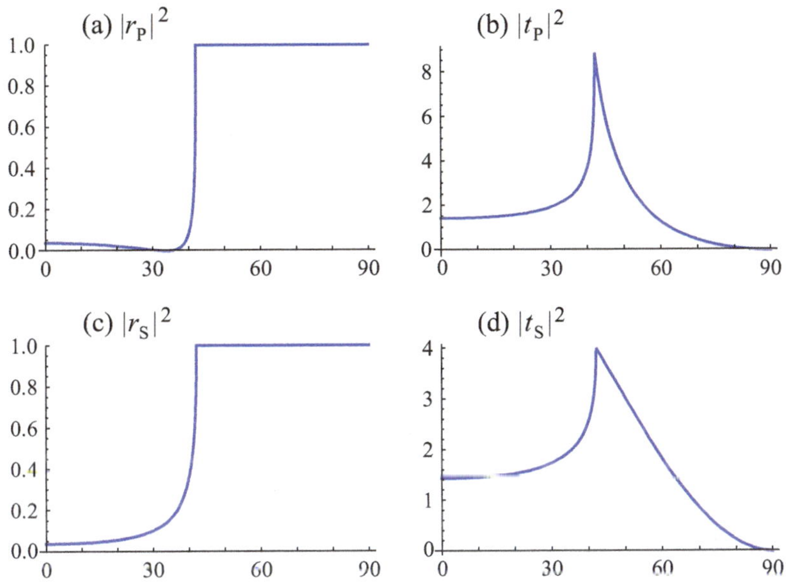

Fig. 2.8 Square of the Fresnel coefficients (**a**) $|r_P|^2$, (**b**) $|t_P|^2$, (**c**) $|r_S|^2$, (**d**) $|t_S|^2$ for the glass-air interface. Other conditions are same as in Fig. 2.7

complex at $\theta_i > \theta_c$ where the total reflection takes place. The square of the Fresnel coefficients, which correspond to the intensity ratio, are also plotted in Fig. 2.8 in the similar manner. The intensity ratio of reflected waves is unity in the condition of total reflection, $|r_P|^2 = |r_S|^2 = 1$ at $\theta_i > \theta_c$, while the transmitted waves attenuate. It is remarkable in Fig. 2.8 that the intensity of transmitted waves $|t_P|^2$, $|t_S|^2$ exhibits a sharp maximum at $\theta_i = \theta_c$ in both P and S polarizations. This enhancement near the critical angle is utilized to augment the sensitivity of interfacial spectroscopy.

A.3 Fresnel Factors for Three-Layer Model

Here we derive the Fresnel factor $\boldsymbol{F}^{i \to j}$ in Eq. (2.18) for the three-layer model.

A.3.1 Fields and Wavevectors

For the three layer model, the incident light with the wave vector $\boldsymbol{k}_I$ and the frequency ω gives rise to four other waves, $\boldsymbol{k}_R$, $\boldsymbol{k}_{T'}$, $\boldsymbol{k}_{R'}$ and $\boldsymbol{k}_T$, as illustrated in Fig. 2.9.

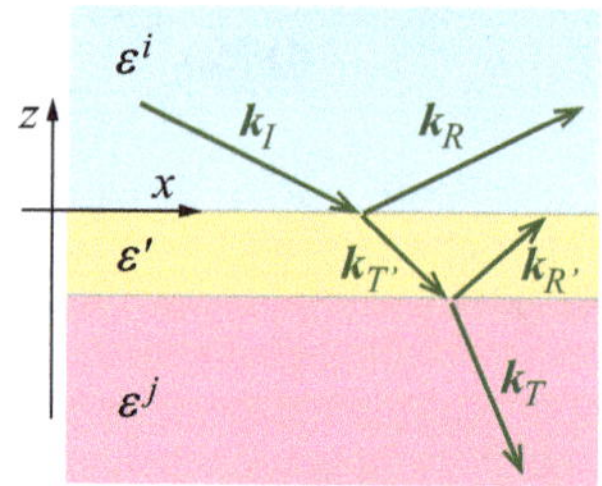

Fig. 2.9 Light waves associated to the incident light $\boldsymbol{k}_I$ in the three layer model

Each electromagnetic field is presented as

$$\boldsymbol{E}_X(\boldsymbol{r}, \omega, t) = \boldsymbol{E}_X \exp(i\boldsymbol{k}_X \cdot \boldsymbol{r} - i\omega t) + c.c. \tag{2.62}$$

$$\begin{aligned} \boldsymbol{H}_X(r, \omega, t) &= \frac{1}{K}(\boldsymbol{k}_X \times \boldsymbol{E}_X) \exp(i\boldsymbol{k}_X \cdot \boldsymbol{r} - i\omega t) + c.c \\ &= \boldsymbol{H}_X \exp(i\boldsymbol{k}_X \cdot \boldsymbol{r} - i\omega t) + c.c. \end{aligned} \tag{2.63}$$

where $X = I, R, T', R'$, or T. $K = \omega/c$ is the angular wavenumber of the light in vacuo. The wavevector $\boldsymbol{k}_X$ and the coefficients $\boldsymbol{E}_X$, $\boldsymbol{H}_X$ for each wave X are summarized as follows.

$$\begin{array}{lccc}
 & \boldsymbol{k}_X & \boldsymbol{E}_X & \boldsymbol{H}_X \\
I: & \boldsymbol{k}_I = \begin{pmatrix} k_x \\ 0 \\ -q^i \end{pmatrix}, & \boldsymbol{E}_I = \begin{pmatrix} \dfrac{q^i}{n^i K} I_P \\ I_S \\ \dfrac{k_x}{n^i K} I_P \end{pmatrix}, & \boldsymbol{H}_I = \dfrac{1}{K} \begin{pmatrix} q^i I_S \\ -n^i K I_P \\ k_x I_S \end{pmatrix}, \\
R: & \boldsymbol{k}_R = \begin{pmatrix} k_x \\ 0 \\ q^i \end{pmatrix}, & \boldsymbol{E}_R = \begin{pmatrix} \dfrac{q^i}{n^i K} R_P \\ R_S \\ -\dfrac{k_x}{n^i K} R_P \end{pmatrix}, & \boldsymbol{H}_R = \dfrac{1}{K} \begin{pmatrix} -q^i R_S \\ n^i K R_P \\ k_x R_S \end{pmatrix}, \\
T': & \boldsymbol{k}_{T'} = \begin{pmatrix} k_x \\ 0 \\ -q' \end{pmatrix}, & \boldsymbol{E}_{T'} = \begin{pmatrix} \dfrac{q'}{n' K} T_P{}' \\ T_S{}' \\ \dfrac{k_x}{n' K} T_P{}' \end{pmatrix}, & \boldsymbol{H}_{T'} = \dfrac{1}{K} \begin{pmatrix} q' T_S{}' \\ -n' K T_P{}' \\ k_x T_S{}' \end{pmatrix}, \\
R': & \boldsymbol{k}_{R'} = \begin{pmatrix} k_x \\ 0 \\ q' \end{pmatrix}, & \boldsymbol{E}_{R'} = \begin{pmatrix} \dfrac{q'}{n' K} R_P{}' \\ R_S{}' \\ -\dfrac{k_x}{n' K} R_P{}' \end{pmatrix}, & \boldsymbol{H}_{R'} = \dfrac{1}{K} \begin{pmatrix} -q' R_S{}' \\ n' K R_P{}' \\ k_x R_S{}' \end{pmatrix},
\end{array}$$

$$T: \quad \boldsymbol{k}_T = \begin{pmatrix} k_x \\ 0 \\ -q^j \end{pmatrix}, \quad \boldsymbol{E}_T = \begin{pmatrix} \dfrac{q^j}{n^j K} T_P \\ T_S \\ \dfrac{k_x}{n^j K} T_P \end{pmatrix}, \quad \boldsymbol{H}_T = \frac{1}{K} \begin{pmatrix} q^j T_S \\ -n^j K T_P \\ k_x T_S \end{pmatrix}, \tag{2.64}$$

where

$$\begin{cases} {k_x}^2 + (q^i)^2 = (n^i K)^2 = \varepsilon^i K^2 \\ {k_x}^2 + (q')^2 = (n' K)^2 = \varepsilon' K^2 \\ {k_x}^2 + (q^j)^2 = (n^j K)^2 = \varepsilon^j K^2 \end{cases} \tag{2.65}$$

and k_x is common among the five waves due to the phase matching condition. In what follows, we determine the relations among the unknown variables, I_P, I_S, R_P, R_S, $T_P{}'$, $T_S{}'$, $R_P{}'$, $R_S{}'$, T_P and T_S in Eq. (2.64) from the boundary conditions. The suffixes P and S signify the variables related to P and S polarizations, respectively.

A.3.2 Boundary Conditions

Here we consider three kinds of boundary conditions among the medium i, interface and medium j for the three-layer model in Fig. 2.9: between (1) i and interface, (2) interface and j, and (3) i and j. At each boundary, the continuity conditions for E_t, D_z, H_t and B_z are formulated with the help of Eq. (2.64).

(1) Medium i and Interface

$\Delta E_t = 0$:

$$\left[E_{I,x} + E_{R,x} = E_{T',x} + E_{R',x}\right]: \quad \frac{q^i}{n^i K}(I_P + R_P) = \frac{q'}{n' K}(T_P{}' + R_P{}') \tag{2.66}$$

$$\left[E_{I,y} + E_{R,y} = E_{T',y} + E_{R',y}\right]: \quad I_S + R_S = T_S{}' + R_S{}' \tag{2.67}$$

$\Delta D_z = 0$:

$$\left[\varepsilon^i (E_{I,z} + E_{R,z}) = \varepsilon' (E_{T',z} + E_{R',z})\right]: \quad \frac{\varepsilon^i k_x}{n^i K}(I_P - R_P) = \frac{\varepsilon' k_x}{n' K}(T_P{}' - R_P{}') \tag{2.68}$$

$\Delta H_t = 0$:

$$\left[H_{I,x} + H_{R,x} = H_{T',x} + H_{R',x}\right]: \quad \frac{1}{K} q^i (I_S - R_S) = \frac{1}{K} q' (T_S{}' - R_S{}') \tag{2.69}$$

$$\left[H_{I,y} + H_{R,y} = H_{T',y} + H_{R',y}\right]: \quad n^i(-I_P + R_P) = n'(-T_P{}' + R_P{}') \tag{2.70}$$

$\Delta B_z = \Delta(\mu H_z) = 0$: where $\mu = 1$ is assumed in the nonmagnetic media.

$$\left[H_{I,z} + H_{R,z} = H_{T',z} + H_{R',z}\right]: \quad \frac{k_x}{K}(I_S + R_S) = \frac{k_x}{K}(T_S{}' + R_S{}') \tag{2.71}$$

In the above six boundary conditions, we notice that Eqs. (2.67) and (2.71), and Eqs. (2.68) and (2.70) are equivalent and thus there exist four independent conditions. These four independent conditions are summarized as follows:

$$\text{From (2.66)}: \quad \frac{q^i}{n^i}(I_P + R_P) = \frac{q'}{n'}(T_P{}' + R_P{}') \tag{2.72}$$

$$\text{From (2.67), (2.71)}: \quad I_S + R_S = T_S{}' + R_S{}' \tag{2.73}$$

$$\text{From (2.69)}: \quad q^i (I_S - R_S) = q' (T_S{}' - R_S{}') \tag{2.74}$$

$$\text{From (2.68), (2.70)}: \quad n^i(-I_P + R_P) = n'(-T_P{}' + R_P{}') \tag{2.75}$$

(2) Interface and Medium j

$\Delta E_t = 0$:

$$\left[E_{T',x} + E_{R',x} = E_{T,x}\right]: \quad \frac{q'}{n'K}(T_P{}' + R_P{}') = \frac{q^j}{n^j K} T_P \tag{2.76}$$

$$\left[E_{T',y} + E_{R',y} = E_{T,y}\right]: \quad T_S{}' + R_S{}' = T_S \tag{2.77}$$

$\Delta D_z = 0$:

$$\left[\varepsilon'(E_{T',z} + E_{R',z}) = \varepsilon^j E_{T,z}\right]: \quad \frac{\varepsilon' k_x}{n'K}(T_P{}' - R_P{}') = \frac{\varepsilon^j k_x}{n^j K} T_P \tag{2.78}$$

$\Delta H_t = 0$:

$$\left[H_{T',x} + H_{R',x} = H_{T,x}\right]: \quad \frac{q'}{K}(T_S{}' - R_S{}') = \frac{q^j}{K} T_S \tag{2.79}$$

$$\left[H_{T',y} + H_{R',y} = H_{T,y}\right]: \quad n'(-T_P{}' + R_P{}') = -n^j T_P \tag{2.80}$$

$\Delta B_z = \Delta(\mu H_z) = 0$:

$$\left[H_{T',z} + H_{R',z} = H_{T,z}\right]: \quad \frac{k_x}{K}(T_S{}' + R_S{}') = \frac{k_x}{K} T_S \tag{2.81}$$

In the above conditions of Eqs. (2.76), (2.77), (2.78), (2.79), (2.80), and (2.81), we have assumed that the phase difference over the thickness of the surface layer is negligible, since the thickness is much smaller than the wavelength. In the above conditions, Eqs. (2.77) and (2.81), and Eqs. (2.78) and (2.80) are equivalent. The four independent conditions are

$$\text{From (2.76)}: \quad \frac{q'}{n'}(T_P{}' + R_P{}') = \frac{q^j}{n^j}T_P \tag{2.82}$$

$$\text{From (2.77), (2.81)}: \quad T_S{}' + R_S{}' = T_S \tag{2.83}$$

$$\text{From (2.79)}: \quad q'(T_S{}' - R_S{}') = q^j T_S \tag{2.84}$$

$$\text{From (2.78), (2.80)}: \quad n'(-T_P{}' + R_P{}') = -n^j T_P \tag{2.85}$$

(3) Media *i* and *j*

$\Delta E_t = 0:$

$$\left[E_{I,x} + E_{R,x} = E_{T,x}\right]: \quad \frac{q^i}{n^i K}(I_P + R_P) = \frac{q^j}{n^j K}T_P \tag{2.86}$$

$$\left[E_{I,y} + E_{R,y} = E_{T,y}\right]: \quad I_S + R_S = T_S \tag{2.87}$$

$\Delta D_z = 0:$

$$\left[\varepsilon^i(E_{I,z} + E_{R,z}) = \varepsilon^j E_{T,z}\right]: \quad \frac{\varepsilon^i k_x}{n^i K}(I_P - R_P) = \frac{\varepsilon^j k_x}{n^j K}T_P \tag{2.88}$$

$\Delta H_t = 0:$

$$\left[H_{I,x} + H_{R,x} = H_{T,x}\right]: \quad \frac{q^i}{K}(I_S - R_S) = \frac{q^j}{K}T_S \tag{2.89}$$

$$\left[H_{I,y} + H_{R,y} = H_{T,y}\right]: \quad n^i(-I_P + R_P) = -n^j T_P \tag{2.90}$$

$\Delta B_z = \Delta(\mu H_z) = 0:$

$$\left[H_{I,z} + H_{R,z} = H_{T,z}\right]: \quad \frac{k_x}{K}(I_S + R_S) = \frac{k_x}{K}T_S \tag{2.91}$$

The boundary conditions of Eqs. (2.86), (2.87), (2.88), (2.89), (2.90), and (2.91) are identical to those for the two-layer model. Therefore, the Fresnel factors to connect the two bulk layers are same as those derived with the two-layer model, irrespective of whether the thin interfacial layer is present. Among the above six conditions, Eqs. (2.87) and (2.91) and Eqs. (2.88) and (2.90) are identical. Therefore, four independent conditions are summarized as

From (2.86) : $$\frac{q^i}{n^i}(I_P + R_P) = \frac{q^j}{n^j} T_P \tag{2.92}$$

From (2.87), (2.91) : $$I_S + R_S = T_S \tag{2.93}$$

From (2.89) : $$q^i (I_S - R_S) = q^j T_S \tag{2.94}$$

From (2.88), (2.90) : $$n^i(-I_P + R_P) = -n^j T_P \tag{2.95}$$

Actually Eqs. (2.92), (2.93), (2.94), (2.95) are derived from Eqs. (2.72), (2.73), (2.74), (2.75) and (2.82), (2.83), (2.84), (2.85). There are a total of eight independent boundary conditions in the three-layer model, which are two sets among Eqs. (2.72), (2.73), (2.74), (2.75), (2.82), (2.83), (2.84), (2.85) and (2.92), (2.93), (2.94), (2.95). We choose Eqs. (2.72), (2.73), (2.74), (2.75) and (2.92), (2.93), (2.94), (2.95) in the following. These conditions are utilized to determine the relations among I_P, I_S, R_P, R_S, $T_P{}'$, $T_S{}'$, $R_P{}'$, $R_S{}'$, T_P and T_S.

A.3.3 Solution of Boundary Equations

Using Eqs. (2.92), (2.93), (2.94), (2.95), following two relations are readily derived:

$$R_P = \frac{\varepsilon^i q^j - \varepsilon^j q^i}{\varepsilon^i q^j + \varepsilon^j q^i} I_P, \tag{2.96}$$

$$R_S = \frac{q^i - q^j}{q^i + q^j} I_S. \tag{2.97}$$

P component By using Eq. (2.96), Eqs. (2.72) and (2.75) are converted to

$$T_P{}' + R_P{}' = \frac{n' q^i}{n^i q'}\left(1 + \frac{\varepsilon^i q^j - \varepsilon^j q^i}{\varepsilon^i q^j + \varepsilon^j q^i}\right) I_P = \frac{2n' n^i q^i q^j}{q'(\varepsilon^i q^j + \varepsilon^j q^i)} I_P, \tag{2.98}$$

$$T_P{}' - R_P{}' = \frac{n^i}{n'}\left(1 - \frac{\varepsilon^i q^j - \varepsilon^j q^i}{\varepsilon^i q^j + \varepsilon^j q^i}\right) I_P = \frac{n^i}{n'} \frac{2\varepsilon^j q^i}{\varepsilon^i q^j + \varepsilon^j q^i} I_P. \tag{2.99}$$

Equations (2.98) and (2.99) allow us to relate the P component of the electric field in the interface layer ($T_P{}'$, $R_P{}'$) to the P component of the incident field (I_P). Using these equations along with Eqs. (2.64), the following relations for the x and z components are derived:

$$E_{T',x} + E_{R',x} = \frac{q'}{n'K}(T_P{}' + R_P{}') = \frac{q'}{n'K} \frac{2n' n^i q^i q^j}{q'(\varepsilon^i q^j + \varepsilon^j q^i)} I_P$$

$$= \frac{q'}{n'K} \frac{2n'n^i q^i q^j}{q'(\varepsilon^i q^j + \varepsilon^j q^i)} \frac{n^i K}{q^i} E_{I,x} = \frac{2\varepsilon^i q^j}{\varepsilon^i q^j + \varepsilon^j q^i} E_{I,x} = F_{xx}^{i\to j} \cdot E_{I,x}, \tag{2.100}$$

$$E_{T'}(z) + E_{R'}(z) = \frac{k_x}{n'K}(T_P{}' - R_P{}') = \frac{k_x}{n'K} \frac{n^i}{n'} \frac{2\varepsilon^j q^i}{(\varepsilon^i q^j + \varepsilon^j q^i)} I_P$$

$$= \frac{k_x}{n'K} \frac{n^i}{n'} \frac{2\varepsilon^j q^i}{(\varepsilon^i q^j + \varepsilon^j q^i)} \frac{n^i K}{k_x} E_{I,z} = \frac{\varepsilon^i}{\varepsilon'} \frac{2\varepsilon^j q^i}{(\varepsilon^i q^j + \varepsilon^j q^i)} E_{I,z} = F_{zz}^{i\to j} \cdot E_{I,z}. \tag{2.101}$$

S component The y component is derived from the S polarization in the same way. Using Eqs. (2.64), (2.73), and (2.97),

$$E_{T',y} + E_{R',y} = T_S{}' + R_S{}' = I_S + R_S$$

$$= \left(1 + \frac{q^i - q^j}{q^i + q^j}\right) I_S = \frac{2q^i}{q^i + q^j} E_{I,y} = F_{yy}^{i\to j} \cdot E_{I,y}. \tag{2.102}$$

Fresnel factor The above equations (2.100), (2.101), and (2.102) derive the Fresnel factor $\boldsymbol{F}^{i\to j}$,

$$\boldsymbol{F}^{i\to j} = \begin{pmatrix} \dfrac{2\varepsilon^i q^j}{\varepsilon^j q^i + \varepsilon^i q^j} & & \\ & \dfrac{2q^i}{q^i + q^j} & \\ & & \dfrac{2\varepsilon^i \varepsilon^j}{\varepsilon'} \dfrac{q^i}{\varepsilon^j q^i + \varepsilon^i q^j} \end{pmatrix} \tag{2.18}$$

The above derivation clarifies that the Fresnel factor $\boldsymbol{F}^{i\to j}$ in Eq. (2.18) for the three-layer model describes the total $(T' + R')$ electric field inside the thin interfacial layer.

Bibliography

1. Born M, Wolf E (1999) Principles of optics, 7th edn. Cambridge University Press, Cambridge
2. Heinz TF (1991) Second-order nonlinear optical effects at surfaces and interfaces. In: Ponath H-E, Stegeman GI (eds) Nonlinear surface electromagnetic phenomena. Elsevier, Amsterdam, pp 353–416
3. Jackson JD (1998) Classical electrodynamics. Wiley, New York
4. Nihonyanagi S, Mondal JA, Yamaguchi S, Tahara T (2013) Structure and dynamics of interfacial water studied by heterodyne-detected vibrational sum-frequency generation. Annu Rev Phys Chem 64:579–603

5. Shen YR (1994) Surface spectroscopy by nonlinear optics In: Hänsch T, Inguscio M (eds) Frontiers in laser spectroscopy: Proceedings of international school of physics, vol CXX. North Holland, Amsterdam, pp 139–165
6. Shen YR (2013) Phase-sensitive sum-frequency spectroscopy. Annu Rev Phys Chem 64:129–150

Chapter 3
Microscopic Expressions of Nonlinear Polarization

Abstract In the SFG processes described in the preceding Chap. 2, material properties were treated as given parameters. This chapter formulates the material properties relevant to the SFG spectroscopy from a microscopic viewpoint. The most important property in the SFG spectra is the frequency-dependent second-order nonlinear susceptibility tensor, $\boldsymbol{\chi}^{(2)}(\Omega, \omega_1, \omega_2)$. We derive $\boldsymbol{\chi}^{(2)}$ by the quantum mechanical perturbation theory of light-matter interactions. In the vibrational SFG spectroscopy, vibrational resonance plays a critical role in the nonlinear response of polarization. We further discuss some basic features of $\boldsymbol{\chi}^{(2)}$, including the relation of its tensor elements to the light polarizations and molecular orientation.

Keywords $\boldsymbol{\chi}^{(2)}$ · Perturbation theory · Vibrational resonance · Molecular orientation

3.1 Density Matrix

In the present discussion on the light-matter interactions, we employ the semiclassical description that the material is treated by quantum mechanics while the light is treated as classical electromagnetic waves which perturb the material states.[1] The quantum states of materials in condensed phase are expressed preferably with the density matrix rather than the wavefunction, for the reasons described below. Before discussing the nonlinear susceptibility, we briefly introduce the density matrix and summarize its merits [1, 13].

[1]The semiclassical theory suffices for properly describing susceptibilities of materials, including the nonlinear ones, as the light fields provide perturbation on the materials states [12].

A. Morita, *Theory of Sum Frequency Generation Spectroscopy*,
Lecture Notes in Chemistry 97, https://doi.org/10.1007/978-981-13-1607-4_3

3.1.1 Definition

First we begin with a simple case that the quantum state of material is expressed by a wavefunction $\psi(r,t)$, which is a proper superposition of energy eigenstates, $\{\phi_n(r)\exp(-i\mathcal{E}_n t/\hbar)\}$;

$$\psi(r,t) = \sum_n \left[c'_n(t) \exp\left(\frac{-i\mathcal{E}_n t}{\hbar} \right) \right] \phi_n(r) = \sum_n c_n(t)\phi_n(r). \tag{3.1}$$

We suppose that the set of eigenstates $\{\phi_n\}$ are orthonormal and the wavefunctions ψ is normalized. The normalization condition for ψ is

$$\sum_n |c_n(t)|^2 = 1. \tag{3.2}$$

The expectation value of an arbitrary physical quantity A at this state ψ is

$$\langle A(t) \rangle = \langle \psi(t)|A|\psi(t) \rangle = \sum_n \sum_m c_n(t)^* c_m(t) \langle n|A|m \rangle . \tag{3.3}$$

Now we introduce the density matrix $\rho(t)$. Its matrix element ρ_{mn} is defined to be

$$\rho_{mn}(t) = c_n(t)^* c_m(t) \tag{3.4}$$

on the basis of the energy eigenfunctions $\{\phi_n\}$. Note the order of suffixes m, n in ρ_{mn} and its Hermitian character, $\rho_{mn} = \rho^*_{nm}$. According to Eqs. (3.3) and (3.4), the expectation value $\langle A(t) \rangle$ could be written using the density matrix ρ by

$$\langle A(t) \rangle = \sum_n \sum_m \rho_{mn} A_{nm} = \mathrm{Tr}\,[\rho A]. \tag{3.5}$$

The diagonal element $\rho_{nn} = |c_n|^2$ means the probability to find the state n. The normalization condition (3.2) is written by

$$\sum_n \rho_{nn} = \mathrm{Tr}\,[\rho] = 1. \tag{3.6}$$

On the other hand, the off-diagonal element ρ_{mn} $(m \neq n)$ represents the coherence between the states m and n, as we discuss later (see Appendix A.1).

The time development of the density matrix ρ is defined from the Schrödinger equation for ψ, $i\hbar \frac{\partial \psi}{\partial t} = \hat{H}\psi$. Substituting the expression of ψ in Eq. (3.1) into this Schrödinger equation, we get the equations for the coefficients $\{c_m(t)\}$,

$$i\hbar \frac{dc_m(t)}{dt} = \sum_n H_{mn} c_n(t). \tag{3.7}$$

Therefore, the time development of $\rho(t)$ is defined by

$$
\begin{aligned}
i\hbar\frac{d\rho_{mn}(t)}{dt} &= i\hbar\left(\frac{dc_n(t)^*}{dt}c_m(t) + c_n(t)^*\frac{dc_m(t)}{dt}\right) \\
&= -\sum_l H_{nl}^* c_l(t)^* c_m(t) + \sum_l c_n(t)^* H_{ml} c_l(t) \\
&= -\sum_l H_{ln}\rho_{ml}(t) + \sum_l H_{ml}\rho_{ln}(t) = [H\rho(t) - \rho(t)H]_{mn} \, . \qquad (3.8)
\end{aligned}
$$

Equation (3.8) is called the Liouville equation.

3.1.2 Features and Advantages

The density matrix ρ offers an alternative means for the wavefunction ψ to represent the quantum state of the system. The merits of employing the density matrix over the wavefunction are manifested when considering (i) statistical ensemble of quantum states, and/or (ii) quantum states of a partial system embedded in a large system, such as a solute molecule in solution. We discuss these two situations in the following.

(i) Ensemble of States

Most of experimental measurements in molecular science usually observe an ensemble of molecules, except for special cases of single molecule measurements. Therefore, ensemble average tends to be involved and plays an important role in interpreting the ordinary experiments from a microscopic viewpoint.

Let us consider an ensemble of quantum states consisting of ψ^j, where the probability of finding the state j is P^j. (Each ψ^j is supposed to be normalized, but need not be orthogonal each other.) In such a situation, the expectation value of a physical quantity A in Eq. (3.3) is given by taking the ensemble average over the states j,

$$
\langle A(t)\rangle \longrightarrow \sum_j^{\text{ensemble}} P^j \left\langle \psi^j(t)|A|\psi^j(t)\right\rangle = \sum_j\sum_n\sum_m P^j c_n^j(t)^* c_m^j(t) \langle n|A|m\rangle \, .
$$

Hereafter the ensemble average is denoted with the overbar. The above formula is accordingly expressed by

$$
\langle A(t)\rangle \longrightarrow \overline{\langle A(t)\rangle} = \overline{\langle\psi(t)|A|\psi(t)\rangle} = \sum_n\sum_m \overline{c_n(t)^* c_m(t)} \langle n|A|m\rangle \, . \qquad (3.9)
$$

We notice that $\overline{\langle A(t)\rangle}$ involves two kinds of average manipulations, the quantum mechanical average for the wavefunction ψ^j and the statistical average over the ensemble j. The definition of the density matrix ρ in Eq. (3.4) is accordingly extended to take account of the statistical ensemble to

$$\rho_{mn}(t) = \sum_{j}^{\text{ensemble}} P^j c_n^j(t)^* c_m^j(t) = \overline{c_n(t)^* c_m(t)}. \tag{3.10}$$

Such ensemble of states is called "mixed state", whereas a state represented with a single wavefunction is called "pure state". The density matrix ρ in Eq. (3.4) account for a pure state, while that in Eq. (3.10) for a mixed state.

Then we extend the formulas of ρ in the preceding subsection to the mixed states. However, it is rather surprising to confirm that the fundamental formulas of ρ are unchanged by the extension of ρ. The following formulas are valid for the mixed states as well as the pure states,

- Eq. (3.5): expectation value $\langle A(t)\rangle$ or $\overline{\langle A(t)\rangle}$,
- Eq. (3.6): normalization condition,
- Eq. (3.8): time development (Liouville equation),

once we replace ρ in Eq. (3.4) with that in Eq. (3.10).

[Problem 3.1] Confirm that Eqs. (3.5), (3.6) and (3.8) are valid for the mixed states as well.

The density matrix allows us to treat pure states and mixed states with the common formalism. This is a remarkable advantage of the density matrix in treating ensemble of states. We also note that it is possible to distinguish whether a given density matrix ρ refers to a mixed state or a pure state by the following criterion,

$$\begin{cases} \mathrm{Tr}\,[\rho^2] = 1 & \text{for pure state,} \\ \mathrm{Tr}\,[\rho^2] < 1 & \text{for mixed state.} \end{cases} \tag{3.11}$$

[Problem 3.2] Prove the criterion Eq. (3.11) to distinguish the pure state and the mixed state. Recall the normalization condition of states and the Schwarz inequality for inner products.

On the other hand, the wavefunction is not as convenient as the density matrix to treat the mixed state, since the mixed state is not represented with a single wavefunction. One cannot take the statistical average for the wavefunctions themselves, since an "ensemble averaged state" $\overline{\psi} = \sum_j P^j \psi^j$ would be a completely different state! For example, the expectation value would be different as follows,

$$\langle\overline{\psi}|A|\overline{\psi}\rangle = \sum_{j,k}^{\text{ensemble}} \sum_{n,m} P^j P^k c_n^j(t)^* c_m^k(t) \langle n|A|m\rangle$$

$$\neq \overline{\langle A\rangle} = \sum_{j}^{\text{ensemble}} \sum_{n,m} P^j c_n^j(t)^* c_m^j(t) \langle n|A|m\rangle .$$

If the mixed state is treated with the wavefunctions ψ^j, one should have to treat the properties (expectation value, time development, etc.) at each wavefunction ψ^j and then to explicitly take the statistical average over the wavefunctions.

(ii) Partial System in Bath

The density matrix is also advantageous in describing a partial system buried in a large system when our main interest is focused on the partial system. Such a situation is quite common in chemistry in condensed phase, e.g. a solute molecule in solution. The wavefunction is hard to account for such situation. This is because treating the wavefunction for the entire system may be tedious, and the wavefunction for a partial system is not well defined when it interacts with the rest (called "bath").

To illustrate such situation, we assume the collective coordinates of a partial system and the bath to be $\boldsymbol{r}$ and $\boldsymbol{R}$, respectively. The partial system has a set of eigenstates $\{\phi_n(\boldsymbol{r})\}$ of the Hamiltonian for the partial system itself. Then the wavefunction of the whole system Ψ can be expressed in principle with the superposition of $\{\phi_n(\boldsymbol{r})\}$ in the same way as Eq. (3.1),

$$\Psi(\boldsymbol{r}, \boldsymbol{R}, t) = \sum_n C_n(\boldsymbol{R}, t)\phi_n(\boldsymbol{r}). \tag{3.12}$$

Treating $\Psi(\boldsymbol{r}, \boldsymbol{R}, t)$ in Eq. (3.12) may be tedious, as the coefficients C_n explicitly depend on the bath coordinates $\boldsymbol{R}$. However, in case that we are interested in a physical quantity for the partial system A, its expectation value is given by

$$\langle A(t)\rangle = \langle\Psi|A|\Psi\rangle = \sum_n \sum_m \langle C_n(\boldsymbol{R}, t)|C_m(\boldsymbol{R}, t)\rangle \langle n|A|m\rangle ,$$

where A is supposed to be a function of the system coordinates $\boldsymbol{r}$. We could take the statistical average over ensemble to get

$$\overline{\langle A(t)\rangle} = \overline{\langle\Psi|A|\Psi\rangle} = \sum_n \sum_m \overline{\langle C_n(\boldsymbol{R}, t)|C_m(\boldsymbol{R}, t)\rangle} \langle n|A|m\rangle . \tag{3.13}$$

Accordingly, we can define the density matrix for the partial system to be

$$\rho_{mn}(t) = \overline{\langle C_n(\boldsymbol{R}, t)|C_m(\boldsymbol{R}, t)\rangle}, \tag{3.14}$$

and thereby Eq. (3.13) becomes equivalent to Eq. (3.5) in terms of the expression of the expectation value. We note that $\rho_{mn}(t)$ in Eq. (3.14) does not include the explicit bath coordinates $\boldsymbol{R}$ any more, as they are averaged out in the right hand side. Equations (3.13) and (3.14) provide a closed description of the partial system within the coordinates $\boldsymbol{r}$. The density matrix for the partial system in Eq. (3.14) is sufficient to describe the physical quantity A for the partial system, with obviating details of the bath wavefunctions C_n.

A remaining issue is to determine the time development of the density matrix thus defined. The time development of the partial system could not be represented by the Liouville equation (3.8) with the Hamiltonian for the partial system H, because the time development is influenced by the interaction with the bath. One could describe the time development using the Hamiltonian and density matrix (or wavefunction) for the whole system, though such a way is not compatible with the merit of using the density matrix formalism. Here we avoid the tedious route to treat the whole system, but rather focus on the time development of the partial system itself in a closed, phenomenological way.

When the partial system interacts with the bath, an outstanding feature arises that the partial system approaches thermal equilibrium. This is a quite general tendency, irrespective of the details of the interaction with the bath. The thermal equilibrium is represented as a mixed state, where the probability of state n is proportional to the Boltzmann factor $\exp(-\mathcal{E}_n/k_B T)$ and the off-diagonal correlation vanishes by random fluctuation. Therefore, the thermal equilibrium is specified by the following steady-state density matrix,

$$\rho_{mn}^{\mathrm{eq}} = \frac{\exp\left(-\mathcal{E}_n/k_B T\right)}{\sum\limits_n \exp\left(-\mathcal{E}_n/k_B T\right)}\,\delta_{mn}, \tag{3.15}$$

where k_B denotes the Boltzmann constant and T the absolute temperature. To incorporate this relaxation in a simple and phenomenological manner, the Liouville equation of (3.8) is extended to

$$i\hbar\frac{d\rho_{mn}(t)}{dt} = [H\rho(t) - \rho(t)H]_{mn} - i\hbar\Gamma_{mn}\left(\rho_{mn}(t) - \rho_{mn}^{\mathrm{eq}}\right), \tag{3.16}$$

where the second term in the right hand side is added to account for the relaxation to the thermal equilibrium. The parameters for relaxation rates Γ_{mn} are real and positive, and also symmetric with respect to the suffixes m, n, $\Gamma_{mn} = \Gamma_{nm}$, due to the Hermitian condition for the density matrix ($\rho_{nm}^* = \rho_{mn}$). The following discussion on the susceptibilities will be derived from the extended Liouville equation of Eq. (3.16).

Finally, we note in passing the validity of Eqs. (3.16) and (3.8). The states m, n in Eq. (3.16) refer to a system in a bath, and the damping parameter Γ_{mn} stems from perturbation from the bath. If the states m, n refer to the entire system (system + bath), the Liouville equation of Eq. (3.16) becomes Eq. (3.8) with obviating Γ_{mn}.

The states for the entire system (system + bath) are hard to treat in practical applications. However, in some occasions, the original formula of Eq. (3.8) is more convenient than Eq. (3.16) to deal with formal properties of the susceptibilities, since the discussion becomes free from the phenomenological parameter. We will encounter such occasions in Sects. 4.3 and 7.4.

3.2 Perturbation Forms of Susceptibilities

3.2.1 *Perturbation Expansion of Density Matrix*

Based on the density matrix introduced above, let us discuss the quantum formulas of susceptibilities [1, 13, 15]. We begin with the general perturbation theory of materials by lights, and apply the formulas to the first-order susceptibility. In what follows, the material in question is arbitrarily chosen, either a single molecule or the interface system of condensed matter, since the following formulas are commonly applicable.

We define the zero-th order state without the perturbation. Before irradiating the light, the material is in the thermal equilibrium state,

$$\rho^{(0)} = \rho^{\mathrm{eq}} = \frac{1}{Q}\exp\left(-\frac{H_0}{k_B T}\right) \quad \left(Q = \mathrm{Tr}\left[\exp\left(-\frac{H_0}{k_B T}\right)\right]\right), \tag{3.17}$$

where H_0 is the Hamiltonian for the material system. (If it is embedded in the bath, H_0 refers to the partial system.) Q is the partition function for the canonical ensemble. The above expression for ρ^{eq} is equivalent to Eq. (3.15), though Eq. (3.15) shows the matrix element on the basis of energy eigenfunctions $\{\phi_n\}$ $(H_0\phi_n = \mathcal{E}_n\phi_n)$.

Then let us turn on the light and allow it to interact with the material. Consequently, the Hamiltonian changes from H_0 to

$$H = H_0 + H' = H_0 - \boldsymbol{\mu}\cdot\boldsymbol{E}(t). \tag{3.18}$$

The second term $H' = -\boldsymbol{\mu}\cdot\boldsymbol{E}(t)$ represents the perturbation, which is the interaction between the electric field $\boldsymbol{E}$ and the dipole moment of the material $\boldsymbol{\mu}$. This expression is based on the electric dipole approximation for the light-matter interaction, on the assumption that the dimension of surface layer is much smaller than the typical wavelength of light. We also note that the light-matter interaction Hamiltonian H' in Eq. (3.18) takes account of only the electric interaction, because the interaction energy with the magnetic field is significantly smaller than with the electric field for nonmagnetic materials (see Appendix A.2).[2] The electric field $\boldsymbol{E}(t)$ in Eq. (3.18) is expressed by the superposition of oscillating fields like Eq. (1.3),

$$E_q(t) = \sum_k E_q(\omega_k)\exp(-i\omega_k t) \qquad (q = x \sim z)$$
$$= E_q(\omega_1)\exp(-i\omega_1 t) + E_q(\omega_2)\exp(-i\omega_2 t) + \cdots + c.c.$$

The perturbation by this electric field alters the density matrix from $\rho^{(0)}$ to ρ,

$$\rho = \rho^{(0)} + \rho^{(1)} + \rho^{(2)} + \cdots .$$

By substituting this perturbation expansion of ρ in the Liouville equation (3.16), we find that each order of perturbation has to satisfy the following equation.

$$\text{0-th: } i\hbar\frac{d\rho_{mn}^{(0)}}{dt} = \left[H_0, \rho^{(0)}\right]_{mn} = 0 \tag{3.19}$$

$$\text{1-st: } i\hbar\frac{d\rho_{mn}^{(1)}}{dt} = \left[H_0, \rho^{(1)}\right]_{mn} + \left[H', \rho^{(0)}\right]_{mn} - i\hbar\Gamma_{mn}\rho_{mn}^{(1)} \tag{3.20}$$

$$\text{2-nd: } i\hbar\frac{d\rho_{mn}^{(2)}}{dt} = \left[H_0, \rho^{(2)}\right]_{mn} + \left[H', \rho^{(1)}\right]_{mn} - i\hbar\Gamma_{mn}\rho_{mn}^{(2)} \tag{3.21}$$

$$\vdots$$

The matrix elements in Eqs. (3.19), (3.20), and (3.21) are represented on the basis set of energy eigenstates that satisfy $H_0 |n\rangle = \mathcal{E}_n |n\rangle$. The zero-th order equation (3.19) means that the original state $\rho^{(0)}$ in thermal equilibrium is a steady state. Since $\rho^{(0)} = \rho^{\text{eq}}$ in Eq. (3.15) is diagonal, it is expressed by $\rho_{mn}^{(0)} = \rho_n^{(0)}\delta_{mn}$ hereafter.

3.2.2 First-Order Susceptibility

The first-order equation (3.20) is expanded by the energy eigenstates,

$$i\hbar\frac{d\rho_{mn}^{(1)}}{dt} = \left[H_0, \rho^{(1)}\right]_{mn} + \left[H', \rho^{(0)}\right]_{mn} - i\hbar\Gamma_{mn}\rho_{mn}^{(1)}$$
$$= (\mathcal{E}_m - \mathcal{E}_n)\rho_{mn}^{(1)} + \langle m|H'|n\rangle\,(\rho_n^{(0)} - \rho_m^{(0)}) - i\hbar\Gamma_{mn}\rho_{mn}^{(1)}.$$

[2]We extend this treatment to include the interaction with electric quadrupole and magnetic dipole in Chap. 7.

This equation is written using $\omega_{mn} = (\mathcal{E}_m - \mathcal{E}_n)/\hbar$ by

$$\frac{d\rho_{mn}^{(1)}}{dt} + i\omega_{mn}\rho_{mn}^{(1)} + \Gamma_{mn}\rho_{mn}^{(1)} = \frac{i}{\hbar}\langle m|H'|n\rangle(\rho_m^{(0)} - \rho_n^{(0)})$$
$$= e^{-(i\omega_{mn}+\Gamma_{mn})t}\frac{d}{dt}\left\{e^{(i\omega_{mn}+\Gamma_{mn})t}\rho_{mn}^{(1)}(t)\right\},$$

which can be readily integrated to obtain the following form,

$$e^{(i\omega_{mn}+\Gamma_{mn})t}\rho_{mn}^{(1)}(t) = \frac{i}{\hbar}(\rho_m^{(0)} - \rho_n^{(0)})\int_{-\infty}^{t}\langle m|H'(\tau)|n\rangle e^{(i\omega_{mn}+\Gamma_{mn})\tau}d\tau$$
$$= -\frac{i}{\hbar}(\rho_m^{(0)} - \rho_n^{(0)})\int_{-\infty}^{t}\sum_q^{x-z}\langle m|\mu_q|n\rangle\sum_k E_q(\omega_k)e^{-i\omega_k\tau}e^{(i\omega_{mn}+\Gamma_{mn})\tau}d\tau$$
$$= -\frac{i}{\hbar}(\rho_m^{(0)} - \rho_n^{(0)})\sum_q\langle m|\mu_q|n\rangle\sum_k E_q(\omega_k)\frac{e^{(-i\omega_k+i\omega_{mn}+\Gamma_{mn})t}}{-i\omega_k + i\omega_{mn} + \Gamma_{mn}}.$$

In the above integration, the boundary condition at $t \to -\infty$ was set to be $\rho(t) = \rho^{(0)}$, assuming that the system was in thermal equilibrium in the past before irradiated by light. Therefore, we have the following expression for $\rho^{(1)}(t)$,

$$\rho_{mn}^{(1)}(t) = \frac{1}{\hbar}(\rho_m^{(0)} - \rho_n^{(0)})\sum_q\langle m|\mu_q|n\rangle\sum_k E_q(\omega_k)\frac{e^{-i\omega_k t}}{\omega_k - \omega_{mn} + i\Gamma_{mn}} \tag{3.22}$$

Equation (3.22) presents the first-order perturbation of the density matrix by the electric field $\boldsymbol{E}$ of light. The perturbed state results in the induced polarization $\boldsymbol{P}^{(1)}(t)$,

$$P_p^{(1)}(t) = \mathrm{Tr}\left[\mu_p\rho^{(1)}(t)\right] = \sum_{m,n}\langle n|\mu_p|m\rangle\rho_{mn}^{(1)}(t)$$
$$= \frac{1}{\hbar}\sum_{m,n}(\rho_m^{(0)} - \rho_n^{(0)})\langle n|\mu_p|m\rangle\sum_q\langle m|\mu_q|n\rangle\sum_k E_q(\omega_k)\frac{e^{-i\omega_k t}}{\omega_k - \omega_{mn} + i\Gamma_{mn}}$$
$$= \sum_k P_p^{(1)}(\omega_k)e^{-i\omega_k t}. \tag{3.23}$$

The above formula allows us to derive the first-order susceptibility of the material, $\chi^{(1)}(\omega)$, which is defined with the induced polarization by the oscillating electric field $E_q(\omega)$

$$P_p^{(1)}(\omega) = \chi_{pq}^{(1)}(\omega)E_q(\omega) \tag{3.24}$$

Comparing Eqs. (3.23) and (3.24), one gets the form of $\chi^{(1)}(\omega)$ as

$$\chi_{pq}^{(1)}(\omega) = \frac{1}{\hbar}\sum_{m,n}(\rho_m^{(0)} - \rho_n^{(0)})\frac{\langle n|\mu_p|m\rangle\langle m|\mu_q|n\rangle}{\omega - \omega_{mn} + i\Gamma_{mn}}. \tag{3.25}$$

Equation (3.25) is equivalently written in the following,

$$\begin{aligned}\chi_{pq}^{(1)}(\omega) &= \frac{1}{\hbar}\sum_{g,n}\rho_g^{(0)}\left[-\frac{\langle g|\mu_p|n\rangle\langle n|\mu_q|g\rangle}{\omega - \omega_{ng} + i\Gamma_{ng}} + \frac{\langle g|\mu_q|n\rangle\langle n|\mu_p|g\rangle}{\omega + \omega_{ng} + i\Gamma_{ng}}\right]\\ &= \sum_g \rho_g^{(0)}\chi_{pq,g}^{(1)}(\omega).\end{aligned} \tag{3.26}$$

The last expression of Eq. (3.26) allows for the interpretation that $\chi_{pq}^{(1)}(\omega)$ is given by thermal average of the susceptibility at the state g, $\chi_{pq,g}^{(1)}(\omega)$, over the population distribution of g.

3.2.3 Second-Order Susceptibility

The second-order susceptibility $\chi^{(2)}$ is derived from the second-order perturbation of the density matrix $\rho^{(2)}$ in Eq. (3.21), or in the following form,

$$\frac{d\rho_{mn}^{(2)}}{dt} + i\omega_{mn}\rho_{mn}^{(2)} + \Gamma_{mn}\rho_{mn}^{(2)} = -\frac{i}{\hbar}\left[H', \rho^{(1)}\right]_{mn}. \tag{3.27}$$

Substituting $\rho^{(1)}$ of Eq. (3.22) into Eq. (3.27), we get the expression for $\rho^{(2)}$,

$$\begin{aligned}\rho_{mn}^{(2)}(t) = &-\frac{1}{\hbar^2}\sum_l\sum_{j,k}\sum_{p,q}^{x\sim z}\\ &\left\{(\rho_l^{(0)} - \rho_n^{(0)})\frac{\langle m|\mu_p|l\rangle\langle l|\mu_q|n\rangle}{\omega_k - \omega_{ln} + i\Gamma_{ln}} - (\rho_m^{(0)} - \rho_l^{(0)})\frac{\langle m|\mu_q|l\rangle\langle l|\mu_p|n\rangle}{\omega_k - \omega_{ml} + i\Gamma_{ml}}\right\}\\ &\cdot\frac{1}{(\omega_j + \omega_k) - \omega_{mn} + i\Gamma_{mn}}E_p(\omega_j)E_q(\omega_k)e^{-i(\omega_j+\omega_k)t}.\end{aligned} \tag{3.28}$$

[Problem 3.3] Derive $\rho^{(2)}(t)$ in Eq. (3.28).

Therefore, the second-order induced polarization $\boldsymbol{P}^{(2)}(t)$ is

$$P_r^{(2)}(t) = \mathrm{Tr}\left[\mu_r \rho^{(2)}(t)\right] = \sum_{m,n} \langle n|\mu_r|m\rangle\, \rho_{mn}^{(2)}(t)$$

$$= -\frac{1}{\hbar^2}\sum_{m,n}\langle n|\mu_r|m\rangle \sum_{l}\sum_{j,k}\sum_{p,q}^{x-z}\left\{(\rho_l^{(0)} - \rho_n^{(0)})\frac{\langle m|\mu_p|l\rangle\langle l|\mu_q|n\rangle}{\omega_k - \omega_{ln} + i\Gamma_{ln}}\right.$$

$$\left.-(\rho_m^{(0)}-\rho_l^{(0)})\frac{\langle m|\mu_q|l\rangle\langle l|\mu_p|n\rangle}{\omega_k-\omega_{ml}+i\Gamma_{ml}}\right\}\frac{1}{(\omega_j+\omega_k)-\omega_{mn}+i\Gamma_{mn}}E_p(\omega_j)E_q(\omega_k)e^{-i(\omega_j+\omega_k)t}. \tag{3.29}$$

The second-order nonlinear susceptibility $\chi_{pqr}^{(2)}(\Omega, \omega_1, \omega_2)$ is derived from this equation, by considering two incident fields, $E_q(\omega_1)$ and $E_r(\omega_2)$, with specific directions (q, r) and frequencies (ω_1, ω_2). Then the sum frequency polarization along the p direction, $P_p^{(2)}(\Omega = \omega_1 + \omega_2)$, is induced by

$$P_p^{(2)}(\Omega = \omega_1 + \omega_2) = \chi_{pqr}^{(2)}(\Omega, \omega_1, \omega_2)E_q(\omega_1)E_r(\omega_2) \tag{3.30}$$

where p, q, r denote the spatial coordinates $x \sim z$. To compare Eq. (3.30) to (3.29), we find that the induced polarization $P_p^{(2)}(\Omega)$ corresponds to a pair of terms, $E_q(\omega_1)E_r(\omega_2)$ and $E_r(\omega_2)E_p(\omega_1)$, with exchanging the suffixes $(p \leftrightarrow q)$ and $(j \leftrightarrow k)$ simultaneously in Eq. (3.29). Therefore, the expression of $\chi_{pqr}^{(2)}$ becomes

$$\chi_{pqr}^{(2)}(\Omega, \omega_1, \omega_2) = -\frac{1}{\hbar^2}\sum_{m,n}\langle n|\mu_p|m\rangle$$

$$\cdot\sum_l\left[\left\{(\rho_l^{(0)} - \rho_n^{(0)})\frac{\langle m|\mu_q|l\rangle\,\langle l|\mu_r|n\rangle}{\omega_2 - \omega_{ln} + i\Gamma_{ln}} - (\rho_m^{(0)} - \rho_l^{(0)})\frac{\langle m|\mu_r|l\rangle\,\langle l|\mu_q|n\rangle}{\omega_2 - \omega_{ml} + i\Gamma_{ml}}\right\}\right.$$

$$\cdot\frac{1}{\Omega - \omega_{mn} + i\Gamma_{mn}}$$

$$+\left\{(\rho_l^{(0)} - \rho_n^{(0)})\frac{\langle m|\mu_r|l\rangle\,\langle l|\mu_q|n\rangle}{\omega_1 - \omega_{ln} + i\Gamma_{ln}} - (\rho_m^{(0)} - \rho_l^{(0)})\frac{\langle m|\mu_q|l\rangle\,\langle l|\mu_r|n\rangle}{\omega_1 - \omega_{ml} + i\Gamma_{ml}}\right\}$$

$$\left.\cdot\frac{1}{\Omega - \omega_{mn} + i\Gamma_{mn}}\right] \tag{3.31}$$

We should note that the perturbation formula of $\boldsymbol{\chi}^{(2)}$ has a number of variations by different notation of suffixes l, m, n. For example, the following expression is also seen in literature,

$$
\begin{aligned}
\chi^{(2)}_{pqr}(\Omega,\omega_1,\omega_2) &= \frac{1}{\hbar^2}\sum_{g,n,m}\rho_g^{(0)} \\
&\left[\frac{\langle g|\mu_p|n\rangle\langle n|\mu_q|m\rangle\langle m|\mu_r|g\rangle}{(\Omega-\omega_{ng}+i\Gamma_{ng})\,(\boldsymbol{\omega_2-\omega_{mg}+i\Gamma_{mg}})} + \frac{\langle g|\mu_p|n\rangle\langle n|\mu_r|m\rangle\langle m|\mu_q|g\rangle}{(\Omega-\omega_{ng}+i\Gamma_{ng})\,(\omega_1-\omega_{mg}+i\Gamma_{mg})}\right. \\
&+\frac{\langle g|\mu_r|m\rangle\langle m|\mu_q|n\rangle\langle n|\mu_p|g\rangle}{(\Omega+\omega_{ng}+i\Gamma_{ng})\,(\boldsymbol{\omega_2+\omega_{mg}+i\Gamma_{mg}})} + \frac{\langle g|\mu_q|m\rangle\langle m|\mu_r|n\rangle\langle n|\mu_p|g\rangle}{(\Omega+\omega_{ng}+i\Gamma_{ng})\,(\omega_1+\omega_{mg}+i\Gamma_{mg})} \\
&-\frac{\langle g|\mu_r|m\rangle\langle m|\mu_p|n\rangle\langle n|\mu_q|g\rangle}{(\Omega-\omega_{nm}+i\Gamma_{nm})}\left(\frac{1}{\boldsymbol{\omega_2+\omega_{mg}+i\Gamma_{mg}}}+\frac{1}{\omega_1-\omega_{ng}+i\Gamma_{ng}}\right) \\
&\left.-\frac{\langle g|\mu_q|m\rangle\langle m|\mu_p|n\rangle\langle n|\mu_r|g\rangle}{(\Omega-\omega_{nm}+i\Gamma_{nm})}\left(\frac{1}{\boldsymbol{\omega_2-\omega_{ng}+i\Gamma_{ng}}}+\frac{1}{\omega_1+\omega_{mg}+i\Gamma_{mg}}\right)\right] \\
&=\sum_g \rho_g^{(0)}\chi^{(2)}_{pqr,g}(\Omega,\omega_1,\omega_2). \qquad (3.32)
\end{aligned}
$$

This expression is amenable to the similar interpretation with Eq. (3.26) that $\chi^{(2)}_{pqr}$ is given by thermal average of the susceptibility at the state g, $\chi^{(2)}_{pqr,g}$, over the population distribution of g. The factors emphasized with bold fonts are discussed in the following section.

3.3 Properties of $\boldsymbol{\chi}^{(2)}$

The above equation (3.32) for $\boldsymbol{\chi}^{(2)}(\Omega,\omega_1,\omega_2)$ is used to discuss qualitative features of nonlinear susceptibility. We summarize some fundamental properties of $\boldsymbol{\chi}^{(2)}$ in relation to the surface spectroscopy. This section deals with the mechanism of resonance in $\boldsymbol{\chi}^{(2)}$ in Sect. 3.3.1, the relations of $\boldsymbol{\chi}^{(2)}$ to molecular orientation in Sect. 3.3.2, and to light polarizations in Sect. 3.3.3.

3.3.1 Vibrational Resonance

Since $\boldsymbol{\chi}^{(2)}$ governs the SFG signal, its frequency dependence describes the spectral shape. In particular, the dependence of $\boldsymbol{\chi}^{(2)}$ on the infrared frequency ω_2 is essential for interpreting the vibrational SFG spectroscopy. The ω_2 dependence of $\boldsymbol{\chi}^{(2)}$ is mainly attributed to vibrational resonance, as discussed in the following.

Among the terms in the right hand side of Eq. (3.32), some denominators including ω_2, such as $(\boldsymbol{\omega_2-\omega_{mg}+i\Gamma_{mg}})$ and $(\boldsymbol{\omega_2-\omega_{ng}+i\Gamma_{ng}})$, are emphasized in a bold font. These denominators indicate resonance when ω_2 is close to ω_{mg} or ω_{ng}, i.e. $\omega_2 \approx \omega_{mg} = (\mathcal{E}_m-\mathcal{E}_g)/\hbar$ or $\omega_2 \approx \omega_{ng} = (\mathcal{E}_n-\mathcal{E}_g)/\hbar$. Suppose ω_2 is an infrared frequency, the state m (or n) in resonance is usually a vibrationally excited state. Other bold denominators including ω_2 in Eq. (3.32), such as $(\boldsymbol{\omega_2+\omega_{mg}+i\Gamma_{mg}})$, also imply the possible resonance at $\omega_2 \approx -\omega_{mg}$ in case of $\omega_{mg} < 0$ $(\mathcal{E}_m < \mathcal{E}_g)$ that the initial state g is a vibrationally excited state. Therefore, one could categorize the terms in Eq. (3.32) into those including the above-mentioned resonant denominators $((\boldsymbol{\omega_2 - \omega_{mg} + i\Gamma_{mg}}), (\boldsymbol{\omega_2 - \omega_{ng} + i\Gamma_{ng}}), (\boldsymbol{\omega_2 + \omega_{mg} + i\Gamma_{mg}}))$ and the rest. The former is called the vibrationally resonant terms and denoted by $\boldsymbol{\chi}^{(2),\mathrm{res}}$, while the latter is called the nonresonant terms $\boldsymbol{\chi}^{(2),\mathrm{nonres}}$. Accordingly,

$$\chi_{pqr}^{(2)} = \chi_{pqr}^{(2),\mathrm{res}} + \chi_{pqr}^{(2),\mathrm{nonres}}, \tag{3.33}$$

where

$$\chi_{pqr}^{(2),\mathrm{res}}(\Omega, \omega_1, \omega_2) = \frac{1}{\hbar^2} \sum_{g,n,m} \rho_g^{(0)}$$

$$\left[\frac{\langle g|\mu_p|n\rangle\langle n|\mu_q|m\rangle\langle m|\mu_r|g\rangle}{(\Omega - \omega_{ng} + i\Gamma_{ng})\,(\boldsymbol{\omega_2 - \omega_{mg} + i\Gamma_{mg}})} - \frac{\langle g|\mu_q|m\rangle\langle m|\mu_p|n\rangle\langle n|\mu_r|g\rangle}{(\Omega - \omega_{nm} + i\Gamma_{nm})\,(\boldsymbol{\omega_2 - \omega_{ng} + i\Gamma_{ng}})} \right.$$
$$\left. + \frac{\langle g|\mu_r|m\rangle\langle m|\mu_q|n\rangle\langle n|\mu_p|g\rangle}{(\Omega + \omega_{ng} + i\Gamma_{ng})\,(\boldsymbol{\omega_2 + \omega_{mg} + i\Gamma_{mg}})} - \frac{\langle g|\mu_r|m\rangle\langle m|\mu_p|n\rangle\langle n|\mu_q|g\rangle}{(\Omega - \omega_{nm} + i\Gamma_{nm})\,(\boldsymbol{\omega_2 + \omega_{mg} + i\Gamma_{mg}})} \right] \tag{3.34}$$

and

$$\chi_{pqr}^{(2),\mathrm{nonres}}(\Omega, \omega_1, \omega_2) = \frac{1}{\hbar^2} \sum_{g,n,m} \rho_g^{(0)}$$

$$\left[\frac{\langle g|\mu_p|n\rangle\langle n|\mu_r|m\rangle\langle m|\mu_q|g\rangle}{(\Omega - \omega_{ng} + i\Gamma_{ng})\,(\omega_1 - \omega_{mg} + i\Gamma_{mg})} - \frac{\langle g|\mu_r|m\rangle\langle m|\mu_p|n\rangle\langle n|\mu_q|g\rangle}{(\Omega - \omega_{nm} + i\Gamma_{nm})\,(\omega_1 - \omega_{ng} + i\Gamma_{ng})} \right.$$
$$\left. + \frac{\langle g|\mu_q|m\rangle\langle m|\mu_r|n\rangle\langle n|\mu_p|g\rangle}{(\Omega + \omega_{ng} + i\Gamma_{ng})\,(\omega_1 + \omega_{mg} + i\Gamma_{mg})} - \frac{\langle g|\mu_q|m\rangle\langle m|\mu_p|n\rangle\langle n|\mu_r|g\rangle}{(\Omega - \omega_{nm} + i\Gamma_{nm})\,(\omega_1 + \omega_{mg} + i\Gamma_{mg})} \right]. \tag{3.35}$$

$\boldsymbol{\chi}^{(2),\mathrm{res}}$ in Eq. (3.34) is further modified to

$$\chi_{pqr}^{(2),\mathrm{res}}(\Omega, \omega_1, \omega_2)$$
$$= \frac{1}{\hbar^2} \sum_{g,n,m} \rho_g^{(0)} \left[\frac{\langle g|\mu_p|n\rangle\langle n|\mu_q|m\rangle\langle m|\mu_r|g\rangle}{(\Omega - \omega_{ng} + i\Gamma_{ng})\,(\omega_2 - \omega_{mg} + i\Gamma_{mg})} \right.$$
$$\left. - \frac{\langle g|\mu_q|m\rangle\langle m|\mu_p|n\rangle\langle n|\mu_r|g\rangle}{(\Omega - \omega_{nm} + i\Gamma_{nm})\,(\omega_2 - \omega_{ng} + i\Gamma_{ng})} \right]$$

$$
\begin{aligned}
&+\frac{1}{\hbar^2}\sum_{g,n,m}\rho_g^{(0)}\left[\frac{\langle g|\mu_r|m\rangle\langle m|\mu_q|n\rangle\langle n|\mu_p|g\rangle}{(\Omega+\omega_{ng}+i\Gamma_{ng})(\omega_2+\omega_{mg}+i\Gamma_{mg})}\right.\\
&\qquad\left.-\frac{\langle g|\mu_r|m\rangle\langle m|\mu_p|n\rangle\langle n|\mu_q|g\rangle}{(\Omega-\omega_{nm}+i\Gamma_{nm})(\omega_2+\omega_{mg}+i\Gamma_{mg})}\right]\\
&=\frac{1}{\hbar^2}\sum_{g,n,m}\rho_g^{(0)}\left[\frac{\langle g|\mu_p|n\rangle\langle n|\mu_q|m\rangle}{\Omega-\omega_{ng}+i\Gamma_{ng}}-\frac{\langle g|\mu_q|n\rangle\langle n|\mu_p|m\rangle}{\Omega+\omega_{nm}+i\Gamma_{nm}}\right]\frac{\langle m|\mu_r|g\rangle}{\omega_2-\omega_{mg}+i\Gamma_{mg}}\\
&+\frac{1}{\hbar^2}\sum_{g,n,m}\rho_m^{(0)}\left[\frac{\langle g|\mu_q|n\rangle\langle n|\mu_p|m\rangle}{\Omega+\omega_{nm}+i\Gamma_{nm}}-\frac{\langle g|\mu_p|n\rangle\langle n|\mu_q|m\rangle}{\Omega-\omega_{ng}+i\Gamma_{ng}}\right]\frac{\langle m|\mu_r|g\rangle}{\omega_2-\omega_{mg}+i\Gamma_{mg}}\\
&=-\frac{1}{\hbar}\sum_{g,m}\left(\rho_g^{(0)}-\rho_m^{(0)}\right)\frac{\langle g|\alpha_{pq}(\Omega)|m\rangle\langle m|\mu_r|g\rangle}{\omega_2-\omega_{mg}+i\Gamma_{mg}},
\end{aligned}
\tag{3.36}
$$

where the last expression includes the matrix elements of the Raman tensor $\langle g|\alpha_{pq}(\Omega)|m\rangle$ defined by

$$
\langle g|\alpha_{pq}(\Omega)|m\rangle=-\frac{1}{\hbar}\sum_n\left[\frac{\langle g|\mu_p|n\rangle\langle n|\mu_q|m\rangle}{\Omega-\omega_{ng}+i\Gamma_{ng}}-\frac{\langle g|\mu_q|n\rangle\langle n|\mu_p|m\rangle}{\Omega+\omega_{nm}+i\Gamma_{nm}}\right].
\tag{3.37}
$$

The matrix element $\langle g|\alpha_{pq}(\Omega)|m\rangle$ is associated to the transition probability of Raman scattering from the state g to m [2, 8, 10, 11]. The diagonal element at $g=m$, $\langle g|\alpha_{pq}(\Omega)|g\rangle$, is equal to $\chi^{(1)}_{pq,g}(\Omega)$ in Eq. (3.26), the first-order susceptibility at the state g. We also note that the nonresonant term $\boldsymbol{\chi}^{(2),\mathrm{nonres}}$ in Eq. (3.35) can be represented in the same manner to be

$$
\chi_{pqr}^{(2),\mathrm{nonres}}=-\frac{1}{\hbar}\sum_{g,m}\left(\rho_g^{(0)}-\rho_m^{(0)}\right)\frac{\langle g|\alpha_{pr}(\Omega)|m\rangle\langle m|\mu_q|g\rangle}{\omega_1-\omega_{mg}+i\Gamma_{mg}}.
\tag{3.38}
$$

The $\boldsymbol{\chi}^{(2)}$ expressions in Eqs. (3.36) and (3.38) include product of the transition dipole moment $\langle m|\mu|g\rangle$ and the Raman tensor $\langle g|\boldsymbol{\alpha}(\Omega)|m\rangle$ in the numerator. Therefore, the SFG process includes both infrared active ($\langle m|\mu|g\rangle\neq 0$) and Raman active ($\langle g|\boldsymbol{\alpha}|m\rangle\neq 0$) transitions. This conclusion is in accord with the basic selection rule of SFG in Chap. 1 that a system having inversion symmetry has no active vibrational mode to SFG due to the IR-Raman mutual exclusion theorem.

The principal difference between Eqs. (3.36) and (3.38) is seen in ω_2 (infrared) or ω_1 (visible) included in the denomenator. In the vibrational SFG spectroscopy, the vibrational transition could be in resonance with the infrared frequency ω_2. The vibrationally resonance term $\chi^{(2),\mathrm{res}}$ could be interpreted formally as successive transitions by the infrared absorption and anti-Stokes Raman scattering, as schemat-

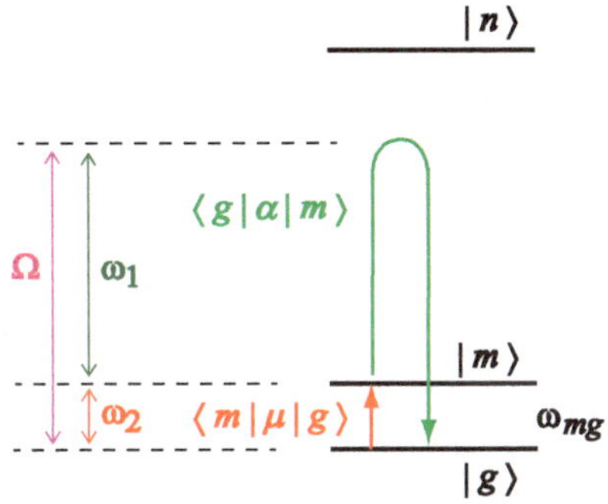

Fig. 3.1 Schematic illustration of vibrationally resonant term $\chi^{(2),\mathrm{res}}$ in the vibrational SFG spectroscopy. ω_2 is in resonance with the energy gap between the vibrational ground $|g\rangle$ and excited $|m\rangle$ states. The intermediate state $|n\rangle$ may be off resonant from ω_1 or Ω

ically illustrated in Fig. 3.1. The main resonance term of $\chi_{pqr}^{(2),\mathrm{res}}$ in Eq. (3.36) has the following spectral lineshape as a function of ω_2,

$$\chi_{pqr}^{(2),\mathrm{res}}(\omega_2) \sim \frac{-C_{mg}}{\omega_2 - \omega_{mg} + i\Gamma_{mg}} = \frac{-C_{mg}\,(\omega_2 - \omega_{mg})}{(\omega_2 - \omega_{mg})^2 + \Gamma_{mg}^2} + i\,\frac{C_{mg}\,\Gamma_{mg}}{(\omega_2 - \omega_{mg})^2 + \Gamma_{mg}^2}, \tag{3.39}$$

where $C_{mg} = (1/\hbar)(\rho_g^{(0)} - \rho_m^{(0)})\langle g|\alpha_{pq}(\Omega)|m\rangle\langle m|\mu_r|g\rangle$. Equation (3.39) is a complex Lorentz function with the central frequency $\omega_2 = \omega_{mg}$ and the width Γ_{mg}, and Fig. 3.2 illustrates the ω_2 dependence of its real and imaginary parts. On the other hand, the nonresonant term $\chi^{(2),\mathrm{nonres}}$ does not show such dependence on ω_2. In a limited frequency range of ω_2 in usual vibrational SFG spectra, the nonresonant term can be approximated as a constant with respect to ω_2. Therefore, $\chi^{(2)}$ is often represented phenomenologically as a superposition of a constant background C^0 and some Lorentz functions in the form,

$$\chi^{(2)}(\omega_2) \approx C^0 - \sum_{k=1}^{N} \frac{C^k}{\omega_2 - \omega^k + i\Gamma^k}. \tag{3.40}$$

Equation (3.40) is often employed in fitting experimental spectra, where the ingredient parameters C^0 and $\{C^k, \omega^k, \Gamma^k\}$ $(k = 1 \sim N)$ are determined so as to reproduce the experimental spectra.

In the above discussion we focused on the vibrational resonance with ω_2 in $\chi^{(2)}(\Omega, \omega_1, \omega_2)$. Conventional applications of vibrational SFG spectroscopy usually employ a fixed visible frequency ω_1 in electronically non-resonant condition. When the frequency Ω is far off the electronic transition energy ($\Omega \ll \omega_{ng}$ in Fig. 3.1), the Raman tensor α_{pq} is well approximated with the polarizability tensor [14] (see Appendix A.3). In the following discussion we mainly deal with vibrationally resonant but electronically non-resonant conditions in the vibrational

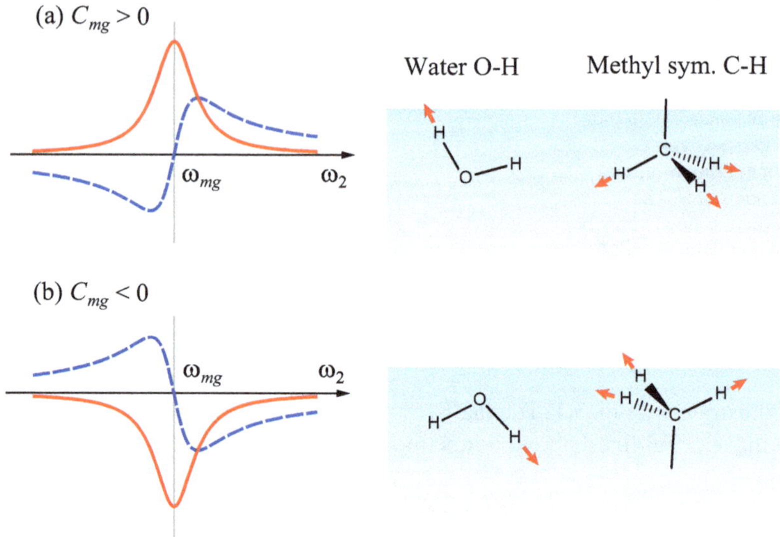

Fig. 3.2 Schematics of Lorentz functions in Eq. (3.39) in the case of (**a**) $C_{mg} > 0$ and (**b**) $C_{mg} < 0$. Real and imaginary parts are written with blue dashed and red solid lines, respectively. Right pictures illustrate corresponding molecular orientations of water and methyl group, as discussed in Sect. 4.2

SFG spectroscopy, though the SFG process involving the electronic resonance has been also discussed [4, 16–18].[3]

3.3.2 Relation to Molecular Orientation

Next we discuss the values of the $\boldsymbol{\chi}^{(2)}$ tensor elements. The $\boldsymbol{\chi}^{(2)}$ tensor can be represented in an arbitrary coordinate system, and the values of the tensor elements vary with the rotation of the system or the coordinate. Here we formulate the relation between the tensor elements and orientation. The relation offers a useful clue to investigate the orientation of molecules at interface by the SFG spectroscopy [9].

Let us recall that the $\boldsymbol{\chi}^{(2)}$ formula of Eq. (3.31) or (3.32) is applicable to either a molecule or the interface system. When ρ and $\boldsymbol{\mu}$ are defined for a single molecule, Eq. (3.31) or (3.32) gives the second-order susceptibility for the molecule, usually called the molecular hyperpolarizability $\boldsymbol{\alpha}^{(2)}(\Omega, \omega_1, \omega_2)$. When ρ and $\boldsymbol{\mu}$ are defined for the interface system, the $\boldsymbol{\chi}^{(2)}$ formula gives the second-order susceptibility of the

[3]The SFG with electronic resonance involves the Raman tensor in electronically resonant condition, and thus related to the resonance Raman scattering. The vibrational SFG spectroscopy including electronic resonance plays an important role in the chiral applications in Chap. 8.

interface. The hyperpolarizability of an individual molecule $\boldsymbol{\alpha}^{(2)}$ and the nonlinear susceptibility of the interface $\boldsymbol{\chi}^{(2)}$ are treated on the same formulation. The relation between $\boldsymbol{\chi}^{(2)}$ and $\boldsymbol{\alpha}^{(2)}$ is discussed as follows. In the following formulas about molecular orientation, the suffixes p, q, r stand for the space-fixed coordinates (x, y, z), while p', q', r' for the molecule-fixed coordinates (ξ, η, ζ).

Two Factors: Density and Orientation

Suppose that the interface system consists of molecules, the second-order susceptibility of the interface $\boldsymbol{\chi}^{(2)}$ is approximated by the sum of the second-order susceptibility of constituent molecules $\boldsymbol{\alpha}^{(2)}$,

$$\chi^{(2)}_{pqr} \approx \sum_{l=1}^{N} \alpha^{(2)}_{l,pqr} = N \cdot \overline{\alpha^{(2)}_{pqr}}, \tag{3.41}$$

where the suffix l stands for the constituent N molecules, and $\overline{\alpha^{(2)}_{pqr}} = \frac{1}{N}\sum_{l=1}^{N} \alpha^{(2)}_{l,pqr}$ is the average of $\boldsymbol{\alpha}^{(2)}$ in the space-fixed coordinate. Equation (3.41) indicates that $\chi^{(2)}_{pqr}$ is proportional to the number of molecules N and the average of hyperpolarizability $\overline{\alpha^{(2)}_{pqr}}$. $\overline{\alpha^{(2)}_{pqr}}$ is the average over molecules with various orientations, and thus an index to the orientational order of the molecules. The two factors, number density and orientational order, are regarded to govern the intensity of SFG signal in a qualitative sense.

We note that Eq. (3.41) expresses the nonlinear polarization of the whole system simply as the assembly of the induced polarizations of molecules. This $\boldsymbol{\chi}^{(2)}$ expression of Eq. (3.41) is widely utilized in qualitative analysis of SFG amplitude. However, this expression neglects dielectric coupling among molecular polarizations, and the underlying approximation can break down in Sect. 9.3 [6, 7]. More accurate treatment of polarization will be provided in Chap. 5.

Rotational Matrix

Equation (3.41) relates the tensor elements $\chi^{(2)}_{pqr}$ and $\alpha^{(2)}_{pqr}$ on the common space-fixed coordinates. However, the molecular property of $\alpha^{(2)}_{pqr}$ is conveniently presented on the molecule-fixed coordinates, while $\chi^{(2)}_{pqr}$ is usually given on the space-fixed coordinates of the experimental geometry. Accordingly, we define the transformation of the hyperpolarizability tensor $\boldsymbol{\alpha}^{(2)}$ from the molecule-fixed coordinates (namely ξ, η, ζ) to the space-fixed ones (x, y, z).

We denote the unit vectors along the molecule-fixed axes ξ, η, ζ by $\boldsymbol{e}_\xi, \boldsymbol{e}_\eta, \boldsymbol{e}_\zeta$, respectively, and the unit vectors along the spaced-fixed axes x, y, z by $\boldsymbol{e}_x, \boldsymbol{e}_y, \boldsymbol{e}_z$.

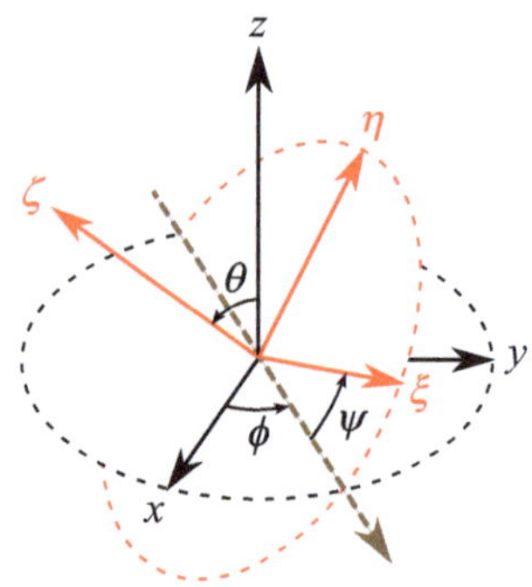

Fig. 3.3 Definition of the Euler angles $\{\phi, \theta, \psi\}$

Then the relation between the two sets of unit vectors is given using the direction cosine matrix $\mathcal{D}$,

$$\begin{pmatrix} \boldsymbol{e}_x \\ \boldsymbol{e}_y \\ \boldsymbol{e}_z \end{pmatrix} = \mathcal{D} \begin{pmatrix} \boldsymbol{e}_\xi \\ \boldsymbol{e}_\eta \\ \boldsymbol{e}_\zeta \end{pmatrix}, \quad \text{where } \mathcal{D} = \begin{pmatrix} (\boldsymbol{e}_x \cdot \boldsymbol{e}_\xi) & (\boldsymbol{e}_x \cdot \boldsymbol{e}_\eta) & (\boldsymbol{e}_x \cdot \boldsymbol{e}_\zeta) \\ (\boldsymbol{e}_y \cdot \boldsymbol{e}_\xi) & (\boldsymbol{e}_y \cdot \boldsymbol{e}_\eta) & (\boldsymbol{e}_y \cdot \boldsymbol{e}_\zeta) \\ (\boldsymbol{e}_z \cdot \boldsymbol{e}_\xi) & (\boldsymbol{e}_z \cdot \boldsymbol{e}_\eta) & (\boldsymbol{e}_z \cdot \boldsymbol{e}_\zeta) \end{pmatrix} \tag{3.42}$$

The matrix $\mathcal{D}$ represents the rotation of the axes, which is given with the Euler angles $\{\phi, \theta, \psi\}$ by[4]

$$\mathcal{D} = \begin{pmatrix} \cos\psi\cos\phi - \cos\theta\sin\phi\sin\psi & -\sin\psi\cos\phi - \cos\theta\sin\phi\cos\psi & \sin\theta\sin\phi \\ \cos\psi\sin\phi + \cos\theta\cos\phi\sin\psi & -\sin\psi\sin\phi + \cos\theta\cos\phi\cos\psi & -\sin\theta\cos\phi \\ \sin\psi\sin\theta & \cos\psi\sin\theta & \cos\theta \end{pmatrix}. \tag{3.43}$$

The present definition of the Euler angles after Goldstein [3] is illustrated in Fig. 3.3, though the way of their definition is not unique. When we express the coordinates of a vector in two ways, by (ξ, η, ζ) in the molecule-fixed coordinates and by (x, y, z) in the space-fixed coordinates, the following relation holds between the two expressions,

$$\begin{pmatrix} x \\ y \\ z \end{pmatrix} = \mathcal{D} \begin{pmatrix} \xi \\ \eta \\ \zeta \end{pmatrix}. \tag{3.44}$$

Comparing Eq. (3.44) to (3.42), we notice that the rotations of coordinates and axes apparently take the same form of transformation. This is in accord with the fact that an arbitrary vector $\boldsymbol{r}$ can be represented in either space-fixed or body-fixed coordinates,

$$\boldsymbol{r} = x\,\boldsymbol{e}_x + y\,\boldsymbol{e}_y + z\,\boldsymbol{e}_z = \xi\,\boldsymbol{e}_\xi + \eta\,\boldsymbol{e}_\eta + \zeta\,\boldsymbol{e}_\zeta,$$

[4] ϕ and ψ in Fig. 3.3 and Eq. (3.43) denote the Euler angles, according to the conventional notation. Distinguish them from the quantum states in this chapter.

which can be converted each other using Eqs. (3.42) and (3.44) as

$$x\,\boldsymbol{e}_x + y\,\boldsymbol{e}_y + z\,\boldsymbol{e}_z = \left(\boldsymbol{e}_x\ \boldsymbol{e}_y\ \boldsymbol{e}_z\right)\begin{pmatrix} x \\ y \\ z \end{pmatrix} = \left(\boldsymbol{e}_\xi\ \boldsymbol{e}_\eta\ \boldsymbol{e}_\zeta\right)\boldsymbol{\mathcal{D}}^T \cdot \boldsymbol{\mathcal{D}}\begin{pmatrix} \xi \\ \eta \\ \zeta \end{pmatrix}$$

$$= \left(\boldsymbol{e}_\xi\ \boldsymbol{e}_\eta\ \boldsymbol{e}_\zeta\right)\begin{pmatrix} \xi \\ \eta \\ \zeta \end{pmatrix} = \xi\,\boldsymbol{e}_\xi + \eta\,\boldsymbol{e}_\eta + \zeta\,\boldsymbol{e}_\zeta .$$

$\boldsymbol{\mathcal{D}}$ in Eq. (3.44) is called the rotational matrix,[5] and the transformation of Eq. (3.44) holds for any vector (first-rank tensor).

Now let us express Eq. (3.41) with the $\boldsymbol{\alpha}^{(2)}$ elements in the molecule-fixed coordinates. The molecular hyperpolarizability $\boldsymbol{\alpha}^{(2)}$ is represented in two ways, $\alpha_{pqr}^{(2),\text{space}}$ in the space-fixed coordinates or $\alpha_{p'q'r'}^{(2),\text{mol}}$ in the molecule-fixed coordinates. As $\boldsymbol{\alpha}^{(2)}$ is a third-rank tensor, the two representations of $\boldsymbol{\alpha}^{(2)}$ elements are related by

$$\alpha_{pqr}^{(2),\text{space}} = \sum_{p'}^{\xi\sim\zeta}\sum_{q'}^{\xi\sim\zeta}\sum_{r'}^{\xi\sim\zeta} \mathcal{D}_{pp'}\mathcal{D}_{qq'}\mathcal{D}_{rr'}\alpha_{p'q'r'}^{(2),\text{mol}} . \tag{3.45}$$

Therefore, $\chi_{pqr}^{(2)}$ in Eq. (3.41) is represented by

$$\chi_{pqr}^{(2)} = \sum_{l}^{\text{molecule}} \alpha_{l,pqr}^{(2)} = \sum_{l}^{\text{molecule}} \sum_{p'}^{\xi\sim\zeta}\sum_{q'}^{\xi\sim\zeta}\sum_{r'}^{\xi\sim\zeta} \mathcal{D}_{l,pp'}\mathcal{D}_{l,qq'}\mathcal{D}_{l,rr'}\alpha_{p'q'r'}^{(2),\text{mol}} , \tag{3.46}$$

using the rotational matrix $\boldsymbol{\mathcal{D}}_l$ specified for the l-th molecule. The average $\overline{\alpha_{pqr}^{(2)}}$ in Eq. (3.41) is accordingly expressed by

$$\overline{\alpha_{pqr}^{(2)}} = \frac{1}{N}\sum_{l=1}^{N}\sum_{p'}^{\xi\sim\zeta}\sum_{q'}^{\xi\sim\zeta}\sum_{r'}^{\xi\sim\zeta} \mathcal{D}_{l,pp'}\mathcal{D}_{l,qq'}\mathcal{D}_{l,rr'}\alpha_{p'q'r'}^{(2),\text{mol}}$$
$$= \sum_{p'}\sum_{q'}\sum_{r'} \overline{\mathcal{D}_{pp'}\mathcal{D}_{qq'}\mathcal{D}_{rr'}}\ \alpha_{p'q'r'}^{(2),\text{mol}} . \tag{3.47}$$

---

[5]The rotation matrix is often introduced with $\boldsymbol{\mathcal{D}}^{-1} = \boldsymbol{\mathcal{D}}^T$, which satisfies

$$\begin{pmatrix} \boldsymbol{e}_\xi \\ \boldsymbol{e}_\eta \\ \boldsymbol{e}_\zeta \end{pmatrix} = \boldsymbol{\mathcal{D}}^T \begin{pmatrix} \boldsymbol{e}_x \\ \boldsymbol{e}_y \\ \boldsymbol{e}_z \end{pmatrix} \quad \text{and} \quad \begin{pmatrix} \xi \\ \eta \\ \zeta \end{pmatrix} = \boldsymbol{\mathcal{D}}^T \begin{pmatrix} x \\ y \\ z \end{pmatrix} .$$

Do not confuse the two definitions, which are transpose each other.

Note that Eq. (3.46) is consistent to the surface sensitivity of $\boldsymbol{\chi}^{(2)}$ mentioned in Sect. 1.1, that an isotropic system has null $\boldsymbol{\chi}^{(2)}$. If a system consists of molecules with fully random isotropic orientation, the average of Eq. (3.47) vanishes, $\overline{\mathcal{D}_{pp'}\mathcal{D}_{qq'}\mathcal{D}_{rr'}} = 0$, which leads to $\boldsymbol{\chi}^{(2)} = 0$.

3.3.3 Tensor Elements of $\boldsymbol{\chi}^{(2)}$ and Polarization

Using the tensor elements of $\boldsymbol{\chi}^{(2)}$ in the space-fixed coordinates, we discuss the relation to polarization combination of SFG measurement. Since $\chi^{(2)}_{pqr}$ is a third-rank tensor, it has $3^3 = 27$ elements in principle. In analyzing experimental SFG spectra, relevant tensor elements of $\chi^{(2)}_{pqr}$ depend on the polarization of lights. There are 8 possible combination of light polarizations in the SFG measurements, SSS, SSP, SPS, PSS, SPP, PSP, PPS, and PPP, as noted in Sect. 2.3. The relation between the experimental configuration and the relevant tensor elements of $\boldsymbol{\chi}^{(2)}$ is the main topic of this subsection.

As discussed in Chap. 2, observed SFG signal is determined with the effective susceptibility $\chi^{(2)}_{\text{eff}}$,

$$\chi^{(2)}_{\text{eff}} = \boldsymbol{e}(\Omega) \cdot \boldsymbol{\chi}^{(2)}(\Omega, \omega_1, \omega_2) : \boldsymbol{e}(\omega_1)\boldsymbol{e}(\omega_2) \tag{2.23}$$

$$= \sum_{p,q,r}^{x\sim z} e_p(\Omega)\chi^{(2)}_{pqr}(\Omega, \omega_1, \omega_2)e_q(\omega_1)e_r(\omega_2).$$

The polarization dependence of SFG signal is represented with three vectors, $\boldsymbol{e}(\Omega)$, $\boldsymbol{e}(\omega_1)$ and $\boldsymbol{e}(\omega_2)$, associated to the polarizations of three lights. These vectors describe the electric fields at the respective frequencies $\omega = \Omega$, ω_1 and ω_2 *inside* the interface, and are determined from the directions of the electric fields in the bulk medium by the Fresnel transformation,

$$\boldsymbol{e}(\omega) = \boldsymbol{F}^{i\to j}(\omega) \cdot \hat{\boldsymbol{e}}^i(\omega). \tag{2.17}$$

Therefore, we treat the directions of the electric fields in the bulk medium $\hat{\boldsymbol{e}}^i(\omega)$ in relation to the experimental measurement.

Then let us consider the experimental configuration of Fig. 2.1, and observe the reflective SFG signal in the medium α. In Fig. 2.1, $\theta^\alpha(\omega_1)$ and $\theta^\alpha(\omega_2)$ denote the incident angles for the visible ω_1 and infrared ω_2 lights, respectively, and $\theta^\alpha(\Omega)$ the emission angle for the sum-frequency Ω light into the medium α. These angles specify the directions of the propagating lights in the medium α. The normalized wavevectors for $\Omega, \omega_1, \omega_2$ are accordingly given as

$$\hat{\boldsymbol{k}}^\alpha(\Omega) = \begin{pmatrix} \sin\theta^\alpha(\Omega) \\ 0 \\ \cos\theta^\alpha(\Omega) \end{pmatrix}, \quad \hat{\boldsymbol{k}}^\alpha(\omega_1) = \begin{pmatrix} \sin\theta^\alpha(\omega_1) \\ 0 \\ -\cos\theta^\alpha(\omega_1) \end{pmatrix}, \quad \hat{\boldsymbol{k}}^\alpha(\omega_2) = \begin{pmatrix} \sin\theta^\alpha(\omega_2) \\ 0 \\ -\cos\theta^\alpha(\omega_2) \end{pmatrix}.$$

Each propagating light has two kinds of polarizations, S and P, as stated in Sect. 2.3 (c1). The directions of the S- and P-polarized electric fields in the medium α are given by the following unit vectors,

- S-polarized:

$$\hat{\boldsymbol{e}}_S^\alpha(\Omega) = \begin{pmatrix} 0 \\ 1 \\ 0 \end{pmatrix}, \quad \hat{\boldsymbol{e}}_S^\alpha(\omega_1) = \begin{pmatrix} 0 \\ 1 \\ 0 \end{pmatrix}, \quad \hat{\boldsymbol{e}}_S^\alpha(\omega_2) = \begin{pmatrix} 0 \\ 1 \\ 0 \end{pmatrix},$$

- P-polarized:

$$\hat{\boldsymbol{e}}_P^\alpha(\Omega)=\begin{pmatrix} -\cos\theta^\alpha(\Omega) \\ 0 \\ \sin\theta^\alpha(\Omega) \end{pmatrix}, \quad \hat{\boldsymbol{e}}_P^\alpha(\omega_1)=\begin{pmatrix} \cos\theta^\alpha(\omega_1) \\ 0 \\ \sin\theta^\alpha(\omega_1) \end{pmatrix}, \quad \hat{\boldsymbol{e}}_P^\alpha(\omega_2)=\begin{pmatrix} \cos\theta^\alpha(\omega_2) \\ 0 \\ \sin\theta^\alpha(\omega_2) \end{pmatrix}.$$

The S-polarized electric field is perpendicular to the xz plane by definition, and three vectors at each frequency, $\hat{\boldsymbol{k}}^\alpha(\omega)$, $\hat{\boldsymbol{e}}_S^\alpha(\omega)$, $\hat{\boldsymbol{e}}_P^\alpha(\omega)$, are orthogonal to each other. For a given combination of light polarizations (SSS, SSP, ..., PPP), $\hat{\boldsymbol{e}}^\alpha(\omega)$ at each frequency ω $(= \Omega, \omega_1, \omega_2)$ is determined, either $\hat{\boldsymbol{e}}_S^\alpha(\omega)$ or $\hat{\boldsymbol{e}}_P^\alpha(\omega)$, in the respective order. Each $\hat{\boldsymbol{e}}^\alpha(\omega)$ $(\omega = \Omega, \omega_1, \omega_2)$ determines $\boldsymbol{e}(\omega)$ by the Fresnel transformation in Eq. (2.17). The set of $\boldsymbol{e}(\Omega)$, $\boldsymbol{e}(\omega_1)$ and $\boldsymbol{e}(\omega_2)$ thus obtained specify $\chi_{\text{eff}}^{(2)}$ in Eq. (2.23) and thereby the relevant tensor element(s) of $\chi_{pqr}^{(2)}(\Omega, \omega_1, \omega_2)$ involved in $\chi_{\text{eff}}^{(2)}$.

We note that all possible combinations of light polarizations do not necessarily provide meaningful information on the interface systems, because independent tensor elements of $\chi_{pqr}^{(2)}$ are much fewer than $3^3 = 27$ in most systems for symmetry reasons. In a typical case that two isotropic bulk phases are in contact with a flat interface, this whole system has $C_{\infty v}$ symmetry with the principal normal z axis. In such case, following seven elements of $\chi_{pqr}^{(2)}$ are non-zero,

$$\chi_{xxz}^{(2)} = \chi_{yyz}^{(2)}, \qquad \chi_{xzx}^{(2)} = \chi_{yzy}^{(2)}, \qquad \chi_{zxx}^{(2)} = \chi_{zyy}^{(2)}, \qquad \chi_{zzz}^{(2)}. \tag{3.48}$$

The other tensor elements vanish under $C_{\infty v}$, since all these elements include x or y odd times. These elements change their sign by the reflection for the yz or xz plane, respectively, though the system should be unchanged for its symmetry $C_{\infty v}$. We also note that the x and y axes are equivalent under $C_{\infty v}$. Consequently, $\chi_{\text{eff}}^{(2)}$ in Eq. (2.23) remains non-zero for only four combinations, SSP, SPS, PSS, and PPP. The amplitudes for the four combinations are represented with the relevant tensor elements and Fresnel factors as follows:

$$\chi_{\text{eff,SSP}}^{(2)} = F_{yy}^{\alpha\to\beta}(\Omega) F_{yy}^{\alpha\to\beta}(\omega_1) F_{zz}^{\alpha\to\beta}(\omega_2) \sin\theta^\alpha(\omega_2) \chi_{yyz}^{(2)}(\Omega, \omega_1, \omega_2), \tag{3.49}$$

$$\chi_{\text{eff,SPS}}^{(2)} = F_{yy}^{\alpha\to\beta}(\Omega) F_{zz}^{\alpha\to\beta}(\omega_1) F_{yy}^{\alpha\to\beta}(\omega_2) \sin\theta^\alpha(\omega_1) \chi_{yzy}^{(2)}(\Omega, \omega_1, \omega_2), \tag{3.50}$$

$$\chi^{(2)}_{\text{eff,PSS}} = F^{\alpha\to\beta}_{zz}(\Omega)F^{\alpha\to\beta}_{yy}(\omega_1)F^{\alpha\to\beta}_{yy}(\omega_2)\sin\theta^\alpha(\Omega)\chi^{(2)}_{zyy}(\Omega,\omega_1,\omega_2), \tag{3.51}$$

$$\begin{aligned}\chi^{(2)}_{\text{eff,PPP}} &= \\ &-F^{\alpha\to\beta}_{xx}(\Omega)F^{\alpha\to\beta}_{xx}(\omega_1)F^{\alpha\to\beta}_{zz}(\omega_2)\cos\theta^\alpha(\Omega)\cos\theta^\alpha(\omega_1)\sin\theta^\alpha(\omega_2)\chi^{(2)}_{xxz}(\Omega,\omega_1,\omega_2)\\ &-F^{\alpha\to\beta}_{xx}(\Omega)F^{\alpha\to\beta}_{zz}(\omega_1)F^{\alpha\to\beta}_{xx}(\omega_2)\cos\theta^\alpha(\Omega)\sin\theta^\alpha(\omega_1)\cos\theta^\alpha(\omega_2)\chi^{(2)}_{xzx}(\Omega,\omega_1,\omega_2)\\ &+F^{\alpha\to\beta}_{zz}(\Omega)F^{\alpha\to\beta}_{xx}(\omega_1)F^{\alpha\to\beta}_{xx}(\omega_2)\sin\theta^\alpha(\Omega)\cos\theta^\alpha(\omega_1)\cos\theta^\alpha(\omega_2)\chi^{(2)}_{zxx}(\Omega,\omega_1,\omega_2)\\ &+F^{\alpha\to\beta}_{zz}(\Omega)F^{\alpha\to\beta}_{zz}(\omega_1)F^{\alpha\to\beta}_{zz}(\omega_2)\sin\theta^\alpha(\Omega)\sin\theta^\alpha(\omega_1)\sin\theta^\alpha(\omega_2)\chi^{(2)}_{zzz}(\Omega,\omega_1,\omega_2).\end{aligned} \tag{3.52}$$

[Problem 3.4] Derive Eqs. (3.49), (3.50), (3.51), and (3.52). Also explain that other combinations of light polarization vanish for the $C_{\infty v}$ interface.

Note that the SSP, SPS, or PSS combination involves one tensor element of $\boldsymbol{\chi}^{(2)}$, whereas the PPP involves a linear combination of four elements of $\boldsymbol{\chi}^{(2)}$.

3.4 Solutions to Problems

3.4.1 *Formulas for Mixed States*

[Problem 3.1] Confirm that Eqs. (3.5), (3.6) and (3.8) are valid for the mixed states as well.

We confirm the three equations using the extended definition of ρ in Eq. (3.10).

1. Eq. (3.5): Expectation value.

$$\langle A(t)\rangle = \text{Tr}\,[\rho A] = \sum_n\sum_m \rho_{mn}(t)\langle n|A|m\rangle \tag{3.5}$$

$$= \sum_n\sum_m\left(\sum_j P^j c^j_n(t)^* c^j_m(t)\right)\langle n|A|m\rangle$$

$$= \sum_j P^j \sum_n\sum_m c^j_n(t)^* c^j_m(t)\langle n|A|m\rangle$$

$$= \sum_j^{\text{ensemble}} P^j\langle\psi^j(t)|A|\psi^j(t)\rangle = \overline{\langle A(t)\rangle}. \tag{3.9}$$

2. Eq. (3.6): Normalization condition.

$$\mathrm{Tr}\,[\rho] = \sum_n \rho_{nn} = 1 \tag{3.6}$$

$$= \sum_n \left(\sum_j P^j c_n^j(t)^* c_n^j(t) \right) = \sum_j P^j \sum_n |c_n^j(t)|^2 = 1,$$

because the probability in the ensemble P^j and each state ψ^j are both normalized to unity,

$$\sum_j P^j = 1 \qquad \text{and} \qquad |\psi^j(t)|^2 = \sum_n |c_n^j(t)|^2 = 1.$$

3. Eq. (3.8): Time development (Liouville equation).

$$i\hbar \frac{d\rho_{mn}(t)}{dt} = [H\rho - \rho H]_{mn} \tag{3.8}$$

$$= i\hbar \sum_j P^j \left(\frac{dc_n^j(t)^*}{dt} c_m^j(t) + c_n^j(t)^* \frac{dc_m^j(t)}{dt} \right)$$

$$= \sum_j P^j \left(-\sum_l H_{nl}^* c_l^j(t)^* c_m^j(t) + \sum_l c_n^j(t)^* H_{ml} c_l^j(t) \right)$$

$$= -\sum_l H_{ln} \sum_j P^j c_l^j(t)^* c_m^j(t) + \sum_l H_{ml} \sum_j P^j c_n^j(t)^* c_l^j(t)$$

$$= -\sum_l H_{ln} \rho_{ml} + \sum_l H_{ml} \rho_{ln} = [H\rho - \rho H]_{mn}\,.$$

3.4.2 Pure and Mixed States

[Problem 3.2] Prove the criterion Eq. (3.11) to distinguish the pure state and the mixed state. Recall the normalization condition of states and the Schwarz inequality for inner products.

First we note the normalization condition for the density matrix ρ,

$$\mathrm{Tr}\,[\rho] = \sum_n \sum_j P^j c_n^{j*} c_n^j = \sum_j P^j \left(\sum_n \left| c_n^j \right|^2 \right) = \sum_j P^j = 1 \qquad (P^j \ge 0).$$

$\mathrm{Tr}\left[\rho^2\right]$ is then calculated using Eq. (3.10),

$$\mathrm{Tr}\left[\rho^2\right] = \sum_{m,l} \rho_{ml}\rho_{lm} = \sum_{m,l}\left(\sum_i P^i c^i_m c^{i*}_l\right)\left(\sum_j P^j c^j_l c^{j*}_m\right)$$
$$= \sum_{i,j} P^i P^j \sum_m c^i_m c^{j*}_m \sum_l c^j_l c^{i*}_l = \sum_{i,j} P^i P^j \left|\sum_m c^i_m c^{j*}_m\right|^2 . \tag{3.53}$$

We invoke the Schwartz inequality,

$$|\boldsymbol{a}\cdot\boldsymbol{b}|^2 \leq |\boldsymbol{a}|^2|\boldsymbol{b}|^2 \tag{3.54}$$

for two arbitrary non-zero vectors $\boldsymbol{a}$ and $\boldsymbol{b}$, where the equal relation holds only for $\boldsymbol{a} = \boldsymbol{b}$. Suppose that $\boldsymbol{a} = \boldsymbol{c}^i$ and $\boldsymbol{b} = \boldsymbol{c}^j$, where $\boldsymbol{c}^i$ and $\boldsymbol{c}^j$ are the sets of coefficients c^i_m and c^j_m, respectively,

$$\boldsymbol{a} = \boldsymbol{c}^i = \begin{pmatrix} c^i_1 \\ c^i_2 \\ \vdots \end{pmatrix}, \qquad \boldsymbol{b} = \boldsymbol{c}^j = \begin{pmatrix} c^j_1 \\ c^j_2 \\ \vdots \end{pmatrix}.$$

Then Eq. (3.54) is written by

$$\left|\sum_m c^i_m c^{j*}_m\right|^2 \leq \left(\sum_m |c^i_m|^2\right)\left(\sum_l |c^j_l|^2\right). \tag{3.55}$$

Applying Eqs. (3.55) to (3.53), the following relation is derived,

$$\mathrm{Tr}\left[\rho^2\right] = \sum_{i,j} P^i P^j \left|\sum_m c^i_m c^{j*}_m\right|^2 \leq \sum_{i,j} P^i P^j \sum_m |c^i_m|^2 \sum_l |c^j_l|^2$$
$$= \left(\sum_i P^i \sum_m |c^i_m|^2\right)^2 = 1. \tag{3.56}$$

The equal relation in Eq. (3.56) is realized only in the case that $c^i_m = c^j_m$ for all m, indicating that the states i and j are identical. This condition should be satisfied between any pair of states in the ensemble. Therefore, the equal relation means the case that the ensemble of states consist of a single identical state (pure state).

3.4.3 Derivation of $\chi^{(2)}$

[Problem 3.3] Derive $\rho^{(2)}(t)$ in Eq. (3.28).

Equation (3.27) is expanded by substituting $\rho^{(1)}$ with Eq. (3.22).

- Left side of Eq. (3.27):

$$\frac{d\rho_{mn}^{(2)}}{dt} + i\omega_{mn}\rho_{mn}^{(2)} + \Gamma_{mn}\rho_{mn}^{(2)} = e^{-(i\omega_{mn}+\Gamma_{mn})t}\frac{d}{dt}\left\{e^{(i\omega_{mn}+\Gamma_{mn})t}\rho_{mn}^{(2)}(t)\right\}$$

- Right side of Eq. (3.27):

$$-\frac{i}{\hbar}\left[H', \rho^{(1)}\right]_{mn}$$

$$= \frac{i}{\hbar}\sum_{l}\sum_{p}^{x-z}\left(\langle m|\mu_p|l\rangle\rho_{ln}^{(1)}(t) - \rho_{ml}^{(1)}(t)\langle l|\mu_p|n\rangle\right)\sum_{j}E_p(\omega_j)e^{-i\omega_j t}$$

$$= \frac{i}{\hbar^2}\sum_{l}\sum_{p,q}^{x-z}\sum_{j,k}\left\{\langle m|\mu_p|l\rangle(\rho_l^{(0)} - \rho_n^{(0)})\langle l|\mu_q|n\rangle\frac{1}{\omega_k - \omega_{ln} + i\Gamma_{ln}}\right.$$

$$\left.-(\rho_m^{(0)} - \rho_l^{(0)})\langle m|\mu_q|l\rangle\frac{1}{\omega_k - \omega_{ml} + i\Gamma_{ml}}\langle l|\mu_p|n\rangle\right\}E_p(\omega_j)E_q(\omega_k)e^{-i(\omega_j+\omega_k)t}.$$

Therefore,

$$\frac{d}{dt}\left\{e^{(i\omega_{mn}+\Gamma_{mn})t}\rho_{mn}^{(2)}(t)\right\}$$

$$= \frac{i}{\hbar^2}e^{(i\omega_{mn}+\Gamma_{mn})t}\sum_{l}\sum_{p,q}^{x-z}\sum_{j,k}\left\{\langle m|\mu_p|l\rangle(\rho_l^{(0)} - \rho_n^{(0)})\langle l|\mu_q|n\rangle\frac{1}{\omega_k - \omega_{ln} + i\Gamma_{ln}}\right.$$

$$\left.-(\rho_m^{(0)} - \rho_l^{(0)})\langle m|\mu_q|l\rangle\frac{1}{\omega_k - \omega_{ml} + i\Gamma_{ml}}\langle l|\mu_p|n\rangle\right\}E_p(\omega_j)E_q(\omega_k)e^{-i(\omega_j+\omega_k)t}.$$

This equation can be integrated by t from $-\infty$ to t, using the following relation,

$$\int_{-\infty}^{t} e^{(i\omega_{mn}+\Gamma_{mn})\tau}e^{-i(\omega_j+\omega_k)\tau}d\tau = \frac{e^{(i\omega_{mn}+\Gamma_{mn})t}e^{-i(\omega_j+\omega_k)t}}{-i(\omega_j+\omega_k) + i\omega_{mn} + \Gamma_{mn}} \qquad (\Gamma_{mn} > 0).$$

Then $\rho^{(2)}$ in Eq. (3.28) is derived,

$$
\begin{aligned}
\rho_{mn}^{(2)}(t) &= \frac{i}{\hbar^2}\sum_l\sum_{j,k}\sum_{p,q}^{x-z}\left\{\langle m|\mu_p|l\rangle\left(\rho_l^{(0)}-\rho_n^{(0)}\right)\langle l|\mu_q|n\rangle\frac{1}{\omega_k-\omega_{ln}+i\Gamma_{ln}}\right.\\
&\left.-\left(\rho_m^{(0)}-\rho_l^{(0)}\right)\langle m|\mu_q|l\rangle\frac{1}{\omega_k-\omega_{ml}+i\Gamma_{ml}}\langle l|\mu_p|n\rangle\right\}E_p(\omega_j)E_q(\omega_k)\frac{e^{-i(\omega_j+\omega_k)t}}{-i(\omega_j+\omega_k)+i\omega_{mn}+\Gamma_{mn}}\\
&= -\frac{1}{\hbar^2}\sum_l\sum_{j,k}\sum_{p,q}^{x-z}\left\{\left(\rho_l^{(0)}-\rho_n^{(0)}\right)\frac{\langle m|\mu_p|l\rangle\langle l|\mu_q|n\rangle}{\omega_k-\omega_{ln}+i\Gamma_{ln}}-\left(\rho_m^{(0)}-\rho_l^{(0)}\right)\frac{\langle m|\mu_q|l\rangle\langle l|\mu_p|n\rangle}{\omega_k-\omega_{ml}+i\Gamma_{ml}}\right\}\\
&\cdot\frac{1}{(\omega_j+\omega_k)-\omega_{mn}+i\Gamma_{mn}}E_p(\omega_j)E_q(\omega_k)e^{-i(\omega_j+\omega_k)t}.
\end{aligned}
\tag{3.28}
$$

3.4.4 *Effective $\chi^{(2)}$ Formula*

[Problem 3.4] Derive Eqs. (3.49), (3.50), (3.51), and (3.52). Also explain that other combinations of light polarization vanish for the $C_{\infty v}$ interface.

Equations (3.49), (3.50), (3.51), and (3.52) are derived from the definition of $\chi_{\text{eff}}^{(2)}$ in Eq. (2.23) and the unit vectors of electric fields below.

- S polarization:

$$
\hat{\boldsymbol{e}}_{\mathrm{S}}^{\alpha}(\Omega)=\hat{\boldsymbol{e}}_{\mathrm{S}}^{\alpha}(\omega_1)=\hat{\boldsymbol{e}}_{\mathrm{S}}^{\alpha}(\omega_2)=\begin{pmatrix}0\\1\\0\end{pmatrix}
$$

- P polarization:

$$
\hat{\boldsymbol{e}}_{\mathrm{P}}^{\alpha}(\Omega)=\begin{pmatrix}-\cos\theta^{\alpha}(\Omega)\\0\\\sin\theta^{\alpha}(\Omega)\end{pmatrix},\quad \hat{\boldsymbol{e}}_{\mathrm{P}}^{\alpha}(\omega_1)=\begin{pmatrix}\cos\theta^{\alpha}(\omega_1)\\0\\\sin\theta^{\alpha}(\omega_1)\end{pmatrix},\quad \hat{\boldsymbol{e}}_{\mathrm{P}}^{\alpha}(\omega_2)=\begin{pmatrix}\cos\theta^{\alpha}(\omega_2)\\0\\\sin\theta^{\alpha}(\omega_2)\end{pmatrix}.
$$

In the SSP case,

$$
\chi_{\text{eff,SSP}}^{(2)}=\sum_{q,r,u}^{x\sim z}e_{\mathrm{S},q}(\Omega)\chi_{qru}^{(2)}(\Omega,\omega_1,\omega_2)e_{\mathrm{S},r}(\omega_1)e_{\mathrm{P},u}(\omega_2)
$$

$$= \sum_{q,r,u}^{x\sim z} \left(F_{qq}^{\alpha\to\beta}(\Omega)\hat{e}_{\mathrm{S},q}^{\alpha}(\Omega)\right) \chi_{qru}^{(2)}(\Omega,\omega_1,\omega_2) \left(F_{rr}^{\alpha\to\beta}(\omega_1)\hat{e}_{\mathrm{S},r}^{\alpha}(\omega_1)\right)$$

$$\left(F_{uu}^{\alpha\to\beta}(\omega_2)\hat{e}_{\mathrm{P},u}^{\alpha}(\omega_2)\right)$$

$$= \left(F_{yy}^{\alpha\to\beta}(\Omega)\right) \chi_{yyx}^{(2)}(\Omega,\omega_1,\omega_2) \left(F_{yy}^{\alpha\to\beta}(\omega_1)\right) \left(F_{xx}^{\alpha\to\beta}(\omega_2)\cos\theta^{\alpha}(\omega_2)\right)$$

$$+ \left(F_{yy}^{\alpha\to\beta}(\Omega)\right) \chi_{yyz}^{(2)}(\Omega,\omega_1,\omega_2) \left(F_{yy}^{\alpha\to\beta}(\omega_1)\right) \left(F_{zz}^{\alpha\to\beta}(\omega_2)\sin\theta^{\alpha}(\omega_2)\right)$$

$$= F_{yy}^{\alpha\to\beta}(\Omega) F_{yy}^{\alpha\to\beta}(\omega_1) F_{zz}^{\alpha\to\beta}(\omega_2)\sin\theta^{\alpha}(\omega_2)\chi_{yyz}^{(2)}(\Omega,\omega_1,\omega_2), \tag{3.49}$$

where the term in a shade box vanishes for the selection rule. The last expression takes account of $\chi_{yyx}^{(2)} = 0$ for the $C_{\infty v}$ interface.

Similar expressions can be derived in the SPS and PSS cases.

$$\chi_{\mathrm{eff,SPS}}^{(2)} = \sum_{q,r,u}^{x\sim z} e_{\mathrm{S},q}(\Omega)\chi_{qru}^{(2)}(\Omega,\omega_1,\omega_2)e_{\mathrm{P},r}(\omega_1)e_{\mathrm{S},u}(\omega_2)$$

$$= F_{yy}^{\alpha\to\beta}(\Omega) F_{zz}^{\alpha\to\beta}(\omega_1) F_{yy}^{\alpha\to\beta}(\omega_2)\sin\theta^{\alpha}(\omega_1)\chi_{yzy}^{(2)}(\Omega,\omega_1,\omega_2), \tag{3.50}$$

$$\chi_{\mathrm{eff,PSS}}^{(2)} = \sum_{q,r,u}^{x\sim z} e_{\mathrm{P},q}(\Omega)\chi_{qru}^{(2)}(\Omega,\omega_1,\omega_2)e_{\mathrm{S},r}(\omega_1)e_{\mathrm{S},u}(\omega_2)$$

$$= F_{zz}^{\alpha\to\beta}(\Omega) F_{yy}^{\alpha\to\beta}(\omega_1) F_{yy}^{\alpha\to\beta}(\omega_2)\sin\theta^{\alpha}(\Omega)\chi_{zyy}^{(2)}(\Omega,\omega_1,\omega_2). \tag{3.51}$$

In the PPP case,

$$\chi_{\mathrm{eff,PPP}}^{(2)} = \sum_{q,r,u}^{x\sim z} e_{\mathrm{P},q}(\Omega)\chi_{qru}^{(2)}(\Omega,\omega_1,\omega_2)e_{\mathrm{P},r}(\omega_1)e_{\mathrm{P},u}(\omega_2)$$

$$= \sum_{q,r,u}^{x\sim z} \left(F_{qq}^{\alpha\to\beta}(\Omega)\hat{e}_{\mathrm{P},q}^{\alpha}(\Omega)\right) \chi_{qru}^{(2)}(\Omega,\omega_1,\omega_2) \left(F_{rr}^{\alpha\to\beta}(\omega_1)\hat{e}_{\mathrm{P},r}^{\alpha}(\omega_1)\right)$$

$$\left(F_{uu}^{\alpha\to\beta}(\omega_2)\hat{e}_{\mathrm{P},u}^{\alpha}(\omega_2)\right)$$

$$= \left(-F_{xx}^{\alpha\to\beta}(\Omega)\cos\theta^{\alpha}(\Omega)\right)\left(F_{xx}^{\alpha\to\beta}(\omega_1)\cos\theta^{\alpha}(\omega_1)\right)\left(F_{xx}^{\alpha\to\beta}(\omega_2)\cos\theta^{\alpha}(\omega_2)\right)\chi_{xxx}^{(2)}$$
$$+ \left(-F_{xx}^{\alpha\to\beta}(\Omega)\cos\theta^{\alpha}(\Omega)\right)\left(F_{xx}^{\alpha\to\beta}(\omega_1)\cos\theta^{\alpha}(\omega_1)\right)\left(F_{zz}^{\alpha\to\beta}(\omega_2)\sin\theta^{\alpha}(\omega_2)\right)\chi_{xxz}^{(2)}$$
$$+ \left(-F_{xx}^{\alpha\to\beta}(\Omega)\cos\theta^{\alpha}(\Omega)\right)\left(F_{zz}^{\alpha\to\beta}(\omega_1)\sin\theta^{\alpha}(\omega_1)\right)\left(F_{xx}^{\alpha\to\beta}(\omega_2)\cos\theta^{\alpha}(\omega_2)\right)\chi_{xzx}^{(2)}$$
$$+ \left(F_{zz}^{\alpha\to\beta}(\Omega)\sin\theta^{\alpha}(\Omega)\right)\left(F_{xx}^{\alpha\to\beta}(\omega_1)\cos\theta^{\alpha}(\omega_1)\right)\left(F_{xx}^{\alpha\to\beta}(\omega_2)\cos\theta^{\alpha}(\omega_2)\right)\chi_{zxx}^{(2)}$$
$$+ \left(-F_{xx}^{\alpha\to\beta}(\Omega)\cos\theta^{\alpha}(\Omega)\right)\left(F_{zz}^{\alpha\to\beta}(\omega_1)\sin\theta^{\alpha}(\omega_1)\right)\left(F_{zz}^{\alpha\to\beta}(\omega_2)\sin\theta^{\alpha}(\omega_2)\right)\chi_{xzz}^{(2)}$$
$$+ \left(F_{zz}^{\alpha\to\beta}(\Omega)\sin\theta^{\alpha}(\Omega)\right)\left(F_{xx}^{\alpha\to\beta}(\omega_1)\cos\theta^{\alpha}(\omega_1)\right)\left(F_{zz}^{\alpha\to\beta}(\omega_2)\sin\theta^{\alpha}(\omega_2)\right)\chi_{zxz}^{(2)}$$

$$
\begin{aligned}
&+ \left(F_{zz}^{\alpha\to\beta}(\Omega)\sin\theta^\alpha(\Omega)\right)\left(F_{zz}^{\alpha\to\beta}(\omega_1)\sin\theta^\alpha(\omega_1)\right)\left(F_{xx}^{\alpha\to\beta}(\omega_2)\cos\theta^\alpha(\omega_2)\right)\chi_{zzx}^{(2)} \\
&+ \left(F_{zz}^{\alpha\to\beta}(\Omega)\sin\theta^\alpha(\Omega)\right)\left(F_{zz}^{\alpha\to\beta}(\omega_1)\sin\theta^\alpha(\omega_1)\right)\left(F_{zz}^{\alpha\to\beta}(\omega_2)\sin\theta^\alpha(\omega_2)\right)\chi_{zzz}^{(2)} \\
&= \left(-F_{xx}^{\alpha\to\beta}(\Omega)\cos\theta^\alpha(\Omega)\right)\left(F_{xx}^{\alpha\to\beta}(\omega_1)\cos\theta^\alpha(\omega_1)\right)\left(F_{zz}^{\alpha\to\beta}(\omega_2)\sin\theta^\alpha(\omega_2)\right)\chi_{xxz}^{(2)} \\
&+ \left(-F_{xx}^{\alpha\to\beta}(\Omega)\cos\theta^\alpha(\Omega)\right)\left(F_{zz}^{\alpha\to\beta}(\omega_1)\sin\theta^\alpha(\omega_1)\right)\left(F_{xx}^{\alpha\to\beta}(\omega_2)\cos\theta^\alpha(\omega_2)\right)\chi_{xzx}^{(2)} \\
&+ \left(F_{zz}^{\alpha\to\beta}(\Omega)\sin\theta^\alpha(\Omega)\right)\left(F_{xx}^{\alpha\to\beta}(\omega_1)\cos\theta^\alpha(\omega_1)\right)\left(F_{xx}^{\alpha\to\beta}(\omega_2)\cos\theta^\alpha(\omega_2)\right)\chi_{zxx}^{(2)} \\
&+ \left(F_{zz}^{\alpha\to\beta}(\Omega)\sin\theta^\alpha(\Omega)\right)\left(F_{zz}^{\alpha\to\beta}(\omega_1)\sin\theta^\alpha(\omega_1)\right)\left(F_{zz}^{\alpha\to\beta}(\omega_2)\sin\theta^\alpha(\omega_2)\right)\chi_{zzz}^{(2)}.
\end{aligned}
\tag{3.52}
$$

The shaded components of $\chi^{(2)}$ also vanish for the $C_{\infty v}$ interface to derive the last expression.

The other combinations of polarization result in null amplitude and yield no SFG signal for the $C_{\infty v}$ interface. The relevant tensor components of $\chi^{(2)}$ included in $\chi_{\text{eff}}^{(2)}$ for the other polarization combinations are listed in the following.

- SPP — $\chi_{yxx}^{(2)}, \chi_{yxz}^{(2)}, \chi_{yzx}^{(2)}, \chi_{yzz}^{(2)}$.
- PSP — $\chi_{xyx}^{(2)}, \chi_{xyz}^{(2)}, \chi_{zyx}^{(2)}, \chi_{zyz}^{(2)}$
- PPS — $\chi_{xxy}^{(2)}, \chi_{xzy}^{(2)}, \chi_{zxy}^{(2)}, \chi_{zzy}^{(2)}$
- SSS — $\chi_{yyy}^{(2)}$

We notice that all these components vanish.

Appendix

A.1 Off-Diagonal Elements of Density Matrix

We have learned in Sect. 3.1 that the density matrix can represent statistical ensemble of states and a pure state in the common formulas. It is instructive to illustrate the distinction between a superposition of quantum states and a statistical ensemble of states. This example is useful to clarify the concept of coherence.

Let us consider two wavefunctions, ϕ_1 and ϕ_2, for example. If the two states are superposed in the quantum sense, the state is represented by a wavefunction,

$$
\psi(t) = c_1(t)\phi_1 + c_2(t)\phi_2 \qquad \left(\text{where } |c_1(t)|^2 + |c_2(t)|^2 = 1\right)
$$

or equivalently by a density matrix

$$
\rho^{\text{pure}}(t) = \begin{pmatrix} c_1c_1^* & c_1c_2^* \\ c_2c_1^* & c_2c_2^* \end{pmatrix}. \tag{3.57}
$$

The above state in Eq. (3.57) is a pure state, where the probabilities of finding the states ϕ_1, ϕ_2 are $P^1 = c_1 c_1^*$, $P^2 = c_2 c_2^*$, respectively. On the other hand, we consider a statistical ensemble consisting of ϕ_1 and ϕ_2 with the probabilities being P^1 and P^2 respectively. This is a mixed state, presented by the following density matrix

$$\rho^{\text{mixed}}(t) = P^1 \begin{pmatrix} 1 & 0 \\ 0 & 0 \end{pmatrix} + P^2 \begin{pmatrix} 0 & 0 \\ 0 & 1 \end{pmatrix} = \begin{pmatrix} c_1 c_1^* & 0 \\ 0 & c_2 c_2^* \end{pmatrix}, \tag{3.58}$$

where $c_1 c_1^* = P^1$ (> 0), $c_2 c_2^* = P^2$ (> 0). Comparing ρ^{pure} and ρ^{mixed}, we see that the diagonal elements are common, indicating that the probabilities of finding ϕ_1, ϕ_2 are the same. However, the off-diagonal elements are distinct between Eqs. (3.57) and (3.58).

The off-diagonal element $\rho_{12} = \overline{c_1 c_2^*}$ implies correlation between the coefficients (c_1 and c_2) of the constituent states (ϕ_1 and ϕ_2). To illustrate the physical meaning of off-diagonal elements, we discuss the following two cases that exhibit no off-diagonal elements. Let us consider an ensemble of states $\{\psi^1, \psi^2, \psi^3, \cdots\}$, where each sample ψ^j is a superposition of two states ($\psi^j = c_1^j \phi_1 + c_2^j \phi_2$) and has a probability of P^j in the ensemble.

Case 1. First case is an extreme one that ψ^j is either ϕ_1 ($c_2^j = 0$) or ϕ_2 ($c_1^j = 0$). Then the diagonal element ρ_{12} vanishes, $\overline{c_1 c_2^*} = \sum_j P^j c_1^j c_2^{j*} = 0$, because either c_1^j or c_2^j is zero in each term of j. This case shows that the off-diagonal element ρ_{12} arises from quantum superposition between ϕ_1 and ϕ_2.

Case 2. The quantum superposition is not a sufficient condition for the off-diagonal elements. We consider another ensemble of $\{\psi^j = c_1^j \phi_1 + c_2^j \phi_2\}$, where the coefficient $c_m^j = |c_m^j| \exp(i\theta_m^j)$ $(m = 1, 2)$ has a definite amplitude $|c_m|$ but a random phase factor θ_m^j $(m = 1, 2)$. Then the ensemble average of the diagonal element has a definite value $|c_m|^2$ while the off-diagonal element vanishes, e.g.

$$\overline{c_1 c_1^*} = \sum_j^{\text{ensemble}} P^j c_1^j c_1^{j*} = \sum_j P^j |c_1| \exp(i\theta_1^j) |c_1| \exp(-i\theta_1^j) = \sum_j P^j |c_1|^2 = |c_1|^2,$$

$$\begin{aligned} \overline{c_1 c_2^*} &= \sum_j^{\text{ensemble}} P^j c_1^j c_2^{j*} = \sum_j P^j |c_1| \exp(i\theta_1^j) |c_2| \exp(-i\theta_2^j) \\ &= |c_1||c_2| \sum_j P^j \exp\{i(\theta_1^j - \theta_2^j)\} = |c_1||c_2| \overline{\exp\{i(\theta_1 - \theta_2)\}} = 0, \end{aligned}$$

because the average of random phase distribution results in cancellation.

From the two cases, we find that the diagonal element is determined by the square of amplitude $|c_m|^2$, irrespective of its phase. On the other hand, the off-diagonal element is sensitive to the relative phase of the two superposition coefficients, c_1

and c_2. If the phases of the two coefficients have no correlation, the off-diagonal element disappears via the ensemble average.

In summary, a finite off-diagonal element ρ_{12} indicates that there involves a definite quantum superposition between the states ϕ_1 and ϕ_2 with some phase relation. In such case the coherence is present between the two states ϕ_1 and ϕ_2.

A.2 Interaction Energy of Nonmagnetic Materials

In Chap. 3, the perturbation Hamiltonian by the irradiated light is given by $H' = -\boldsymbol{\mu} \cdot \boldsymbol{E}(t)$ in Eq. (7.2), which represents the interaction to the electric field $\boldsymbol{E}$. This is based on the assumption that interaction energy with the magnetic field of light is negligible compared to that with the electric field for ordinary nonmagnetic materials. Here we estimate their relative orders of magnitude to justify this assumption.

The interaction energy of material with the magnetic field H is

$$U_m = -mH = -\chi_m H^2,$$

where m is the magnetization, and χ_m is the magnetic susceptibility, which is dimensionless in the cgs Gauss units. (A possible factor 1/2 is omitted for simplicity to estimate the order of magnitude.) Typical values of χ_m for nonmagnetic materials are in the range of $|\chi_m| = 10^{-4} \sim 10^{-6}$ for paramagnetic materials, and $|\chi_m| = 10^{-6} \sim 10^{-7}$ for diamagnetic materials [5]. On the other hand, the interaction energy with the electric field E is analogously presented by

$$U_e = -PE = -\chi_e E^2,$$

where χ_e is the electric susceptibility, also dimensionless in the cgs Gauss units. Typical range of χ_e is in the order $|\chi_e| = 10^{-1} \sim 10^{-2}$. For example, χ_m and χ_e of liquid benzene are roughly estimated to

$$\chi_m \simeq -5.48 \times 10^{-5}\ \mathrm{cm^3/mol} \cdot \frac{0.8765\ \mathrm{g/cm^3}}{78.11\ \mathrm{g/mol}} = -6.1 \times 10^{-7},$$

$$\chi_e \simeq \frac{1}{4\pi}(2.2825 - 1) = 1.0 \times 10^{-1},$$

using the experimental values of molar magnetic susceptibility ($-5.48 \times 10^{-5}\ \mathrm{cm^3/mol}$), density ($0.8765\ \mathrm{g/cm^3}$), molecular weight (78.11 g/mol), and dielectric constant (2.2825) [5].

The light field consists of electric and magnetic fields, whose amplitudes are related to $|E| \simeq |H|$, as seen in Eq. (2.14). Therefore, the ratio of electric and magnetic interactions U_m/U_e is evaluated to

$$\left|\frac{U_m}{U_e}\right| = \left|\frac{\chi_m H^2}{\chi_e E^2}\right| \simeq \left|\frac{\chi_m}{\chi_e}\right| \ll 1$$

This relation confirms that typical interaction energy with magnetic field is much smaller than that with the electric field in ordinary nonmagnetic materials.

A.3 Polarizability Approximation for Raman Tensor

This subsection shows that the Raman tensor of Eq. (3.37) is approximated with the polarizability in electronically off-resonant conditions [10, 11, 14]. The Raman tensor defined in Eq. (3.37) includes the states g, n and m, which denote the initial, intermediate and final states, respectively. These states are represented as products of electronic and vibrational states on the basis of the Born-Oppenheimer approximation,

$$\begin{aligned} |g\rangle &= \left|g^e(\boldsymbol{r}, \boldsymbol{R})\right\rangle \left|g^v(\boldsymbol{R})\right\rangle, \\ |n\rangle &= \left|n^e(\boldsymbol{r}, \boldsymbol{R})\right\rangle \left|n^v(\boldsymbol{R})\right\rangle, \\ |m\rangle &= \left|m^e(\boldsymbol{r}, \boldsymbol{R})\right\rangle \left|m^v(\boldsymbol{R})\right\rangle = \left|g^e\right\rangle \left|m^v\right\rangle, \end{aligned} \tag{3.59}$$

where the superscript e designates the electronic states and v the vibrational states. $\boldsymbol{r}$ and $\boldsymbol{R}$ denote the coordinates for electrons and nuclei, respectively. In the ordinary Raman process illustrated in Fig. 3.4, both the initial state g and the final state m are supposed to be the electronically ground state g^e, while their vibrational states are different. Here we assume that the ground electronic state g^e is unique and not degenerated. The total energy is also represented as the sum of electronic and vibrational energies,

$$\begin{aligned} \mathcal{E}_g &= \mathcal{E}_g^e + \mathcal{E}_g^v, \\ \mathcal{E}_n &= \mathcal{E}_n^e + \mathcal{E}_n^v, \\ \mathcal{E}_m &= \mathcal{E}_m^e + \mathcal{E}_m^v = \mathcal{E}_g^e + \mathcal{E}_m^v. \end{aligned} \tag{3.60}$$

We substitute Eqs. (3.59) and (3.60) into the expression of Raman tensor. *If the excitation energy* $\hbar\Omega$ *is off resonant and thus the condition* $\left|\left(\mathcal{E}_n^e - \mathcal{E}_g^e\right) - \hbar\Omega\right| \gg \left|\mathcal{E}_n^v - \mathcal{E}_g^v\right|$ *is satisfied* (see Fig. 3.4), then the denominators of Eq. (3.37) are approximated to be

$$* \frac{1}{\Omega - \omega_{ng} + i\Gamma_{ng}} = \left[\Omega - \frac{\left(\mathcal{E}_n^e - \mathcal{E}_g^e\right) + \left(\mathcal{E}_n^v - \mathcal{E}_g^v\right)}{\hbar} + i\Gamma_{ng}\right]^{-1}$$

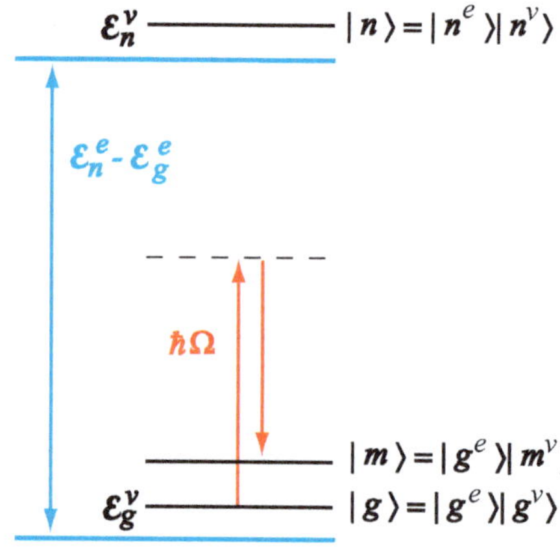

Fig. 3.4 Raman process in off-resonant condition

$$\cong \left[\Omega - \frac{\mathcal{E}_n^e - \mathcal{E}_g^e}{\hbar} + i\Gamma_{ng}\right]^{-1} = \frac{1}{\Omega - \omega_{ng}^e + i\Gamma_{ng}}, \tag{3.61}$$

$$* \frac{1}{\Omega + \omega_{mn} + i\Gamma_{mn}} \cong \frac{1}{\Omega + \omega_{nm}^e + i\Gamma_{nm}} = \frac{1}{\Omega + \omega_{ng}^e + i\Gamma_{ng}}. \tag{3.62}$$

Using these approximations of Eqs. (3.61) and (3.62),[6] the Raman tensor element of Eq. (3.37) is represented by

$$\begin{aligned}
\langle g|\alpha_{pq}(\Omega)|m\rangle \cong &-\sum_{n^e}\sum_{n^v}\left[\frac{\langle g^v|\langle g^e|\mu_p|n^e\rangle|n^v\rangle\langle n^v|\langle n^e|\mu_q|m^e\rangle|m^v\rangle}{\Omega - \omega_{ng}^e + i\Gamma_{ng}}\right.\\
&\left.-\frac{\langle g^v|\langle g^e|\mu_q|n^e\rangle|n^v\rangle\langle n^v|\langle n^e|\mu_p|m^e\rangle|m^v\rangle}{\Omega + \omega_{ng}^e + i\Gamma_{ng}}\right]\\
&= \left\langle g^v\left|-\sum_{n^e}\left[\frac{\langle g^e|\mu_p|n^e\rangle\langle n^e|\mu_q|g^e\rangle}{\Omega - \omega_{ng}^e + i\Gamma_{ng}} - \frac{\langle g^e|\mu_q|n^e\rangle\langle n^e|\mu_p|g^e\rangle}{\Omega + \omega_{ng}^e + i\Gamma_{ng}}\right]\right|m^v\right\rangle\\
&= \langle g^v(\boldsymbol{R})|\alpha_{pq}(\Omega, \boldsymbol{R})|m^v(\boldsymbol{R})\rangle,
\end{aligned} \tag{3.63}$$

where the completeness condition $\sum_{n^v}|n^v\rangle\langle n^v| = 1$ has been adopted. The final expression of $\alpha_{pq}(\Omega, \boldsymbol{R})$ is no longer an explicit function of the electronic coordinate $\boldsymbol{r}$, as $\boldsymbol{r}$ is already integrated out in Eq. (3.63). $\alpha_{pq}(\Omega, \boldsymbol{R})$ means the polarizability of the electronic ground state at the frequency Ω and given nuclear

[6]Equations (3.61) and (3.62) also involve the approximation of the dampling factor Γ for electronic states. Actually the damping factor Γ is insignificant here in the electronically off-resonant conditions, and often neglected in the present discussion.

coordinates $\boldsymbol{R}$. When Ω is far off resonant from the electronic excited state (i.e. $\hbar\Omega \ll \mathcal{E}_n^e - \mathcal{E}_g^e$), the dispersion of α can be neglected and $\alpha_{pq}(\Omega, \boldsymbol{R})$ is further approximated with the static polarizability, $\alpha_{pq}(\Omega, \boldsymbol{R}) \cong \alpha_{pq}(\Omega = 0, \boldsymbol{R})$.

Bibliography

1. Boyd RW (2003) Nonlinear optics. Academic, Amsterdam
2. Dirac PAM (1927) The quantum theory of dispersion. Prof R Soc Lond A114:710–728
3. Goldstein H, Poole C, Safko J (2001) Classical mechanics, 3rd edn. Pearson, San Francisco
4. Hayashi M, Lin SH, Raschke MB, Shen YR (2002) A molecular theory for doubly resonant IR-UV-vis sum-frequency generation. J Phys Chem A 106:2271–2282
5. Haynes WM (ed) (2012) CRC handbook of chemistry and physics, 93rd edn. CRC Press, Boca Raton
6. Ishiyama T, Morita A (2006) Intermolecular correlation effect in sum frequency generation spectroscopy of electrolyte aqueous solution. Chem Phys Lett 431:78–82
7. Ishiyama T, Morita A (2007) Molecular dynamics study of gas-liquid aqueous sodium halide interfaces II. analysis of vibrational sum frequency generation spectra. J Phys Chem C 111:738–748
8. Kramers HA, Heisenberg W (1925) Üder die streuung von strahlung durch atome. Z Phys 31:681–708
9. Lambert AG, Davies PB, Neivandt DJ (2005) Implementing the theory of sum frequency generation vibrational spectroscopy: a tutorial review. Appl Spec Rev 40:103–145
10. Long DA (1977) Raman spectroscopy. McGraw-Hill, New York
11. Long DA (2002) The raman effect. Wiley, New York
12. Loudon R (2000) The quantum theory of light, 3rd edn. Oxford University Press, Oxford
13. Mukamel S (1995) Principles of nonlinear optical spectroscopy. Oxford University Press, New York
14. Placzek G (1934) Rayleigh-streuung und Raman-effekt. Handbuch der Radiologie 6:205–374
15. Shen YR (1984) The principle of nonlinear optics. Wiley, New York
16. Weiss PA, Silverstein DW, Jensen L (2014) Non-condon effects on the doubly resonant sum frequency generation of rhodamine 6G. J Phys Chem Lett 5:329–335
17. Zaleśny R, Bartkowiak W, Champagne B (2003) Ab initio calculations of doubly resonant sum-frequency generation second-order polarizabilities of LiH. Chem Phys Lett 380:549–555
18. Zheng R-H, Wei W-M, Jing Y-Y, Liu H, Shi Q (2013) Theoretical study of doubly resonant sum-frequency vibrational spectroscopy for 1,1-bi-2-naphthol molecules on water surface. J Phys Chem C 117:11117–11123

Chapter 4
Two Computational Schemes of $\chi^{(2)}$

Abstract Now we provide the computational schemes of SFG spectroscopy on the basis of the microscopic theory described in the preceding Chap. 3. The theme of this chapter is to define two methods of calculating $\chi^{(2)}$ spectra, via energy representation and time-dependent representation. These methods can be utilized to calculate the SFG spectra by molecular dynamics (MD) simulation. These methods connect the formal theory of $\chi^{(2)}$ to actual spectra of interfaces that consist of molecules, and thus open new routes of SFG analysis with the aid of MD simulation.

Keywords Energy representation · Polarization analysis · Time correlation function

4.1 Energy Representation

Now we develop a computational scheme of SFG spectroscopy on the basis of the perturbation formula of $\chi^{(2)}$, which is capable of describing experimental SFG spectra of actual interfaces.

Equation (3.46) provides a recipe to construct $\chi^{(2)}$ from the molecular hyperpolarizability $\boldsymbol{\alpha}_l^{(2),\mathrm{mol}}$ and the rotational matrix $\mathcal{D}_l$ of constituent molecules. The molecular hyperpolarizability $\boldsymbol{\alpha}^{(2)}$ is represented as the second-order susceptibility of a single molecule as discussed in Sect. 3.2, and it consists of the vibrational resonant term $\boldsymbol{\alpha}^{(2),\mathrm{res}}$ and nonresonant term $\boldsymbol{\alpha}^{(2),\mathrm{nonres}}$ after Eq. (3.33), $\boldsymbol{\alpha}^{(2)} = \boldsymbol{\alpha}^{(2),\mathrm{res}} + \boldsymbol{\alpha}^{(2),\mathrm{nonres}}$. The former term $\boldsymbol{\alpha}^{(2),\mathrm{res}}$ is represented in the same way as Eq. (3.36) by

$$\alpha_{pqr}^{(2),\mathrm{res}}(\Omega, \omega_1, \omega_2) = -\frac{1}{\hbar}\sum_{g,m}\left(\rho_g^{(0)} - \rho_m^{(0)}\right)\frac{\langle g|\alpha_{pq}(\Omega)|m\rangle\langle m|\mu_r|g\rangle}{\omega_2 - \omega_{mg} + i\Gamma_{mg}}, \tag{4.1}$$

where α_{pq} and μ_r refer to polarizability and dipole moment of a *single molecule*, respectively. In the following we further introduce some approximations to Eq. (4.1) for convenience of practical analysis of SFG spectra.

A. Morita, *Theory of Sum Frequency Generation Spectroscopy*,
Lecture Notes in Chemistry 97, https://doi.org/10.1007/978-981-13-1607-4_4

Then we introduce following four assumptions, (i)–(iv).

(i) The vibrational mode(s) in question are intramolecular vibrations, such as O-H or C-H stretching. The vibrational SFG spectroscopy deals with these vibrations in most cases.
(ii) The energy gap of the vibrational mode $\hbar\omega_{mg}$ is considerably larger than the thermal energy $k_B T$. Thus the thermal population of the state g is dominated by the vibrational ground state, $\rho_g^{(0)} \gg \rho_m^{(0)}$.
(iii) The vibrational mode a in question is treated as a harmonic oscillator. Consequently, the matrix elements $\langle g|\boldsymbol{\alpha}|m\rangle$ and $\langle m|\boldsymbol{\alpha}|g\rangle$ in Eq. (4.1) remain non-zero only when the state m is the first excited vibrational state of the mode a. This is in accord with the conventional selection rule of Raman scattering or infrared absorption. The non-zero matrix elements are given by [2]

$$\left\langle g|\hat{Y}|m\right\rangle = \left\langle m|\hat{Y}|g\right\rangle = \left\langle 1|\hat{Y}|0\right\rangle = \sqrt{\frac{\hbar}{2m_a\omega_a}}\left(\frac{\partial Y}{\partial Q_a}\right), \tag{4.2}$$

where Y denotes either $\boldsymbol{\alpha}$ or $\boldsymbol{\mu}$. ω_a is the harmonic frequency of the mode a. Q_a is the normal mode coordinate, and m_a is its reduced mass. (m_a depends on the definition of Q_a, and is often set to unity by properly defining Q_a.)
(iv) In the electronically off-resonant condition that $\hbar\Omega$ is far off the electronic excitation energy (see Fig. 3.1), the dispersion in the Raman tensor $\boldsymbol{\alpha}(\Omega)$ by Ω can be neglected and hence $\boldsymbol{\alpha}(\Omega)$ is regarded as the static polarizability tensor.

By employing the above four assumptions, Eq. (4.1) becomes

$$\alpha_{pqr}^{(2),\mathrm{res}}(\omega_2) = -\sum_a^{\mathrm{mode}} \frac{1}{2m_a\omega_a}\left(\frac{\partial \alpha_{pq}}{\partial Q_a}\right)\left(\frac{\partial \mu_r}{\partial Q_a}\right)\frac{1}{\omega_2 - \omega_a + i\Gamma_a}, \tag{4.3}$$

where the suffix a refers to the intramolecular vibrational mode(s) to be investigated by SFG, and Γ_a denotes Γ_{mg} in Eq. (4.1). This expression of $\boldsymbol{\alpha}^{(2)}$ allows for modeling frequency dependence of the hyperpolarizability. In Eq. (4.3), the derivative quantities $(\partial\alpha_{pq}/\partial Q_a)$, $(\partial\mu_r/\partial Q_a)$ and the frequency ω_a for the mode a can be obtained by quantum chemical calculations. We can thereby evaluate $\alpha_{pqr}^{(2),\mathrm{res}}$ in Eq. (4.3) in the molecule-fixed coordinates.

Using the $\alpha_{pqr}^{(2),\mathrm{res}}(\omega_2)$ thus defined, the nonlinear susceptibility of the interface $\boldsymbol{\chi}^{(2)}$ is constructed with the help of MD calculation. The $\boldsymbol{\chi}^{(2)}$ expression of Eq. (3.46) includes the rotation matrix $\mathcal{D}_l$ of l-th molecule. The orientation of constituent molecules at the interface can be sampled by MD simulation for the interface system. At each time step of MD simulation, the rotation matrix $\mathcal{D}_l$ for the l-th molecule is obtained from the instantaneous configuration of the l-th molecule and used to convert the hyperpolarizability tensor of the molecule $\boldsymbol{\alpha}_l^{(2),\mathrm{res}}$ from the molecule-fixed coordinates to the space-fixed coordinates. The contributions of all molecules in the interface system are summed to determine

$\chi^{(2),\mathrm{res}}$. The instantaneous value of $\chi^{(2),\mathrm{res}}$ at each time step is statistically averaged over ensemble of MD snapshots. The averaged value of $\chi^{(2),\mathrm{res}}$ is thereby obtained,

$$\chi_{pqr}^{(2),\mathrm{res}} = \overline{\sum_{l}^{\mathrm{molecule}} \sum_{p'}^{\xi\sim\zeta} \sum_{q'}^{\xi\sim\zeta} \sum_{r'}^{\xi\sim\zeta} \mathcal{D}_{l,pp'} \mathcal{D}_{l,qq'} \mathcal{D}_{l,rr'} \alpha_{l,p'q'r'}^{(2),\mathrm{res}}}. \tag{4.4}$$

Equations (4.3) and (4.4) provide a route to evaluate the resonant term $\chi^{(2),\mathrm{res}}$ through quantum chemical and molecular dynamics calculations.[1] On the other hand, the nonresonant term $\chi^{(2),\mathrm{nonres}}$ is regarded as a constant over the ω_2 frequency range of SFG measurements. In practical analysis of SFG spectra, it could be treated as an empirical parameter to fit the experimental spectra. Then $\chi^{(2)} = \chi^{(2),\mathrm{res}} + \chi^{(2),\mathrm{nonres}}$ is obtained by Eq. (3.33). The above procedure illustrates the essential idea to calculate $\chi^{(2)}$ on the basis of the energy representation.

Advantages and disadvantages The above analysis method of SFG spectra based on the energy representation of $\chi^{(2)}$ modeling has been proposed by Morita and Hynes [15], and first applied to the water-vapor interface. This method has been subsequently applied and developed by other researchers [4, 21, 22, 27]. The advantages of the energy representation of $\chi^{(2)}$ modeling are summarized as follows.

1. The model of molecular hyperpolarizability is based on quantum chemical calculations of constituent molecules of the interface. Chemical specificity of the molecules is readily reflected in the calculation of SFG spectra.
2. The calculated $\chi^{(2)}$ in Eq. (4.4) is straightforwardly decomposed into and assigned to molecular species, since Eq. (4.4) represents $\chi^{(2)}$ as the summation of molecular contributions. The effect of molecular orientation on the $\chi^{(2)}$ tensor is also readily analyzed [8, 9].
3. Computational cost of MD simulation is relatively modest, compared to the time-dependent representation of $\chi^{(2)}$ modeling in Sect. 4.3. The MD simulation is used for sampling the molecular orientations at interfaces, and the molecular orientation is readily investigated using conventional force fields.

The above features indicate that the energy representation of $\chi^{(2)}$ modeling provides a convenient tool to interpret experimental spectra from molecular properties and orientation with modest cost of computation.

On the other hand, this method involves some difficulties as follows.

1. Accurate molecular modeling is hard to be generalized in condensed environment. This is particularly the case for hydrogen-bonding systems, because the perturbation on the vibrational frequency and transition matrix elements in Eq. (4.3) is substantial.

[1] The molecular hyperpolarizability $\alpha_l^{(2),\mathrm{res}}$ in Eq. (4.4) includes the suffix l to allow for distinguishing different species.

2. Equation (4.4) constructs $\boldsymbol{\chi}^{(2)}$ with a simple sum of molecular hyperpolarizabilities. This feature has a merit for decomposition analysis as mentioned in Advantage 2 above, though it is hard to incorporate the effects of intermolecular couplings. For example, the vibrational modes Q_a are largely perturbed via intra- and inter-molecular couplings, and may depend on the environment. Vibrational correlation among neighboring molecules can have a substantial effect on $\boldsymbol{\chi}^{(2)}$ in strongly interacting system [11, 12]. The local field effect arises from intermolecular dielectric couplings.
3. The above $\boldsymbol{\chi}^{(2)}$ model is not capable of determining the damping term Γ_a in Eq. (4.3). It arises from dephasing of the vibrations, and has to be treated as empirical parameters. The motional effect of molecular orientation on $\boldsymbol{\chi}^{(2)}$ is omitted [26], as discussed in Sect. 4.4.

In summary, the $\boldsymbol{\chi}^{(2)}$ model of Eqs. (4.3) and (4.4) is useful to interpret experimental spectra in a qualitatively plain manner, though it is rather difficult to make this modeling method a predictive tool of experimental spectra. The $\boldsymbol{\chi}^{(2)}$ model of Eqs. (4.3) and (4.4) may not be convenient to accurately reflect the molecular behaviors in condensed phase, including the perturbation of molecular properties, vibrational couplings and dephasing. For example, the perturbation on the molecule hyperpolarizability $\boldsymbol{\alpha}^{(2),\mathrm{res}}$ in Eq. (4.3), generally called "solvent effect" in quantum chemistry, is critical in the SFG spectrum of water. The first paper by Morita and Hynes [15] made elaborated modeling of the solvent effect of water, to represent the substantial perturbation on the transition dipole $(\partial\mu/\partial Q_a)$, transition polarizability $(\partial\alpha/\partial Q_a)$, and the frequency ω_a of water [18]. However, such elaborated modeling hinders from applying it to general interfaces beside the pure water. It is therefore desirable to develop an alternative method which allows for more rigorous and general modeling of $\boldsymbol{\chi}^{(2)}$ and SFG spectroscopy. The need of extending the $\boldsymbol{\chi}^{(2)}$ modeling spurred us to develop another method based on the time-dependent representation [16, 17]. That method is described in Sect. 4.3 below.

4.2 Examples of $\chi^{(2)}$ Tensor and Orientation

The above $\boldsymbol{\chi}^{(2)}$ model in Sect. 4.1 is actually quite useful to discuss the qualitative relation between $\chi^{(2)}$ tensor and molecular orientation for specific molecular species. Here we discuss the relation in some typical examples of O-H and C-H stretching vibrations, the two most commonly measured vibrational bands by the SFG spectroscopy. Further details of computational analysis will be discussed in the application chapters of 9 and 10.

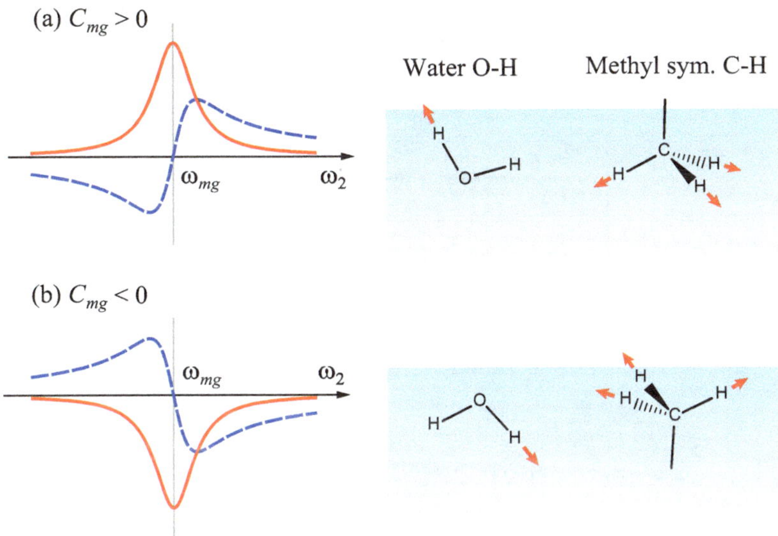

Fig. 4.1 Schematics of Lorentz functions in Eq. (3.39) in the case of (**a**) $C_{mg} > 0$ and (**b**) $C_{mg} < 0$. Real and imaginary parts are written with blue dashed and red solid lines, respectively. (Same as Fig. 3.2 in Chap. 3)

4.2.1 O-H Stretching

The O-H stretching vibration appears in $3000 \sim 3800$ cm^{-1} region, and is typically seen in water and alcohols. Here we focus on a single O-H bond for simplicity, and discuss the qualitative relation between the orientation of O-H bond and the lineshape of $\chi^{(2)}$ spectra, as illustrated in Fig. 4.1. In actual condensed phase, the O-H bonds tend to form hydrogen bonds, and the vibrational modes are coupled and delocalized. Therefore, the relation illustrated here is an ideal picture and offers the basis toward treating more realistic O-H bond vibrations.

Let us consider a local vibration of the O-H$_1$ bond of a water molecule, illustrated along the ζ axis in the panel of Table 4.1, and introduce the local mode coordinate Q to be the O-H$_1$ distance. Then the calculated derivatives of dipole and polarizability with respect to Q in Eq. (4.2) are displayed in Table 4.1. The table shows that the derivatives $(\partial\mu_{r'}/\partial Q)$ and $(\partial\alpha_{p'q'}/\partial Q)$ are dominated by the main elements along the ζ axis in the molecule-fixed coordinates, i.e. $(\partial\mu_\zeta/\partial Q)$ and $(\partial\alpha_{\zeta\zeta}/\partial Q)$, respectively. Accordingly, we roughly approximate $\alpha^{(2),\mathrm{mol}}$ by considering only the product of the main elements for the local O-H vibration,

$$\alpha^{(2),\mathrm{mol}}_{p'q'r'} = \frac{1}{2m\omega}\left(\frac{\partial\alpha_{p'q'}}{\partial Q}\right)\left(\frac{\partial\mu_{r'}}{\partial Q}\right)\frac{-1}{\omega_2-\omega+i\Gamma} \tag{4.5}$$

Table 4.1 Calculated derivatives of dipole $\partial\mu_{r'}/\partial Q$ and polarizability $\partial\alpha_{p'q'}/\partial Q$ of a water molecule by B3LYP/d-aug-cc-pVTZ [15]. Units: atomic units. The coordinate Q is $Q = r_{\text{O-H}_1} - r^{\text{eq}}_{\text{O-H}}$, where $r^{\text{eq}}_{\text{O-H}} = 0.9575$ Å. Right picture illustrates the configuration of water in the molecule-fixed coordinates (ξ, η, ζ)

$\partial\mu_{r'}/\partial Q$	$r' = \xi$	η	ζ
	-0.058	0	0.157
$\partial\alpha_{p'q'}/\partial Q$	$q' = \xi$	η	ζ
$p' = \xi$	1.539	0	-0.163
η	0	1.656	0
ζ	-0.163	0	7.200

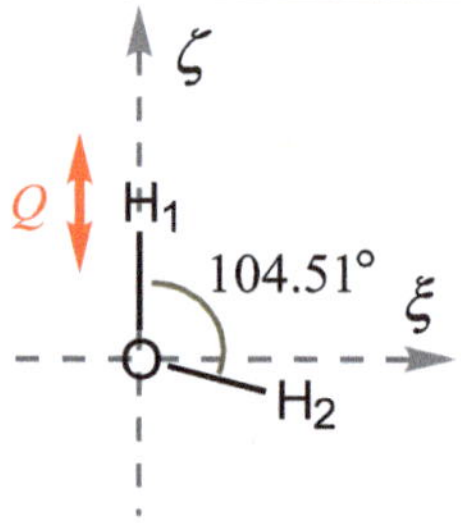

$$\approx \begin{cases} \dfrac{-C_{\zeta\zeta\zeta}}{\omega_2 - \omega + i\Gamma} & (p'q'r') = (\zeta\zeta\zeta) \\ 0 & (p'q'r') = \text{otherwise} \end{cases} \tag{4.6}$$

where

$$C_{\zeta\zeta\zeta} = \frac{1}{2m\omega}\left(\frac{\partial\alpha_{\zeta\zeta}}{\partial Q}\right)\left(\frac{\partial\mu_\zeta}{\partial Q}\right). \tag{4.7}$$

Equation (4.5) or (4.6) is a Lorentz function in the same form as in Eq. (3.39). Therefore, its real and imaginary spectra as a function of ω_2 are illustrated in Fig. 4.1a, noting that $C_{\zeta\zeta\zeta} > 0$ is obtained with the values in Table 4.1.

Sign of Im[$\chi^{(2)}$] Then we discuss the O-H bond that orients to an arbitrary direction in the space-fixed coordinate. $\alpha^{(2),\text{space}}$ is related to $\alpha^{(2),\text{mol}}$ by

$$\alpha^{(2),\text{space}}_{pqr} = \sum_{p'}^{\xi\sim\zeta}\sum_{q'}^{\xi\sim\zeta}\sum_{r'}^{\xi\sim\zeta} \mathcal{D}_{pp'}\mathcal{D}_{qq'}\mathcal{D}_{rr'}\alpha^{(2),\text{mol}}_{p'q'r'}, \tag{3.45}$$

$$= -\frac{1}{2m\omega}\left(\frac{\partial\alpha_{pq}}{\partial Q}\right)\left(\frac{\partial\mu_r}{\partial Q}\right)\frac{1}{\omega_2 - \omega + i\Gamma},$$

and the imaginary part is

$$\text{Im}\left[\alpha^{(2),\text{space}}_{pqr}\right] = \frac{1}{2m\omega}\left(\frac{\partial\alpha_{pq}}{\partial Q}\right)\left(\frac{\partial\mu_r}{\partial Q}\right)\frac{\Gamma}{(\omega_2 - \omega)^2 + \Gamma^2}. \tag{4.8}$$

Using the assumption of Eq. (4.6), Im[$\alpha^{(2),\text{space}}_{yyz}$] for the SSP polarization is represented with the help of Eqs. (3.45) and (3.43),

$$\text{Im}\left[\alpha^{(2)}_{yyz}\right] \approx \text{Im}\left[\mathcal{D}_{y\zeta}\mathcal{D}_{y\zeta}\mathcal{D}_{z\zeta}\frac{-C_{\zeta\zeta\zeta}}{\omega_2 - \omega + i\Gamma}\right] = \cos^2\phi\,\sin^2\theta\,\cos\theta \cdot C_{\zeta\zeta\zeta}\frac{\Gamma}{(\omega_2 - \omega)^2 + \Gamma^2}, \tag{4.9}$$

where ϕ, θ, ψ are the Euler angles in Fig. 3.3, and the superscript "space" is omitted.

We notice in Eq. (4.9) that the sign of Im[$\alpha^{(2)}_{yyz}$] is determined by the sign of $\cos\theta$, as the other factors are always positive. By its definition, θ is the tilt angle of the local O-H$_1$ bond (ζ axis) from the surface normal (z axis) in the space-fixed coordinate (see Fig. 3.3). Therefore, if the O-H$_1$ bond points upward ($\cos\theta > 0$), Im[$\alpha^{(2)}_{yyz}$] spectrum is positive (Fig. 4.1a). On the other hand, if the O-H$_1$ bond points downward ($\cos\theta < 0$), Im[$\alpha^{(2)}_{yyz}$] spectrum is negative (Fig. 4.1b). This qualitative relation between the O-H orientation and the sign of Im[$\alpha^{(2)}_{yyz}$] is useful to interpret the sign of O-H band of Im[$\chi^{(2)}_{yyz}$] spectrum in the SSP polarization.

4.2.2 C-H Stretching

The C-H stretching vibration usually appears in 2800–3000 cm^{-1} region. The C-H vibrations are widely seen in organic molecules, particularly in alkyl moieties such as methyl (CH_3-) and methylene (-CH_2-) groups. The relation between molecular orientation and $\chi^{(2)}$ tensor elements has been formulated in details [7–9, 23]. Here we briefly present essential formulations in the case of methyl C-H symmetric stretching mode. Further discussion of the C-H bands including other modes will be given in Chap. 10 with the help of MD simulation.

To make the qualitative relation clear, we assume the (pseudo) C_{3v} symmetry for the methyl group. The symmetry allows us to simplify the vibrational analysis of the local methyl C-H vibrations. (Note that such symmetry may not rigorously hold for actual methyl groups, such as in methanol or ethanol [10, 24, 25].) Therefore, we treat the C-H mode of acetonitrile as an ideal C_{3v} molecule. Table 4.2 displays the calculated derivatives of dipole moment and polarizability of acetonitrile with respect to the methyl C-H symmetric stretching mode in the molecule-fixed coordinates [19]. The dipole derivative $\partial\mu_{r'}/\partial q_1$ has a non-zero element along the $r' = \zeta$ axis (molecular principal axis) with respect to the C-H symmetric stretching, and the polarizability derivative $\partial\alpha_{p'q'}/\partial q_1$ has non-zero diagonal elements for $(p'q') = (\xi\xi) = (\eta\eta)$ and $(p'q') = (\zeta\zeta)$. We further notice that the dipole derivative $\partial\mu_\zeta/\partial q_1$ is negative, and the ratio of the two independent elements of the polarizability derivative,

$$R = \frac{(\partial\alpha_{\xi\xi}/\partial q_1)}{(\partial\alpha_{\zeta\zeta}/\partial q_1)} = \frac{(\partial\alpha_{\eta\eta}/\partial q_1)}{(\partial\alpha_{\zeta\zeta}/\partial q_1)} \tag{4.10}$$

is estimated to be 0.80 in Table 4.2.

Sign of Im[$\chi^{(2)}$] First, we provide qualitative discussion about the molecular orientation and the sign of the $\alpha^{(2),\text{space}}_{yyz}$ element for the SSP polarization. The sign of Im[$\alpha^{(2),\text{space}}_{yyz}$] is determined by the product of $(\partial\alpha_{yy}/\partial q_1)$ and $(\partial\mu_z/\partial q_1)$ in the space-fixed coordinates. The latter is determined by the molecular orientation as

$$\left(\frac{\partial\mu_z}{\partial q_1}\right) = \sum_{r'}^{\xi\sim\zeta} \mathcal{D}_{zr'}\left(\frac{\partial\mu_{r'}}{\partial q_1}\right) = -0.005970\cdot\cos\theta,$$

Table 4.2 Calculated derivatives of dipole $\partial\mu_{r'}/\partial q_1$ and polarizability $\partial\alpha_{p'q'}/\partial q_1$ of acetonitrile by B3LYP/aug-cc-pVTZ [19]. Units: atomic units. q_1 denotes the normal mode coordinate of C-H symmetric stretching defined for the stretching direction with unit reduced mass. Right picture illustrate the configuration of acetonitrile in the molecule-fixed coordinates (ξ, η, ζ)

$\partial\mu_{r'}/\partial q_1$	$r' = \xi$	η	ζ
	0	0	-0.005970
$\partial\alpha_{p'q'}/\partial q_1$	$q' = \xi$	η	ζ
$p' = \xi$	0.2967	0	0
η	0	0.2967	0
ζ	0	0	0.3731

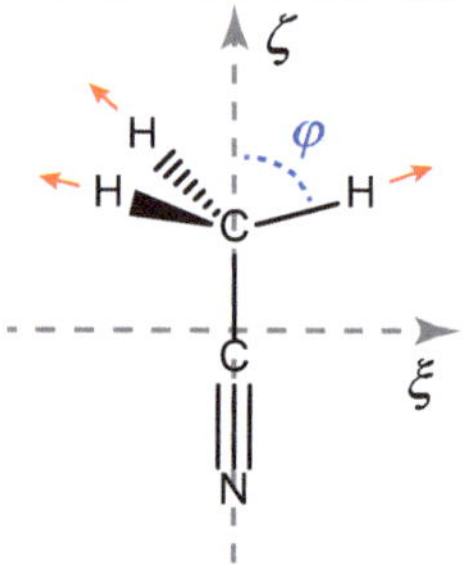

which indicates that the z element $(\partial\mu_z/\partial q_1)$ is negative when $\cos\theta > 0$ (upward orientation of the methyl group) while positive when $\cos\theta < 0$ (downward). On the other hand, the polarizability derivative $(\partial\alpha_{yy}/\partial q_1)$ is relatively insensitive to the orientation, since the polarizability derivative is not much anisotropic.[2] Therefore, the sign of $\alpha_{yyz}^{(2),\text{space}}$ is determined by the tilt angle θ of the methyl axis. The negative $\mathrm{Im}[\chi_{yyz}^{(2)}]$ band of the methyl C-H stretching band is indicative of the upward orientation of the methyl group, while positive $\mathrm{Im}[\chi_{yyz}^{(2)}]$ band is indicative of the downward orientation. These features are illustrated also in Fig. 4.1.

Polarization ratios In the experimental SFG measurements, the ratios of different $\chi_{pqr}^{(2)}$ tensor elements are utilized to investigate the molecular orientation at interfaces. Here we formulate the ratios of different $\chi^{(2)}$ elements, $B = \chi_{yyz}^{(2)}/\chi_{yzy}^{(2)}$ and $C = \chi_{zzz}^{(2)}/\chi_{yyz}^{(2)}$, in relation to the methyl orientation. These polarization ratios B and C can be evaluated experimentally by using the combinations of SSP, SPS and PPP polarizations.

Let us suppose that the $\chi_{pqr}^{(2)}$ elements are given by Eq. (3.41), $\chi_{pqr}^{(2)} \approx N \cdot \overline{\alpha_{pqr}^{(2)}}$. Accordingly, B and C are represented by

$$B = \frac{\chi_{yyz}^{(2)}}{\chi_{yzy}^{(2)}} = \frac{N \cdot \overline{\alpha_{yyz}^{(2)}}}{N \cdot \overline{\alpha_{yzy}^{(2)}}} = \frac{\overline{\alpha_{yyz}^{(2)}}}{\overline{\alpha_{yzy}^{(2)}}}, \qquad C = \frac{\chi_{zzz}^{(2)}}{\chi_{yyz}^{(2)}} = \frac{N \cdot \overline{\alpha_{zzz}^{(2)}}}{N \cdot \overline{\alpha_{yyz}^{(2)}}} = \frac{\overline{\alpha_{zzz}^{(2)}}}{\overline{\alpha_{yyz}^{(2)}}}. \tag{4.11}$$

---

[2]If the polarizability derivative tensor is approximated with a spherical tensor ($R \approx 1$) by $\partial\alpha_{p'q'}/\partial q_1 \approx c\,\delta_{p'q'}$, the yy element $(\partial\alpha_{yy}/\partial q_1)$ is shown to be independent of the transformation,

$$\left(\frac{\partial\alpha_{yy}}{\partial q_1}\right) = \sum_{p'}^{\xi\sim\zeta}\sum_{q'}^{\xi\sim\zeta} \mathcal{D}_{yp'}\mathcal{D}_{yq'} \left(\frac{\partial\alpha_{p'q'}}{\partial q_1}\right) \approx \sum_{p'}^{\xi\sim\zeta}\sum_{q'}^{\xi\sim\zeta} \mathcal{D}_{yp'}\mathcal{D}_{yq'}\, c\,\delta_{p'q'} = c\ (>0).$$

Since the anisotropy is not large ($R \simeq 0.8$ in Table 4.2), the qualitatively similar trend should hold for the acetonitrile molecule.

Equation (4.11) shows that the ratios B and C are free from the number of molecules N, as it is cancelled. $\overline{\alpha_{pqr}^{(2)}}$ in the space-fixed coordinates is represented by

$$\overline{\alpha_{pqr}^{(2)}} = \sum_{p'}^{\xi\sim\zeta}\sum_{q'}^{\xi\sim\zeta}\sum_{r'}^{\xi\sim\zeta} \overline{\mathcal{D}_{pp'}\mathcal{D}_{qq'}\mathcal{D}_{rr'}}\,\alpha_{p'q'r'}^{(2),\mathrm{mol}}, \tag{3.47}$$

which includes the molecular hyperpolarizability tensor $\alpha_{p'q'r'}^{(2),\mathrm{mol}}$ in the molecule-fixed coordinates (ξ, η, ζ) and the rotational matrix $\mathcal{D}$. $\alpha_{p'q'r'}^{(2),\mathrm{mol}}$ in Eq. (3.47) is given in the analogous form with Eq. (4.5),

$$\alpha_{p'q'r'}^{(2),\mathrm{mol}} = \frac{1}{2m\omega}\left(\frac{\partial \alpha_{p'q'}}{\partial q_1}\right)\left(\frac{\partial \mu_{r'}}{\partial q_1}\right)\frac{-1}{\omega_2 - \omega + i\Gamma}, \tag{4.12}$$

and the derivatives of dipole and polarizability in the molecule-fixed coordinates take the following form,

$$\left(\frac{\partial \boldsymbol{\mu}}{\partial q_1}\right) = \begin{pmatrix} 0 \\ 0 \\ \mu \end{pmatrix}, \quad \left(\frac{\partial \boldsymbol{\alpha}}{\partial q_1}\right) = \begin{pmatrix} R\alpha & 0 & 0 \\ 0 & R\alpha & 0 \\ 0 & 0 & \alpha \end{pmatrix}, \tag{4.13}$$

as a consequence of the C_{3v} symmetry. The values of μ, R, and α in Eq. (4.13) are obtained from Table 4.2.

On the other hand, the rotational matrix $\boldsymbol{\mathcal{D}}$ in Eq. (3.47) is represented using the three Euler angles $\{\phi, \theta, \psi\}$ by Eq. (3.43). The average manipulation in Eq. (3.47) can be simplified by exploiting symmetries. The azimuthal angle ϕ should be uniformly distributed due to the macroscopic C_∞ symmetry of the interface, and the distribution of the rotational angle ψ around the principal axis of the molecule is also assumed to be uniform for the methyl group with the C_{3v} symmetry. Consequently, only the tilt angle θ remains to define the molecular orientation after the distributions for the other two angles ϕ and ψ are averaged out. Therefore, the two ratios B and C for different $\chi_{pqr}^{(2)}$ elements are represented in the following form [19, 20, 23, 28].

$$B = \frac{\chi_{yyz}^{(2)}}{\chi_{yzy}^{(2)}} = \frac{(1+R)\,\overline{\cos\theta} - (1-R)\,\overline{\cos^3\theta}}{(1-R)(\overline{\cos\theta} - \overline{\cos^3\theta})} \tag{4.14}$$

$$C = \frac{\chi_{zzz}^{(2)}}{\chi_{yyz}^{(2)}} = \frac{2\{R\,\overline{\cos\theta} + (1-R)\,\overline{\cos^3\theta}\}}{(1+R)\,\overline{\cos\theta} - (1-R)\,\overline{\cos^3\theta}} \tag{4.15}$$

[Problem 4.1] Derive Eqs. (4.14) and (4.15) on the basis of the above assumptions.

Equations (4.14) and (4.15) are used to the polarization analysis of SFG measurements. These equations include only three parameters, R, $\overline{\cos\theta}$ and $\overline{\cos^3\theta}$, and thus suitable to simple modeling of the polarization analysis. R is estimated by various means, such as the depolarization ratio of the Raman scattering [14], ab initio or density functional theory calculations, and simple geometric argument of molecules. If one further assumes a relation of $\overline{\cos\theta}$ and $\overline{\cos^3\theta}$, one could derive the molecular orientation θ from the B and/or C ratios. The simplest assumption would be $\overline{\cos^3\theta} \approx (\overline{\cos\theta})^3$. Then the tilt angle is estimated from B and C in Eqs. (4.14) and (4.15), respectively, to be

$$\overline{\cos\theta} \approx \left[\frac{B-(1+R)/(1-R)}{B-1}\right]^{1/2} \quad \text{or} \quad \overline{\cos\theta} \approx \left[\frac{(1+R)C-2R}{(1-R)(C+2)}\right]^{1/2}. \tag{4.16}$$

We note that the assumption $\overline{\cos^3\theta} \approx (\overline{\cos\theta})^3$ is valid when the tilt angle θ is well determined (or the distribution of θ is sufficiently narrow) at the interface. We further discuss the polarization analysis with the help of MD simulation in Chap. 10.

4.3 Time-Dependent Representation

4.3.1 Time Correlation Function

This section presents an alternative expression of $\chi^{(2)}$ based on the time correlation function, instead of using the perturbation formula in the preceding sections. According to the theory of statistical mechanics, physical quantities associated to the response to external perturbation are generally expressed using time correlation functions [13, 14].

$\chi^{(2),\mathrm{res}}$ in Eq. (3.36) is converted to an equivalent formula with time correlation function as follows,

$$\chi^{(2),\mathrm{res}}_{pqr}(\Omega,\omega_1,\omega_2) = -\frac{1}{\hbar}\sum_{g,m}\left(\rho_g^{(0)}-\rho_m^{(0)}\right)\langle g|\alpha_{pq}|m\rangle\frac{\langle m|\mu_r|g\rangle}{\omega_2-\omega_{mg}+i\Gamma_{mg}} \tag{3.36}$$

$$= \frac{i}{\hbar}\int_0^\infty dt\sum_{g,m}\left(\rho_g^{(0)}-\rho_m^{(0)}\right)\langle g|\alpha_{pq}|m\rangle\langle m|\mu_r|g\rangle\exp\left[i(\omega_2-\omega_{mg}+i\Gamma_{mg})t\right]$$

$$= \frac{i}{\hbar}\int_0^\infty dt\sum_{g,m}\left(\rho_g^{(0)}-\rho_m^{(0)}\right)\langle g|\alpha_{pq}(t)|m\rangle\langle m|\mu_r|g\rangle\exp(i\omega_2 t)$$

$$= \frac{i}{\hbar}\int_0^\infty dt\sum_{g,m}\rho_g^{(0)}\left(\langle g|\alpha_{pq}(t)|m\rangle\langle m|\mu_r|g\rangle-\langle g|\mu_r|m\rangle\langle m|\alpha_{pq}(t)|g\rangle\right)\exp(i\omega_2 t)$$

$$= \frac{i}{\hbar}\int_0^\infty dt \sum_g \rho_g^{(0)} \langle g|\alpha_{pq}(t)\mu_r - \mu_r\alpha_{pq}(t)|g\rangle \exp(i\omega_2 t)$$

$$= \frac{i}{\hbar}\int_0^\infty dt \overline{\langle \alpha_{pq}(t)\mu_r - \mu_r\alpha_{pq}(t)\rangle} \exp(i\omega_2 t). \tag{4.17}$$

In the above derivation from the second line to third, we adopted the Heisenberg picture to regard the Raman tensor α as a time dependent quantity,

$$\langle g|\alpha_{pq}(t)|m\rangle = \langle g|\alpha_{pq}|m\rangle \exp\left[i(-\omega_{mg} + i\Gamma_{mg})t\right]. \tag{4.18}$$

This relation is validated by the equivalence of Schrödinger and Heisenberg pictures, as discussed in the following. In general, time dependence of the expectation value of an arbitrary physical quantity A is represented in two ways,

$$\langle A(t)\rangle = \sum_{g,m} A_{gm}\rho_{mg}(t) = \sum_{g,m} A_{gm}(t)\rho_{mg}, \tag{4.19}$$

by Schrödinger and Heisenberg pictures, respectively. The former regards the state (density matrix ρ) is a function of time, while the latter regards the physical quantity A as time dependent. In the Schrödinger picture, the time development of the density matrix ρ is given by the Liouville equation (3.16),

$$\begin{aligned} i\hbar\frac{d\rho_{mg}(t)}{dt} &= [H_0\rho - \rho H_0]_{mg} - i\hbar\Gamma_{mg}\left(\rho_{mg}(t) - \rho_{mg}^{\text{eq}}\right) \\ &= \left(\mathcal{E}_m - \mathcal{E}_g\right)\rho_{mg}(t) - i\hbar\Gamma_{mg}\left(\rho_{mg}(t) - \rho_{mg}^{\text{eq}}\right), \end{aligned}$$

where the time development is driven by the Hamiltonian $H = H_0$ (with no perturbation) and the states g and m are eigenstates of H_0. Therefore, the off-diagonal element $\rho_{mg}(m \neq g)$ is given as a solution,

$$\rho_{mg}(t) = \rho_{mg}\exp\left[i(-\omega_{mg} + i\Gamma_{mg})t\right],$$

which satisfies the proper boundary condition $\rho_{mg}(t) \to 0$ at $t \to \infty$ ($\Gamma_{mg} > 0$). Consequently, $\langle A(t)\rangle$ in Eq. (4.19) is presented by

$$\langle A(t)\rangle = \sum_{g,m} A_{gm}\rho_{mg}(t) = \sum_{g,m} A_{gm}\rho_{mg}\exp\left[i(-\omega_{mg} + i\Gamma_{mg})t\right] = \sum_{g,m} A_{gm}(t)\rho_{mg}.$$

This solution leads to the equivalent Heisenberg picture by

$$A_{gm}(t) = \langle g|A(t)|m\rangle = A_{gm}\exp\left[i(-\omega_{mg} + i\Gamma_{mg})t\right]. \tag{4.20}$$

This relation elucidates Eq. (4.18) by setting $A = \alpha_{pq}$. We note again that the above discussion about the time evolution is based on the Hamiltonian H_0, *not* including the perturbation. The diagonal elements ($g = m$) in Eq. (4.18) do not contribute to Eq. (4.17), due to the vanishing factor $(\rho_g^{(0)} - \rho_m^{(0)}) = 0$ for $g = m$.

In summary, Eq. (4.17) gives an equivalent expression of $\boldsymbol{\chi}^{(2),\mathrm{res}}$ to Eq. (3.36). It indicates that the vibrational resonant term of the nonlinear susceptibility $\boldsymbol{\chi}^{(2),\mathrm{res}}$ can be represented by the Fourier-Laplace transformation of the time correlation function between the Raman tensor α_{pq} and the dipole moment μ_r for the interface system. It is a rigorous expression of $\boldsymbol{\chi}^{(2),\mathrm{res}}$ on the basis of quantum mechanics.

4.3.2 *Classical Analogue*

The time correlation formula of Eq. (4.17) allows for an alternative computational scheme of $\boldsymbol{\chi}^{(2),\mathrm{res}}$. The time correlation functions are utilized to evaluate various properties of statistical mechanics [13], and amenable to be computed by molecular dynamics (MD) simulation [1, 6]. In order to use this formula with MD simulation, however, we should obtain a classical version of Eq. (4.17) since usual MD simulations are carried out on the basis of classical mechanics. Deriving the classical expression is the theme of this subsection.

$\boldsymbol{\chi}^{(2),\mathrm{res}}$ in Eq. (4.17) is expressed by the Fourier-Laplace transformation of the time correlation function $\mathcal{F}(t)$,

$$\chi_{pqr}^{(2),\mathrm{res}}(\Omega, \omega_1, \omega_2) = \int_0^\infty dt \exp(i\omega_2 t)\, \mathcal{F}(t), \tag{4.21}$$

where

$$\mathcal{F}(t) = \frac{i}{\hbar} \overline{\langle \alpha_{pq}(t)\mu_r - \mu_r\alpha_{pq}(t)\rangle} = \frac{i}{\hbar} \overline{\langle \delta\alpha_{pq}(t)\delta\mu_r - \delta\mu_r\delta\alpha_{pq}(t)\rangle} \tag{4.22}$$

In this expression $\delta\alpha_{pq}(t) = \alpha_{pq}(t) - \overline{\langle\alpha_{pq}\rangle}$ and $\delta\mu_r = \mu_r - \overline{\langle\mu_r\rangle}$ denote the displacements from the average values. Equation (4.22) is apparently a quantum mechanical expression, since it includes $\hbar$ and a commutation relation $[\alpha_{pq}(t), \mu_r] = \alpha_{pq}(t)\mu_r - \mu_r\alpha_{pq}(t) \neq 0$. This form is not amenable to the classical limit, since both $\hbar$ and the commutation relation would go to zero and thus $\mathcal{F}(t) \to 0/0$. To obtain a classical analogue, we change Eq. (4.21) to an equivalent form with the canonical time correlation function [13].

A related function $\mathcal{G}(t)$ is introduced using the canonical time correlation function,

$$\begin{aligned}\mathcal{G}(t) &= \left\langle \alpha_{pq}(t) - \alpha_{pq}^\circ;\ \mu_r - \mu_r^\circ \right\rangle \\ &= \frac{1}{\beta}\int_0^\beta d\lambda\, \overline{\left\langle \exp(\lambda\mathcal{H}) \left(\alpha_{pq}(t) - \alpha_{pq}^\circ\right) \exp(-\lambda\mathcal{H}) \left(\mu_r - \mu_r^\circ\right)\right\rangle},\end{aligned} \tag{4.23}$$

where $\beta = 1/(k_B T)$, and $\mathcal{H}$ is the Hamiltonian for the entire system (partial system and bath). $\boldsymbol{\alpha}^\circ$ and $\boldsymbol{\mu}^\circ$ are the diagonal part of $\boldsymbol{\alpha}$ and $\boldsymbol{\mu}$, respectively, on the basis of energy eigenstates.[3] $\langle A; B\rangle$ denotes the canonical time correlation function defined as follows,

$$\langle A; B\rangle = \frac{1}{\beta}\int_0^\beta d\lambda\ \overline{\langle \exp(\lambda\mathcal{H})A\exp(-\lambda\mathcal{H})B\rangle}. \tag{4.24}$$

[Problem 4.2] Prove $\langle A; B\rangle = \langle B; A\rangle$ for arbitrary operators A and B in Eq. (4.24). This indicates that the operators in the canonical correlation function are commutative like a classical one.

Then we find the relation between $\mathcal{F}(t)$ and $\mathcal{G}(t)$. In the following derivation we define $\mathcal{H}$ and ρ for the entire system (partial system and bath). The Liouville equation is formally given using $\mathcal{H}$ by

$$i\hbar\frac{\partial\rho(t)}{\partial t} = \mathcal{H}\rho(t) - \rho(t)\mathcal{H}. \tag{3.8}$$

This equation for the entire system involves no damping term, as we mentioned in Sect. 3.1. Using $\mathcal{H}$ and $\rho(t)$, $\mathcal{F}(t)$ in Eq. (4.22) is written by

$$\begin{aligned}\mathcal{F}(t) &= \frac{i}{\hbar}\overline{\langle\delta\alpha_{pq}(t)\delta\mu_r - \delta\mu_r\delta\alpha_{pq}(t)\rangle}\\ &= \frac{i}{\hbar}\mathrm{Tr}\Bigg[\rho\Bigg\{\exp\left(\frac{i\mathcal{H}t}{\hbar}\right)\delta\alpha_{pq}\exp\left(\frac{-i\mathcal{H}t}{\hbar}\right)\delta\mu_r - \delta\mu_r\exp\left(\frac{i\mathcal{H}t}{\hbar}\right)\delta\alpha_{pq}\\ &\quad\exp\left(\frac{-i\mathcal{H}t}{\hbar}\right)\Bigg\}\Bigg],\end{aligned} \tag{4.25}$$

where $\exp(-i\mathcal{H}t/\hbar)$ is the time development operator. $\mathcal{G}(t)$ in Eq. (4.23) is expanded using the eigenstates of the entire system, m, n, by

[3] Therefore, the matrix elements for an arbitrary operator A° $(= \alpha^\circ_{pq}$ or $\mu^\circ_r)$ are represented using energy eigenstates m, n of the entire system as

$$\langle m|A^\circ|n\rangle = \begin{cases}\langle m|A|n\rangle & (m = n)\\ 0 & (m \neq n)\end{cases}.$$

This is equivalent to the long-time average of $\langle m|A|n\rangle$,

$$\langle m|A^\circ|n\rangle = \lim_{T\to\infty}\frac{1}{T}\int_0^T dt\ \langle m|A(t)|n\rangle = \lim_{T\to\infty}\frac{1}{T}\int_0^T dt\ \exp\left(\frac{i(\mathcal{E}_m - \mathcal{E}_n)t}{\hbar}\right)\langle m|A|n\rangle.$$

Note that A° is commutative with $\mathcal{H}$, and thus $A^\circ(t) = \exp(i\mathcal{H}t/\hbar)A^\circ\exp(-i\mathcal{H}t/\hbar) = A^\circ$.

$$\mathcal{G}(t) = \frac{1}{\beta}\int_0^\beta d\lambda \overline{\langle \exp(\lambda\mathcal{H})\left(\alpha_{pq}(t) - \alpha^\circ_{pq}\right)\exp(-\lambda\mathcal{H})(\mu_r - \mu^\circ_r)\rangle}$$

$$= \frac{1}{\beta}\int_0^\beta d\lambda\ \mathrm{Tr}\Big[\rho\ \exp(\lambda\mathcal{H})\exp\left(\frac{i\mathcal{H}t}{\hbar}\right)\left(\alpha_{pq} - \alpha^\circ_{pq}\right)\exp\left(\frac{-i\mathcal{H}t}{\hbar}\right)$$

$$\exp(-\lambda\mathcal{H})\left(\mu_r - \mu^\circ_r\right)\Big]$$

$$= \frac{1}{\beta}\int_0^\beta d\lambda \sum_m\sum_n \frac{\exp(-\beta\mathcal{E}_m)}{Q}\exp(\lambda\mathcal{E}_m)\exp\left(\frac{i\mathcal{E}_m t}{\hbar}\right)\left(\langle m|\alpha_{pq}|n\rangle - \langle m|\alpha^\circ_{pq}|n\rangle\right)$$

$$\cdot \exp\left(-\frac{i\mathcal{E}_n t}{\hbar}\right)\exp(-\lambda\mathcal{E}_n)\left(\langle n|\mu_r|m\rangle - \langle n|\mu^\circ_r|m\rangle\right)$$

$$= \frac{1}{\beta}\sum_m\sum_{n(\neq m)} \frac{\exp(-\beta\mathcal{E}_m)}{Q}\exp\left(\frac{i\mathcal{E}_m t}{\hbar}\right)\langle m|\alpha_{pq}|n\rangle\exp\left(\frac{-i\mathcal{E}_n t}{\hbar}\right)\langle n|\mu_r|m\rangle$$

$$\cdot\frac{\exp\left(\beta(\mathcal{E}_m - \mathcal{E}_n)\right) - 1}{\mathcal{E}_m - \mathcal{E}_n}$$

$$= \frac{1}{Q}\sum_m\sum_{n(\neq m)} \frac{\exp(-\beta\mathcal{E}_m) - \exp(-\beta\mathcal{E}_n)}{-\beta(\mathcal{E}_m - \mathcal{E}_n)}\exp\left(\frac{i\mathcal{E}_m t}{\hbar}\right)\langle m|\delta\alpha_{pq}|n\rangle$$

$$\exp\left(\frac{-i\mathcal{E}_n t}{\hbar}\right)\langle n|\delta\mu_r|m\rangle, \tag{4.26}$$

where $\mathcal{E}_m, \mathcal{E}_n$ are the energy eigenvalues for the states m, n. $Q = \sum_m \exp(-\beta\mathcal{E}_m)$ is the partition function for the entire system. During the derivation in Eq. (4.26), the diagonal terms ($m = n$) of $A - A^\circ$ ($A = \alpha_{pq}$ or μ_r) vanish, and $\langle m|\delta A|n\rangle = \langle m|A - \overline{\langle A\rangle}|n\rangle = \langle m|A|n\rangle$ is employed for $m \neq n$.

Now Eq. (4.21) is expressed with $\mathcal{G}(t)$ instead of $\mathcal{F}(t)$ using Eqs. (4.25) and (4.26) in the following way,

$$\chi^{(2),\mathrm{res}}_{pqr}(\Omega, \omega_1, \omega_2) = \int_0^\infty dt\exp(i\omega_2 t)\,\mathcal{F}(t) \tag{4.21}$$

$$= \frac{i}{\hbar}\int_0^\infty dt\ \exp(i\omega_2 t)\sum_m\sum_n\frac{\exp(-\beta\mathcal{E}_m)}{Q}\Big\{\exp\left(\frac{i\mathcal{E}_m t}{\hbar}\right)\langle m|\delta\alpha_{pq}|n\rangle$$

$$\exp\left(\frac{-i\mathcal{E}_n t}{\hbar}\right)\langle n|\delta\mu_r|m\rangle$$

$$-\ \langle m|\delta\mu_r|n\rangle\exp\left(\frac{i\mathcal{E}_n t}{\hbar}\right)\langle n|\delta\alpha_{pq}|m\rangle\exp\left(\frac{-i\mathcal{E}_m t}{\hbar}\right)\Big\}$$

$$= \frac{i}{\hbar}\int_0^\infty dt\ \exp(i\omega_2 t)\sum_m\sum_{n(\neq m)}\frac{\exp(-\beta\mathcal{E}_m) - \exp(-\beta\mathcal{E}_n)}{Q}$$

$$
\begin{aligned}
&\quad \cdot \exp\left(\frac{i\mathcal{E}_m t}{\hbar}\right)\langle m|\delta\alpha_{pq}|n\rangle \exp\left(\frac{-i\mathcal{E}_n t}{\hbar}\right)\langle n|\delta\mu_r|m\rangle \\
&= -\frac{1}{Q}\sum_m \sum_{n(\neq m)} \frac{\exp(-\beta\mathcal{E}_m)-\exp(-\beta\mathcal{E}_n)}{\mathcal{E}_m-\mathcal{E}_n}\langle m|\delta\alpha_{pq}|n\rangle\langle n|\delta\mu_r|m\rangle \\
&\quad - i\omega_2\int_0^\infty dt\ \exp(i\omega_2 t)\cdot\frac{1}{Q}\sum_m\sum_{n(\neq m)}\frac{\exp(-\beta\mathcal{E}_m)-\exp(-\beta\mathcal{E}_n)}{\mathcal{E}_m-\mathcal{E}_n} \\
&\quad \cdot \exp\left(\frac{i\mathcal{E}_m t}{\hbar}\right)\langle m|\delta\alpha_{pq}|n\rangle \exp\left(\frac{-i\mathcal{E}_n t}{\hbar}\right)\langle n|\delta\mu_r|m\rangle \\
&= \beta\mathcal{G}(0) + i\omega_2\beta\int_0^\infty dt\ \mathcal{G}(t)\exp(i\omega_2 t). \qquad (4.27)
\end{aligned}
$$

From the third line to the fourth line, integration by parts is performed.

Equation (4.27) is amenable to the classical limit by replacing the canonical correlation function $\mathcal{G}(t)$ in Eq. (4.23) with the classical time correlation function [13],

$$
\mathcal{G}(t) = \left\langle \alpha_{pq}(t) - \alpha^\circ_{pq};\ \mu_r - \mu^\circ_r \right\rangle \quad \longrightarrow \quad \mathcal{G}_{\rm cl}(t) = \left\langle \delta\alpha_{pq}(t)\ \delta\mu_r \right\rangle_{\rm cl}.
$$

In the classical limit, the operators in Eq. (4.23) become commutable and the long-time average of $\boldsymbol{\alpha}$ and $\boldsymbol{\mu}$ correspond to the ensemble average. Therefore, $\boldsymbol{\chi}^{(2),\rm res}$ in Eq. (4.27) finds a classical analogue to be

$$
\begin{aligned}
\chi^{(2),\rm res}_{pqr}(\Omega,\omega_1,\omega_2) &\simeq \beta\mathcal{G}_{\rm cl}(0) + i\omega_2\beta\int_0^\infty dt\mathcal{G}_{\rm cl}(t)\exp(i\omega_2 t) \\
&= \frac{1}{k_B T}\left\langle\delta\alpha_{pq}\delta\mu_r\right\rangle_{\rm cl} + \frac{i\omega_2}{k_B T}\int_0^\infty dt\ \left\langle\delta\alpha_{pq}(t)\delta\mu_r\right\rangle_{\rm cl}\exp(i\omega_2 t). \qquad (4.28)
\end{aligned}
$$

This classical expression of $\boldsymbol{\chi}^{(2),\rm res}$ in Eq. (4.28) is utilized in calculating the SFG spectroscopy by MD simulation. Hereafter we omit the subscript "cl" and the overbar in the classical time correlation function unless otherwise noted.[4] Accordingly, Eq. (4.28) could be expressed as

$$
\begin{aligned}
\chi^{(2),\rm res}_{pqr}(\Omega,\omega_1,\omega_2) &= \frac{1}{k_B T}\left\langle\delta\alpha_{pq}\delta\mu_r\right\rangle + \frac{i\omega_2}{k_B T}\int_0^\infty dt\left\langle\delta\alpha_{pq}(t)\delta\mu_r\right\rangle\exp(i\omega_2 t) \\
&= \frac{1}{k_B T}\left(\left\langle\alpha_{pq}\mu_r\right\rangle - \left\langle\alpha_{pq}\right\rangle\left\langle\mu_r\right\rangle\right) + \frac{i\omega_2}{k_B T}\int_0^\infty dt\left(\left\langle\alpha_{pq}(t)\mu_r\right\rangle - \left\langle\alpha_{pq}\right\rangle\left\langle\mu_r\right\rangle\right)\exp(i\omega_2 t) \\
&= \frac{1}{k_B T}\left\langle\alpha_{pq}\mu_r\right\rangle + \frac{i\omega_2}{k_B T}\int_0^\infty dt\left\langle\alpha_{pq}(t)\mu_r\right\rangle\exp(i\omega_2 t). \qquad (4.29)
\end{aligned}
$$

[4] In the classical mechanics, the distinction between quantum average $\langle A\rangle$ and statistical average $\overline{A}$ in Sect. 3.1 disappears. Accordingly the classical time correlation function $\langle A(t)B\rangle_{\rm cl}$ is equivalent to $\overline{A(t)B}$.

Finally, we make the following three notes on the above derivation of the classical analogue.

Note 1. In actual application to interpret vibrational spectra, we are interested in the frequency dependence of $\boldsymbol{\chi}^{(2)}(\omega_2)$. In such cases, the second term of Eq. (4.29) is often regarded as the vibrational resonant term,

$$\chi_{pqr}^{(2),\mathrm{res}}(\omega_2) = \frac{i\omega_2}{k_B T}\int_0^\infty dt\,\langle\alpha_{pq}(t)\mu_r\rangle \exp(i\omega_2 t). \tag{4.30}$$

On the other hand, the first term of Eq. (4.29) is constant over the frequency ω_2, and thus it could be effectively regarded as a part of the nonresonant background $\boldsymbol{\chi}^{(2),\mathrm{nonres}}$.

Note 2. The present classical form of Eq. (4.30) can be derived in an alternative manner using the harmonic oscillator model [3, 17]. The harmonic oscillators are exactly soluble both by quantum and classical mechanics, and thus allow for finding the correspondence of quantum and classical descriptions.

Note 3. The classical formula of $\boldsymbol{\chi}^{(2)}$ does not necessarily reproduce its quantitative amplitude. This issue is related to the attempts to seek a proper quantum correction factor for time correlation functions [5]. The spectral lineshapes are reliable though, as it is insensitive to the details of the quantum correction factor over the frequency range of ω_2. Therefore, Eq. (4.29) or (4.30) is useful enough for the lineshape analysis of the vibrational spectra.

Equation (4.30) is the fundamental formula to calculate $\boldsymbol{\chi}^{(2),\mathrm{res}}$ by MD simulation. In the MD calculations of Eq. (4.30), the evaluation of polarization properties in the interface system is critically important. We will discuss these polarization properties in the following Chaps. 5 and 6.

4.4 Motional Effect on $\chi^{(2)}$

The time correlation formula of $\boldsymbol{\chi}^{(2)}$ derived above is able to incorporate the dynamical effects on $\boldsymbol{\chi}^{(2)}$, including the vibrational couplings and dephasing, since the dynamics of molecules are naturally reflected in the time correlation function $\langle\alpha(t)\mu\rangle$. This is an advantage of the time correlation formula of $\boldsymbol{\chi}^{(2)}$, while the $\boldsymbol{\chi}^{(2)}$ model by the energy representation in Sect. 4.1 is not suitable to account for the dynamical effects. Here we discuss the motional effect of orientation on $\chi^{(2)}$ on the basis of the time-dependent representation [26].

4.4.1 Relation of Two $\chi^{(2)}$ Models

For this purpose, we clarify the relation and difference of the two $\chi^{(2)}$ modeling on the basis of energy representation and time-dependent representation. The energy representation model of $\chi^{(2)}$ in Eqs. (4.3) and (4.4) can be derived from the time-dependent representation in Eq. (4.29) with employing some assumptions.

Since Eq. (4.29) refers to the nonlinear susceptibility $\chi^{(2)}$ of the whole interface system, $\boldsymbol{\alpha}$ and $\boldsymbol{\mu}$ in this equation also stand for the polarizability and dipole of the whole system. Thus $\boldsymbol{\alpha}$ and $\boldsymbol{\mu}$ of the whole interface system are expressed as the sum of molecular quantities,

$$\alpha_{pq} \approx \sum_{l}^{\text{molecule}} \sum_{p'}^{\xi\sim\zeta} \sum_{q'}^{\xi\sim\zeta} \mathcal{D}_{l,pp'} \mathcal{D}_{l,qq'} \alpha_{l,p'q'}, \tag{4.31}$$

$$\mu_r \approx \sum_{k}^{\text{molecule}} \sum_{r'}^{\xi\sim\zeta} \mathcal{D}_{k,rr'} \mu_{k,r'}, \tag{4.32}$$

where $\alpha_{l,p'q'}$ and $\mu_{k,p'}$ denote the polarizability tensor of l-th molecule and the dipole vector of k-th molecule, respectively, in the molecule-fixed coordinates. We note that Eqs. (4.31) and (4.32) are regarded as a crude approximation to neglect the local field effect. The local field effect will be discussed in details in Chap. 5, and thereby the above formulas will be refined in that chapter.

By substituting Eqs. (4.31) and (4.32) into Eq. (4.29), the following approximate expression of $\chi^{(2)}$ is obtained,

$$\chi_{pqr}^{(2),\text{res}}(\omega_2) = \frac{1}{k_B T}\langle \alpha_{pq}\mu_r\rangle + \frac{i\omega_2}{k_B T}\int_0^\infty dt \left\langle \alpha_{pq}(t)\mu_r(0)\right\rangle \exp(i\omega_2 t) \tag{4.29}$$

$$\begin{aligned}
&\approx \frac{1}{k_B T}\left\langle \sum_{l}^{\text{molecule}} \sum_{p',q'}^{\xi\sim\zeta} \mathcal{D}_{l,pp'}\mathcal{D}_{l,qq'}\alpha_{l,p'q'} \cdot \sum_{k}^{\text{molecule}} \sum_{r'}^{\xi\sim\zeta} \mathcal{D}_{k,rr'}\mu_{k,r'} \right\rangle \\
&+ \frac{i\omega_2}{k_B T}\int_0^\infty dt \left\langle \sum_{l}^{\text{molecule}} \sum_{p',q'}^{\xi\sim\zeta} \mathcal{D}_{l,pp'}(t)\mathcal{D}_{l,qq'}(t)\alpha_{l,p'q'}(t) \cdot \sum_{k}^{\text{molecule}} \sum_{r'}^{\xi\sim\zeta} \mathcal{D}_{k,rr'}(0)\mu_{k,r'}(0) \right\rangle \exp(i\omega_2 t) \\
&\approx \sum_{l}^{\text{molecule}} \sum_{p',q',r'}^{\xi\sim\zeta} \frac{1}{k_B T}\left\langle \mathcal{D}_{l,pp'}\mathcal{D}_{l,qq'}\alpha_{l,p'q'} \cdot \mathcal{D}_{l,rr'}\mu_{l,r'}\right\rangle \\
&+ \sum_{l}^{\text{molecule}} \sum_{p',q',r'}^{\xi\sim\zeta} \frac{i\omega_2}{k_B T}\int_0^\infty dt \left\langle \mathcal{D}_{l,pp'}(t)\mathcal{D}_{l,qq'}(t)\alpha_{l,p'q'}(t) \cdot \mathcal{D}_{l,rr'}(0)\mu_{l,r'}(0)\right\rangle \exp(i\omega_2 t).
\end{aligned} \tag{4.33}$$

From the second line to the third, we have employed another approximation to neglect the cross correlations between different molecules ($l \neq k$). This approximation will be critically examined with some examples of aqueous electrolyte systems in Sect. 9.3.

Equation (4.33) shows explicitly that the time correlation function includes the rotational matrices $\mathcal{D}(t)$ as a function of t, and thereby the orientational motion of molecules. In a case that the orientational motion is slow and thus regarded to be fixed during the decay time of the time correlation function, Eq. (4.33) becomes

$$\chi_{pqr}^{(2),\mathrm{res}}(\omega_2)$$

$$\approx \overline{\sum_{l}^{\text{molecule}} \sum_{p',q',r'}^{\xi\sim\zeta} \mathcal{D}_{l,pp'}\mathcal{D}_{l,qq'}\mathcal{D}_{l,rr'} \left\{ \frac{1}{k_B T}\left\langle \alpha_{l,p'q'}\mu_{l,r'}\right\rangle + \frac{i\omega_2}{k_B T}\int_0^\infty dt \left\langle \alpha_{l,p'q'}(t)\mu_{l,r'}(0)\right\rangle \exp(i\omega_2 t)\right\}} \tag{4.34}$$

(fixed orientational motion)

where the average in the curly bracket is taken first before the average over the orientation.[5] In Eq. (4.34), the dynamics in the time correlation function is governed by intramolecular vibrations. We represent the molecular vibrations with normal mode(s) Q_a. Thus the time evolution of a physical quantity Y $(=\alpha,\mu)$ is driven by the normal mode(s),

$$Y(t) = Y(0) + \sum_a^{\text{mode}} \left(\frac{\partial Y}{\partial Q_a}\right) Q_a(t) + \cdots, \tag{4.35}$$

and the vibration of mode a is given with a damped harmonic oscillator,

$$\langle Q_a(t)Q_a\rangle = \langle Q_a^2\rangle \exp(-\Gamma_a t)\cos(\omega_a t) = \frac{k_B T}{m_a\omega_a^2}\exp(-\Gamma_a t)\cos(\omega_a t). \tag{4.36}$$

Then the curly bracket in Eq. (4.34) including the Fourier-Laplace transform is written as

$$\begin{aligned}
&\frac{1}{k_B T}\langle\alpha\mu\rangle + \frac{i\omega}{k_B T}\int_0^\infty dt\,\langle\alpha(t)\mu(0)\rangle\exp(i\omega t)\\
&\approx \frac{1}{k_B T}\langle\alpha\mu\rangle + \frac{i\omega}{k_B T}\int_0^\infty dt\left\{\langle\alpha\mu\rangle + \sum_a^{\text{mode}}\left(\frac{\partial\alpha}{\partial Q_a}\right)\left(\frac{\partial\mu}{\partial Q_a}\right)\langle Q_a(t)Q_a(0)\rangle\right\}\exp(i\omega t)\\
&= \frac{i\omega}{k_B T}\sum_a\left(\frac{\partial\alpha}{\partial Q_a}\right)\left(\frac{\partial\mu}{\partial Q_a}\right)\frac{k_B T}{m_a\omega_a^2}\int_0^\infty dt\,\exp(-\Gamma_a t)\cos(\omega_a t)\,\exp(i\omega t)\\
&= -\sum_a\left(\frac{\partial\alpha}{\partial Q_a}\right)\left(\frac{\partial\mu}{\partial Q_a}\right)\frac{\omega}{2m_a\omega_a^2}\left(\frac{1}{\omega-\omega_a+i\Gamma_a}+\frac{1}{\omega+\omega_a+i\Gamma_a}\right)
\end{aligned}$$

[5]Here we denote time correlation function by angle bracket $\langle A\rangle$ and orientational average by overbar $\overline{A}$. In the classical mechanics, both notations $\langle A\rangle$ and $\overline{A}$ indicate the statistical average and thus they are essentially equivalent in this context.

$$\approx -\sum_a \frac{1}{2m_a\omega_a}\left(\frac{\partial\alpha}{\partial Q_a}\right)\left(\frac{\partial\mu}{\partial Q_a}\right)\frac{1}{\omega-\omega_a+i\Gamma_a}, \tag{4.37}$$

where the last expression takes the resonant term with the mode a by the rotating wave approximation. Using the result of Eq. (4.37), $\chi_{pqr}^{(2),\mathrm{res}}(\omega_2)$ in Eq. (4.34) becomes

$$\chi_{pqr}^{(2),\mathrm{res}}(\omega_2) \approx \overline{\sum_l^{\text{molecule}} \sum_{p',q',r'}^{\xi\sim\zeta} \mathcal{D}_{l,pp'}\mathcal{D}_{l,qq'}\mathcal{D}_{l,rr'}\left\{-\sum_a^{\text{mode}} \frac{1}{2m_a\omega_a}\left(\frac{\partial\alpha_{l,p'q'}}{\partial Q_a}\right)\left(\frac{\partial\mu_{l,r'}}{\partial Q_a}\right)\frac{1}{\omega_2-\omega_a+i\Gamma_a}\right\}}. \tag{4.38}$$

Equation (4.38) coincides with the $\chi^{(2),\mathrm{res}}$ expression of Eqs. (4.3) and (4.4) in Sect. 4.1 on the basis of energy representation.

4.4.2 *Slow Limit and Fast Limit*

The above derivation of Eqs. (4.3) and (4.4) helps justifying the $\chi^{(2),\mathrm{res}}$ formula of energy representation as well as manifesting the approximations involved in Eqs. (4.3) and (4.4). One important approximation is that the molecular orientation is fixed during the correlation time of the vibrations. This approximation of static orientation has been used to derive Eq. (4.34) in the above discussion. If the molecular orientation changes within the correlation time, the decay profile of the time correlation function is influenced by the orientational motion.

To discuss this motional effect, we could assume the other extreme case that the orientational motion is fast enough in comparison with the correlation time. Then the rotational average is taken first, and Eq. (4.33) should be

$$\chi_{pqr}^{(2),\mathrm{res}}(\omega_2) \approx \overline{\sum_l^{\text{molecule}} \left\{\frac{1}{k_BT}\left\langle\overline{\alpha_{l,pq}}\cdot\overline{\mu_{l,r}}\right\rangle + \frac{i\omega_2}{k_BT}\int_0^\infty dt\left\langle\overline{\alpha_{l,pq}(t)}\cdot\overline{\mu_{l,r}(0)}\right\rangle\exp(i\omega_2 t)\right\}}. \tag{4.39}$$

(fast orientational motion)

In this limit of fast orientational motion, the orientationally averaged values of α and μ,

$$\overline{\alpha_{l,pq}(t)} = \overline{\sum_{p',q'}^{\xi\sim\zeta} \mathcal{D}_{l,pp'}(t)\mathcal{D}_{l,qq'}(t)\alpha_{l,p'q'}(t)} \quad \text{and} \quad \overline{\mu_{l,r}(t)} = \overline{\sum_{r'}^{\xi\sim\zeta} \mathcal{D}_{l,rr'}(t)\mu_{l,r'}(t)},$$

are included in the time correlation function in Eq. (4.39). Wei and Shen argued that the SPS spectra are particularly sensitive to the motional effect of molecular orientation [26]. This is because the $\chi^{(2)}_{yzy}$ tensor component associated to the SPS spectra includes α_{yz} and μ_y components, and the off-diagonal element of polarizability $\overline{\alpha_{yz}}$ is readily reduced by the orientational average.

The realistic situations fall between the two extreme cases of Eqs. (4.34) and (4.39). We note that the motional effect of molecular orientation is naturally incorporated in the MD calculation of $\boldsymbol{\chi}^{(2)}$ by the time correlation function formula.

4.5 Solutions to Problems

4.5.1 Polarization Ratios

[Problem 4.1] Derive Eqs. (4.14) and (4.15) on the basis of the above assumptions.

The ratios B and C are presented using Eq. (3.41) by

$$B = \frac{\chi^{(2)}_{yyz}}{\chi^{(2)}_{yzy}} = \frac{N\cdot\overline{\alpha^{(2)}_{yyz}}}{N\cdot\overline{\alpha^{(2)}_{yzy}}} = \frac{\overline{\alpha^{(2)}_{yyz}}}{\overline{\alpha^{(2)}_{yzy}}}, \qquad C = \frac{\chi^{(2)}_{zzz}}{\chi^{(2)}_{yyz}} = \frac{N\cdot\overline{\alpha^{(2)}_{zzz}}}{N\cdot\overline{\alpha^{(2)}_{yyz}}} = \frac{\overline{\alpha^{(2)}_{zzz}}}{\overline{\alpha^{(2)}_{yyz}}}.$$

Therefore, we formulate $\overline{\alpha^{(2)}_{yyz}}$, $\overline{\alpha^{(2)}_{yzy}}$, and $\overline{\alpha^{(2)}_{zzz}}$ using Eqs. (3.43), (3.47) and (4.13) as follows.

$$\begin{aligned}
\overline{\alpha^{(2)}_{yyz}} &= \sum_{p'}\sum_{q'}\sum_{r'}\overline{\mathcal{D}_{yp'}\mathcal{D}_{yq'}\mathcal{D}_{zr'}}\left(\frac{\partial\alpha_{p'q'}}{\partial q_1}\right)\left(\frac{\partial\mu_{r'}}{\partial q_1}\right)\\
&= \overline{\mathcal{D}_{y\xi}\mathcal{D}_{y\xi}\mathcal{D}_{z\zeta}}\ R\alpha\mu + \overline{\mathcal{D}_{y\eta}\mathcal{D}_{y\eta}\mathcal{D}_{z\zeta}}\ R\alpha\mu + \overline{\mathcal{D}_{y\zeta}\mathcal{D}_{y\zeta}\mathcal{D}_{z\zeta}}\ \alpha\mu\\
&= \overline{(\cos\psi\sin\phi + \cos\theta\cos\phi\sin\psi)^2(\cos\theta)}\ R\alpha\mu\\
&\quad + \overline{(-\sin\psi\sin\phi + \cos\theta\cos\phi\cos\psi)^2(\cos\theta)}\ R\alpha\mu\\
&\quad + \overline{(-\sin\theta\cos\phi)^2(\cos\theta)}\ \alpha\mu\\
&= \overline{(\cos\theta\sin^2\phi + \cos^3\theta\cos^2\phi)}\ R\alpha\mu + \overline{\cos\theta\sin^2\theta\cos^2\psi}\ \alpha\mu\\
&= (\overline{\cos\theta} + \overline{\cos^3\theta})\ \frac{R\alpha\mu}{2} + (\overline{\cos\theta} - \overline{\cos^3\theta})\ \frac{\alpha\mu}{2},
\end{aligned} \tag{4.40}$$

$$\overline{\alpha^{(2)}_{yzy}} = \sum_{p'}\sum_{q'}\sum_{r'}\overline{\mathcal{D}_{yp'}\mathcal{D}_{zq'}\mathcal{D}_{yr'}}\left(\frac{\partial\alpha_{p'q'}}{\partial q_1}\right)\left(\frac{\partial\mu_{r'}}{\partial q_1}\right)$$

$$
\begin{aligned}
&= \overline{\mathcal{D}_{y\xi}\mathcal{D}_{z\xi}\mathcal{D}_{y\zeta}}\ R\alpha\mu + \overline{\mathcal{D}_{y\eta}\mathcal{D}_{z\eta}\mathcal{D}_{y\zeta}}\ R\alpha\mu + \overline{\mathcal{D}_{y\zeta}\mathcal{D}_{z\zeta}\mathcal{D}_{y\zeta}}\ \alpha\mu \\
&= \overline{(\cos\psi\sin\phi + \cos\theta\cos\phi\sin\psi)(\sin\psi\sin\theta)(-\sin\theta\cos\phi)}\ R\alpha\mu \\
&\quad + \overline{(-\sin\psi\sin\phi + \cos\theta\cos\phi\cos\psi)(\cos\psi\sin\theta)(-\sin\theta\cos\phi)}\ R\alpha\mu \\
&\quad + \overline{(-\sin\theta\cos\phi)(\cos\theta)(-\sin\theta\cos\phi)}\ \alpha\mu \\
&= \overline{(-\cos\theta\sin^2\theta\cos^2\phi - 2\sin^2\theta\cos\psi\sin\psi\cos\phi\sin\phi)}\ R\alpha\mu \\
&\quad \overline{+\cos\theta\sin^2\theta\cos^2\phi}\ \alpha\mu \\
&= -(\overline{\cos\theta} - \overline{\cos^3\theta})\ \frac{R\alpha\mu}{2} + (\overline{\cos\theta} - \overline{\cos^3\theta})\ \frac{\alpha\mu}{2},
\end{aligned} \tag{4.41}
$$

$$
\begin{aligned}
\overline{\alpha^{(2)}_{zzz}} &= \sum_{p'}\sum_{q'}\sum_{r'} \overline{\mathcal{D}_{zp'}\mathcal{D}_{zq'}\mathcal{D}_{zr'}} \left(\frac{\partial\alpha_{p'q'}}{\partial q_1}\right)\left(\frac{\partial\mu_{r'}}{\partial q_1}\right) \\
&= \overline{\mathcal{D}_{z\xi}\mathcal{D}_{z\xi}\mathcal{D}_{z\zeta}}\ R\alpha\mu + \overline{\mathcal{D}_{z\eta}\mathcal{D}_{z\eta}\mathcal{D}_{z\zeta}}\ R\alpha\mu + \overline{\mathcal{D}_{z\zeta}\mathcal{D}_{z\zeta}\mathcal{D}_{z\zeta}}\ \alpha\mu \\
&= \overline{(\sin\psi\sin\theta)^2(\cos\theta)}\ R\alpha\mu + \overline{(\cos\psi\sin\theta)^2(\cos\theta)}\ R\alpha\mu + \overline{(\cos^3\theta)}\ \alpha\mu \\
&= \overline{\cos\theta\sin^2\theta}\ R\alpha\mu + \overline{\cos^3\theta}\ \alpha\mu \\
&= (\overline{\cos\theta} - \overline{\cos^3\theta})\ R\alpha\mu + \overline{\cos^3\theta}\ \alpha\mu.
\end{aligned} \tag{4.42}
$$

In the above derivation of Eqs. (4.40), (4.41), and (4.42), the average over ϕ and ψ is carried out with the assumption of uniform distribution. Using the results of Eqs. (4.40), (4.41), and (4.42), B and C are given by

$$
\begin{aligned}
B = \frac{\chi^{(2)}_{yyz}}{\chi^{(2)}_{yzy}} = \frac{\overline{\alpha^{(2)}_{yyz}}}{\overline{\alpha^{(2)}_{yzy}}} &= \frac{(\overline{\cos\theta} + \overline{\cos^3\theta})\ \frac{R\alpha\mu}{2} + (\overline{\cos\theta} - \overline{\cos^3\theta})\ \frac{\alpha\mu}{2}}{-(\overline{\cos\theta} - \overline{\cos^3\theta})\ \frac{R\alpha\mu}{2} + (\overline{\cos\theta} - \overline{\cos^3\theta})\ \frac{\alpha\mu}{2}} \\
&= \frac{(1+R)\ \overline{\cos\theta} - (1-R)\ \overline{\cos^3\theta}}{(1-R)(\overline{\cos\theta} - \overline{\cos^3\theta})},
\end{aligned} \tag{4.14}
$$

$$
\begin{aligned}
C = \frac{\chi^{(2)}_{zzz}}{\chi^{(2)}_{yyz}} = \frac{\overline{\alpha^{(2)}_{zzz}}}{\overline{\alpha^{(2)}_{yyz}}} &= \frac{(\overline{\cos\theta} - \overline{\cos^3\theta})\ R\alpha\mu + \overline{\cos^3\theta}\ \alpha\mu}{(\overline{\cos\theta} + \overline{\cos^3\theta})\ \frac{R\alpha\mu}{2} + (\overline{\cos\theta} - \overline{\cos^3\theta})\ \frac{\alpha\mu}{2}} \\
&= \frac{2\{R\ \overline{\cos\theta} + (1-R)\ \overline{\cos^3\theta}\}}{(1+R)\ \overline{\cos\theta} - (1-R)\ \overline{\cos^3\theta}}.
\end{aligned} \tag{4.15}
$$

We notice that the final results of B and C in Eqs. (4.14) and (4.15) do not include N, α and μ.

4.5.2 Canonical Time Correlation Function

[Problem 4.2] Prove $\langle A; B\rangle = \langle B; A\rangle$ for arbitrary operators A and B in Eq. (4.24). This indicates that the operators in the canonical correlation function are commutative like a classical one.

$\langle A; B\rangle$ is

$$\begin{aligned}\langle A; B\rangle &= \frac{1}{\beta}\int_0^\beta d\lambda\ \overline{\langle \exp(\lambda\mathcal{H})A\exp(-\lambda\mathcal{H})B\rangle} \qquad (4.24)\\ &= \frac{1}{\beta}\int_0^\beta d\lambda\ \mathrm{Tr}\left[\frac{\exp(-\beta\mathcal{H})}{Q}\exp(\lambda\mathcal{H})A\exp(-\lambda\mathcal{H})B\right],\end{aligned}$$

where Q is the partition function. The variable λ in the above integral is transformed to $\lambda' = \beta - \lambda$. Consequently,

$$\begin{aligned}\langle A; B\rangle &= \frac{-1}{\beta Q}\int_\beta^0 d\lambda'\ \mathrm{Tr}\left[\exp(-\lambda'\mathcal{H})A\exp\{(\lambda'-\beta)\mathcal{H}\}B\right]\\ &= \frac{1}{\beta Q}\int_0^\beta d\lambda'\ \mathrm{Tr}\left[\exp(-\beta\mathcal{H})\exp(\lambda'\mathcal{H})B\exp(-\lambda'\mathcal{H})A\right]\\ &= \frac{1}{\beta}\int_0^\beta d\lambda'\ \overline{\langle \exp(\lambda'\mathcal{H})B\exp(-\lambda'\mathcal{H})A\rangle}\\ &= \langle B; A\rangle\,.\end{aligned}$$

Bibliography

1. Allen MP, Tildesley DJ (1987) Computer simulation of liquids. Clarendon Press, Oxford
2. Atkins PW, Friedman RS (2010) Molecular quantum mechanics, 5th edn. Oxford University Press, Oxford
3. Bader JS, Berne BJ (1994) Quantum and classical relaxation rates from classical simulations. J Chem Phys 100:8359–8366
4. Buch V, Tarbuck T, Richmond GL, Groenzin H, Li I, Shultz MJ (2007) Sum frequency generation surface spectra of ice, water, and acid solution investigated by an exciton model. J Chem Phys 127:204710
5. Egorov SA, Everitt KF, Skinner JL (1999) Quantum dynamics and vibrational relaxation. J Phys Chem A 103:9494–9499
6. Frenkel D, Smit B (1996) Understanding molecular simulation. Academic, New York
7. Gan W, Wu B-H, Zhang Z, Guo Y, Wang H-F (2007) Vibrational spectra and molecular orientation with experimental configuration analysis in surface sum frequency generation (SFG). J Phys Chem C 111:8716–8725

8. Hirose C, Akamatsu A, Domen K (1992) Formulas for the analysis of surface sum-frequency generation spectrum by CH stretching modes of methyl and methylene groups. J Chem Phys 96:997–1004
9. Hirose C, Hiroyoshi Y, Akamatsu N, Domen K (1993) Orientation analysis by simulation of vibrational sum frequency generation spectrum: CH stretching bands of the methyl group. J Phys Chem 97:10064–10069
10. Ishihara T, Ishiyama T, Morita A (2015) Surface structure of methanol/water solutions via sum-frequency orientational analysis and molecular dynamics simulation. J Phys Chem C 119:9879–9889
11. Ishiyama T, Morita A (2006) Intermolecular correlation effect in sum frequency generation spectroscopy of electrolyte aqueous solution. Chem Phys Lett 431:78–82
12. Ishiyama T, Morita A (2007) Molecular dynamics study of gas-liquid aqueous sodium halide interfaces II. analysis of vibrational sum frequency generation spectra. J Phys Chem C 111:738–748
13. Kubo R, Toda M, Hashitsume N (1991) Statistical mechanics II: nonequilibrium statistical mechanics, 2nd edn. Springer, Berlin
14. McQuarrie DA (2000) Statistical mechanics, 2nd edn. University of Science Books, Sausalito
15. Morita A, Hynes JT (2000) A theoretical analysis of the sum frequency generation spectrum of the water surface. Chem Phys 258:371–390
16. Morita A, Hynes JT (2002) A theoretical analysis of the sum frequency generation spectrum of the water surface II. Time-dependent approach. J Phys Chem B 106:673–685
17. Morita A, Ishiyama T (2008) Recent progress in theoretical analysis of vibrational sum frequency generation spectroscopy. Phys Chem Chem Phys 10:5801–5816
18. Pimentel GC, McClellan AL (1960) The Hydrogen Bond. W. H. Freemann, San Francisco
19. Saito K, Peng Q, Qiao L, Wang L, Joutsuka T, Ishiyama T, Ye S, Morita A (2017) Theoretical and experimental examination on SFG polarization analysis at acetonitrile-water solution surfaces. Phys Chem Chem Phys 19:8941–8961
20. Shultz MJ, Bisson P, Groenzin H, Li I (2010) Multiplexed polarization spectroscopy: measuring surface hyperpolarizability orientation. J Chem Phys 133:054702
21. Walker DS, Hore DK, Richmond GL (2006) Understanding the population, coordination, and orientation of water species contributing to the nonlinear optical spectroscopy of the vapor-water interface through molecular dynamics simulations. J Phys Chem B 110:20451–20459
22. Walker DS, Richmond GL (2007) Understanding the effects of hydrogen bonding at the vapor-water interface: vibrational sum frequency spectroscopy of H_2O/HOD/D_2O mixtures studied using molecular dynamics simulations. J Phys Chem C 111:8321–8330
23. Wang H-F, Gan W, Lu R, Rao Y, Wu B-H (2005) Quantitative spectral and orientational analysis in surface sum frequency generation vibrational spectroscopy (SFG-VS). Int Rev Phys Chem 24:191–256
24. Wang L, Ishiyama T, Morita A (2017) Theoretical investigation of C-H vibrational spectroscopy. 1. Modeling of methyl and methylene groups of ethanol with different conformers. J Phys Chem A 121:6687–6700
25. Wang L, Ishiyama T, Morita A (2017) Theoretical investigation of C-H vibrational spectroscopy. 2. Unified assignment method of IR, Raman and SFG spectra of ethanol. J Phys Chem A 121:6701–6712
26. Wei X, Shen YR (2001) Motional effect in surface sum-frequency vibrational spectroscopy. Phys Rev Lett 86:4799
27. Yeh YL, Zhang C, Held H, Mebel AM, Wei X, Lin SH, Shen Y (2001) Structure of the acetone liquid/vapor interface. J Chem Phys 114:1837–1843
28. Zhang D, Gutow J, Eisenthal KB (1994) Vibrational spectra, orientations, and phase transitions in long-chain amphiphiles at the air/water interface: probing the head and tail groups by sum frequency generation. J Phys Chem 98:13729–13734

Chapter 5
Molecular Theory of Local Field

Abstract In applying the microscopic SFG theory to actual interfaces, polarization properties of condensed dielectric media have to be properly described. These properties in uniform bulk media are often described with a phenomenological model, though the detailed microscopic treatment becomes indispensable at inhomogeneous interfaces. This chapter summarizes the molecular theory of polarization properties and offer a general scheme to calculate them by molecular simulation. These properties can be determined in principle from molecular properties and their interactions. This theory is of general significance to light-matter interactions, and also an integral part in the computational analysis of SFG.

Keywords Local field correction factor · Effective polarizability · Interfacial dielectric constant

The preceding chapter has given two representations of the second-order nonlinear susceptibility $\boldsymbol{\chi}^{(2)}$, using the polarizability $\boldsymbol{\alpha}$ and the dipole moment $\boldsymbol{\mu}$ of the interface system. In this chapter, we argue how the polarizability and dipole moment of the interface system are described from microscopic configuration of molecules. In the following discussion, the polarizability and dipole moment of the *whole interface system* are denoted with the capital letters by $\boldsymbol{A}$ and $\boldsymbol{M}$, respectively, to distinguish them from molecular properties. In the condensed phase, the polarization properties such as $\boldsymbol{A}$ and $\boldsymbol{M}$ are affected by electrostatic interactions among constituent molecules. The electrostatic intermolecular interactions mutually polarize the constituent molecules, and consequently, the polarizability or dipole moment of the whole system do not become the simple sum of polarizability or dipole moment of non-interacting molecules. The reliable description of $\boldsymbol{\chi}^{(2)}$ should take account of this effect of polarization coupling.

The coupling of polarizations is the origin of local field effect in a dielectric medium, i.e. the local electric field at a molecule in the condensed phase is influenced by the electrostatic interactions from neighboring molecules and thereby deviates from the external field [3]. In the classical theory of dielectrics, the local field effect is often treated by the simple Lorentz model or its analogues for a bulk

A. Morita, *Theory of Sum Frequency Generation Spectroscopy*,
Lecture Notes in Chemistry 97, https://doi.org/10.1007/978-981-13-1607-4_5

medium [3]. Such classical models deal with the homogeneous bulk medium as a dielectric continuum. In the inhomogeneous environment of interface, however, the local field is more complicated than that in the homogeneous medium, and the theory of dielectric continuum model should become less reliable. Here we discuss the local field effect at the interface fully from a molecular point of view [6, 7]. The microscopic theory of local field can accurately describe the effects of polarization coupling in arbitrary environment, and is straightforwardly applicable to MD simulation.

In Sect. 5.1 we formulate the general microscopic expressions of dipole moment and polarizability for the interface system. In Sect. 5.2 these expressions are applied to the nonlinear susceptibility $\boldsymbol{\chi}^{(2)}$ with fully incorporating the local field effect. The local field effect in inhomogeneous environment is relevant to the dielectric constant at interfaces ε'. Thus, we discuss this issue of ε' from the microscopic viewpoint in Sect. 5.3.

In the following discussion, molecular polarization is represented with the polarizability $\boldsymbol{\alpha}$ and dipole $\boldsymbol{\mu}$ of constituent molecules in a general manner. The effect of local field can be described with polarizable MD simulations in general, irrespective of the modeling method of polarization. To implement the present theory in practical MD simulation, however, one has to resort to a certain kind of polarizable molecular model. There are a number of kinds of polarizable models, such as the point dipole [1, 2, 11], fluctuating charge [8, 10, 13], and Drude oscillator [5, 14, 15], which have their respective ways to represent the electronic polarization of molecules. The following theory of local field captures the essential mechanism of polarization interactions and is applicable to the various polarizable model, though slight modifications may be required for some models to represent the polarization properties. We present an example of the Charge Response Kernel (CRK) model in Sect. 6.4.

5.1 Local Field Correction Factor

Self-consistent polarization We discuss the polarization at an arbitrary microscopic configuration of molecules in the interface system. Let us suppose that the i-th molecule is located at $\boldsymbol{r}(i)$ at an instantaneous moment, whose permanent dipole moment vector and polarizability tensor are denoted by $\boldsymbol{\mu}^0(i)$ and $\alpha(i)$, respectively. The polarization of molecules interact each other, and consequently the total (permanent + induced) dipole moment of the i-th molecule $\boldsymbol{\mu}(i)$ and the electric field at the i-th molecule $\boldsymbol{E}(i)$ are determined in the following form,

$$\mu_p(i) = \mu_p^0(i) + \sum_q^{x-z} \alpha_{pq}(i) E_q(i), \tag{5.1}$$

$$E_p(i) = E_p^0(i) - \sum_{j(\neq i)} \sum_q^{x-z} T_{pq}(ij) \mu_q(j), \tag{5.2}$$

where the suffixes p, q denote the space-fixed coordinates, x, y, or z. The first term of the right-hand side of Eq. (5.1) means the permanent dipole moment of the i-th molecule, and the second term the induced dipole. $\boldsymbol{E}(i)$ in Eq. (5.2) consists of two terms. The field generated by the dipoles of neighboring molecules is represented in the second term of the right-hand side of Eq. (5.2), while the first term $\boldsymbol{E}^0(i)$ accounts for the other sources of field than the dipole. $T(ij)$ is called the dipole-dipole coupling tensor, which describes the electric field at $\boldsymbol{r}(i)$ generated by a dipole moment at $\boldsymbol{r}(j)$. The explicit form of $T_{pq}(ij)$ is[1]

$$T_{pq}(ij) = \frac{\delta_{pq}}{r(ij)^3} - \frac{3r_p(ij)r_q(ij)}{r(ij)^5}, \tag{5.3}$$

where $r(ij) = |\boldsymbol{r}(ij)| = |\boldsymbol{r}(i) - \boldsymbol{r}(j)|$. Equations (5.1) and (5.2) define the coupled relation between the dipole moment $\boldsymbol{\mu}(i)$ and the electric field $\boldsymbol{E}(i)$,[2] and accordingly both quantities should be solved simultaneously in a self-consistent manner. In polarizable MD simulation, instantaneous polarizations of constituent molecules are determined at each time step t self-consistently in this way. We also note that $\boldsymbol{E}^0$ in Eq. (5.2) includes the externally imposed field as well as the intermolecular interactions other than the dipole-dipole coupling.

Solutions of $\boldsymbol{\mu}$ and $\boldsymbol{E}$ Since Eqs. (5.1) and (5.2) are linear, the coupled equations can be solved analytically. Equations (5.1) and (5.2) are written in the matrix form,

$$\begin{aligned} \boldsymbol{\mu} &= \boldsymbol{\mu}^0 + \boldsymbol{\alpha}\boldsymbol{E}, \\ \boldsymbol{E} &= \boldsymbol{E}^0 - \boldsymbol{T}\boldsymbol{\mu}. \end{aligned} \tag{5.4}$$

In Eq (5.4), $\boldsymbol{\mu}$, $\boldsymbol{\mu}^0$, $\boldsymbol{E}$, $\boldsymbol{E}^0$ are $3N$-dimensional vectors, with N being the number of molecules. They are represented by

$$\boldsymbol{\mu} = \begin{pmatrix} \boldsymbol{\mu}(1) \\ \boldsymbol{\mu}(2) \\ \vdots \\ \boldsymbol{\mu}(N) \end{pmatrix} = \begin{pmatrix} \mu_x(1) \\ \mu_y(1) \\ \mu_z(1) \\ \mu_x(2) \\ \vdots \\ \mu_z(N) \end{pmatrix}, \qquad \boldsymbol{\mu}^0 = \begin{pmatrix} \boldsymbol{\mu}^0(1) \\ \boldsymbol{\mu}^0(2) \\ \vdots \\ \boldsymbol{\mu}^0(N) \end{pmatrix} = \begin{pmatrix} \mu_x^0(1) \\ \mu_y^0(1) \\ \mu_z^0(1) \\ \mu_x^0(2) \\ \vdots \\ \mu_z^0(N) \end{pmatrix},$$

[1] Some other literature employs the reverse sign for the dipole-dipole tensor [4, 9].

[2] Equations (5.1) and (5.2) account for the dipole-dipole interaction to describe the local field, and neglect the retardation of the electromagnetic interaction. The latter is relevant to the radiation and is treated in a separate manner (see Sect. 5.3).

$$\boldsymbol{E} = \begin{pmatrix} \boldsymbol{E}(1) \\ \boldsymbol{E}(2) \\ \vdots \\ \boldsymbol{E}(N) \end{pmatrix} = \begin{pmatrix} E_x(1) \\ E_y(1) \\ E_z(1) \\ E_x(2) \\ \vdots \\ E_z(N) \end{pmatrix}, \quad \boldsymbol{E}^0 = \begin{pmatrix} \boldsymbol{E}^0(1) \\ \boldsymbol{E}^0(2) \\ \vdots \\ \boldsymbol{E}^0(N) \end{pmatrix} = \begin{pmatrix} E_x^0(1) \\ E_y^0(1) \\ E_z^0(1) \\ E_x^0(2) \\ \vdots \\ E_z^0(N) \end{pmatrix}, \tag{5.5}$$

where the components $\boldsymbol{\mu}(i)$, $\boldsymbol{\mu}^0(i)$, $\boldsymbol{E}(i)$, $\boldsymbol{E}^0(i)$ for the i-th molecule are three-dimensional vectors. On the other hand, $\boldsymbol{\alpha}$ and $\boldsymbol{T}$ are $3N \times 3N$ matrices, whose components are written in the following form,

$$\boldsymbol{\alpha} = \begin{pmatrix} \boldsymbol{\alpha}(1) & \boldsymbol{0} & \cdots & \boldsymbol{0} \\ \boldsymbol{0} & \boldsymbol{\alpha}(2) & & \\ \vdots & & \ddots & \\ \boldsymbol{0} & & & \boldsymbol{\alpha}(N) \end{pmatrix}, \quad \boldsymbol{T} = \begin{pmatrix} \boldsymbol{0} & \boldsymbol{T}(12) & \cdots & \boldsymbol{T}(1N) \\ \boldsymbol{T}(21) & \boldsymbol{0} & & \boldsymbol{T}(2N) \\ \vdots & & \ddots & \vdots \\ \boldsymbol{T}(N1) & \boldsymbol{T}(N2) & \cdots & \boldsymbol{0} \end{pmatrix}. \tag{5.6}$$

In this Eq. (5.6), $\boldsymbol{\alpha}$ is symmetric since $\boldsymbol{\alpha}(i)$ denotes the 3×3 symmetric polarizability tensor of the i-th molecule. $\boldsymbol{T}$ is also symmetric as $\boldsymbol{T}(ij)$ is a symmetric 3×3 matrix. Then Eq. (5.4) is formally solved as follows.

$$\boldsymbol{E} = \boldsymbol{E}^0 - \boldsymbol{T}\boldsymbol{\mu} = \boldsymbol{E}^0 - \boldsymbol{T}\left(\boldsymbol{\mu}^0 + \boldsymbol{\alpha}\boldsymbol{E}\right), \tag{5.7}$$

$$\boldsymbol{\mu} = \boldsymbol{\mu}^0 + \boldsymbol{\alpha}\boldsymbol{E} = \boldsymbol{\mu}^0 + \boldsymbol{\alpha}\left(\boldsymbol{E}^0 - \boldsymbol{T}\boldsymbol{\mu}\right). \tag{5.8}$$

Therefore,

$$\boldsymbol{E} = [\boldsymbol{1} + \boldsymbol{T}\boldsymbol{\alpha}]^{-1}\left(\boldsymbol{E}^0 - \boldsymbol{T}\boldsymbol{\mu}^0\right) = \boldsymbol{g}\left(\boldsymbol{E}^0 - \boldsymbol{T}\boldsymbol{\mu}^0\right), \tag{5.9}$$

$$\boldsymbol{\mu} = [\boldsymbol{1} + \boldsymbol{\alpha}\boldsymbol{T}]^{-1}\left(\boldsymbol{\mu}^0 + \boldsymbol{\alpha}\boldsymbol{E}^0\right) = \boldsymbol{g}^T\left(\boldsymbol{\mu}^0 + \boldsymbol{\alpha}\boldsymbol{E}^0\right) = \boldsymbol{g}^T\boldsymbol{\mu}^{0\prime}, \tag{5.10}$$

where $\boldsymbol{\mu}^{0\prime} = \boldsymbol{\mu}^0 + \boldsymbol{\alpha}\boldsymbol{E}^0$. $\boldsymbol{g}$ is a $3N \times 3N$ matrix defined by

$$\boldsymbol{g} = [\boldsymbol{1} + \boldsymbol{T}\boldsymbol{\alpha}]^{-1} \qquad \left(\text{or} \quad \boldsymbol{g}^T = [\boldsymbol{1} + \boldsymbol{\alpha}\boldsymbol{T}]^{-1}\right). \tag{5.11}$$

Equations (5.9) and (5.10) manifest the physical meaning of $\boldsymbol{g}$ and $\boldsymbol{g}^T$. Equation (5.9) indicates that $\boldsymbol{g}$ is the factor to modify the electric field from $\boldsymbol{E}^0 - \boldsymbol{T}\boldsymbol{\mu}^0$ to $\boldsymbol{E}$,

$$\left(\boldsymbol{E}^0 - \boldsymbol{T}\boldsymbol{\mu}^0\right) \xrightarrow{\boldsymbol{g}} \boldsymbol{E}.$$

The left side consists of the "external" field $\boldsymbol{E}^0$ and the electric field from permanent dipoles of neighboring molecules $-\boldsymbol{T}\boldsymbol{\mu}^0$. $\boldsymbol{E}^0 - \boldsymbol{T}\boldsymbol{\mu}^0$ is a hypothetical electric field in Eq. (5.7) with the polarizability switched off, $\boldsymbol{\alpha} = 0$. In the realistic case, however, the dipole moments of molecules are affected each other, and as a consequence, the electric field changes from $\boldsymbol{E}^0 - \boldsymbol{T}\boldsymbol{\mu}^0$ to $\boldsymbol{E}$. $\boldsymbol{g}$ is therefore considered to be a microscopic definition of the local field correction factor, and deviates from unity when the constituent molecules are polarizable.

In the same vein, Eq. (5.10) illustrates the role of $\boldsymbol{g}^T$, which changes the dipole moments of molecules from $\boldsymbol{\mu}^{0'}$ to $\boldsymbol{\mu}$,

$$\left(\boldsymbol{\mu}^0 + \boldsymbol{\alpha}\boldsymbol{E}^0\right) = \boldsymbol{\mu}^{0'} \xrightarrow{\boldsymbol{g}^T} \boldsymbol{\mu}.$$

The left side $\boldsymbol{\mu}^{0'}$ stands for the dipole moment consisting of the permanent dipole $\boldsymbol{\mu}^0$ and the induced one by the external field $\boldsymbol{\alpha}\boldsymbol{E}^0$. Accordingly, $\boldsymbol{\mu}^{0'}$ does not include the dipole-dipole interaction among molecular polarizations, and $\boldsymbol{g}^T$ takes account of the dipole-dipole coupling effect on the polarizations. To summarize, both factors $\boldsymbol{g}$ and $\boldsymbol{g}^T$ account for the effect of the self-consistent coupling among polarizations. $\boldsymbol{g}$ affects on the electric field while $\boldsymbol{g}^T$ on the dipole moment. Note the distinction between $\boldsymbol{g}$ and $\boldsymbol{g}^T$ in the transpose relation.

Dipole *M* Using the solutions of $\boldsymbol{E}$ and $\boldsymbol{\mu}$ in Eqs. (5.9) and (5.10), we can derive the dipole moment and polarizability of the whole system. The dipole moment of the whole system $\boldsymbol{M}$ is given by the sum of total (permanent + induced) dipole moments of constituent molecules,

$$\boldsymbol{M} = \sum_{i=1}^{N} \boldsymbol{\mu}(i). \tag{5.12}$$

$\boldsymbol{\mu}(i)$ is a part of the $3N$-dimensional vector $\boldsymbol{\mu}$ in Eq. (5.5). Equation (5.12) is expanded in the component representation using Eq. (5.10) by

$$\begin{aligned} M_p &= \sum_{i=1}^{N} \mu_p(i) = \sum_{i,j}^{N} \sum_{q}^{x\sim z} g_{pq}^T(ij) \left(\mu_q^0(j) + \sum_{r}^{x\sim z} \alpha_{qr}(j) E_r^0(j) \right) \\ &= \sum_{i,j}^{N} \sum_{q}^{x\sim z} g_{pq}^T(ij) \mu_q^{0'}(j) = \sum_{i,j} \sum_{q}^{x\sim z} \mu_q^{0'}(j) g_{qp}(ji) = \sum_{j=1}^{N} \sum_{q}^{x\sim z} \mu_q^{0'}(j) f_{qp}(j). \end{aligned} \tag{5.13}$$

$\boldsymbol{g}(ij)$ and $\boldsymbol{g}^T(ij)$ are 3×3 matrices, which are parts of the $3N \times 3N$ matrices of $\boldsymbol{g}$ and $\boldsymbol{g}^T$, respectively,

$$\boldsymbol{g} = \begin{pmatrix} \boldsymbol{g}(11) & \boldsymbol{g}(12) & \cdots & \boldsymbol{g}(1N) \\ \boldsymbol{g}(21) & \boldsymbol{g}(22) & & \\ \vdots & & \ddots & \\ \boldsymbol{g}(N1) & & & \boldsymbol{g}(NN) \end{pmatrix}, \quad \boldsymbol{g}^T = \begin{pmatrix} \boldsymbol{g}^T(11) & \boldsymbol{g}^T(12) & \cdots & \boldsymbol{g}^T(1N) \\ \boldsymbol{g}^T(21) & \boldsymbol{g}^T(22) & & \boldsymbol{g}^T(2N) \\ \vdots & & \ddots & \vdots \\ \boldsymbol{g}^T(N1) & \boldsymbol{g}^T(N2) & \cdots & \boldsymbol{g}^T(NN) \end{pmatrix}.$$

In the last expression of Eq. (5.13), $\boldsymbol{f}(i)$ is a 3×3 matrix for each i as,

$$f_{pq}(i) = \sum_{j=1}^{N} g_{pq}(ij). \tag{5.14}$$

$\boldsymbol{f}(i)$ indicates the correction factor for the dipole moment of the i-th molecule.

Polarizability *A* The polarizability of the whole system $\boldsymbol{A}$ is defined as the derivative of the dipole moment $\boldsymbol{M}$ with respect to the external field. Here we introduce the generalized effective polarizability of $3N \times 3N$ tensor, $\boldsymbol{\alpha}^{\text{eff}} = \partial\boldsymbol{\mu}/\partial\boldsymbol{E}^0$, as the derivative of $\boldsymbol{\mu}$ with respect to the electric field $\boldsymbol{E}^0$. It is derived by differentiating Eq. (5.10) as

$$\boldsymbol{\alpha}^{\text{eff}} = \frac{\partial\boldsymbol{\mu}}{\partial\boldsymbol{E}^0} = \boldsymbol{g}^T\boldsymbol{\alpha} = [\mathbf{1} + \boldsymbol{\alpha}\boldsymbol{T}]^{-1}\boldsymbol{\alpha} = \boldsymbol{\alpha}[\mathbf{1} + \boldsymbol{T}\boldsymbol{\alpha}]^{-1} = \boldsymbol{\alpha}\boldsymbol{g}. \tag{5.15}$$

$\boldsymbol{\alpha}^{\text{eff}}$ is written by

$$\boldsymbol{\alpha}^{\text{eff}} = \begin{pmatrix} \boldsymbol{\alpha}^{\text{eff}}(11) & \boldsymbol{\alpha}^{\text{eff}}(12) & \cdots & \boldsymbol{\alpha}^{\text{eff}}(1N) \\ \boldsymbol{\alpha}^{\text{eff}}(21) & \boldsymbol{\alpha}^{\text{eff}}(22) & & \\ \vdots & & \ddots & \vdots \\ \boldsymbol{\alpha}^{\text{eff}}(N1) & & \cdots & \boldsymbol{\alpha}^{\text{eff}}(NN) \end{pmatrix},$$

using the 3×3 component matrices $\boldsymbol{\alpha}^{\text{eff}}(ij) = \partial\boldsymbol{\mu}(i)/\partial\boldsymbol{E}^0(j)$. $\boldsymbol{\alpha}^{\text{eff}}(ij)$ is the derivative of the dipole moment of the i-th molecule with respect to the electric field at the j-th molecule. Note that $\boldsymbol{\alpha}^{\text{eff}}(ij)$ may have finite values for $i \neq j$ due to the interaction between induced polarizations, in contrast to $\boldsymbol{\alpha}$ in Eq. (5.6).

Suppose that the external field at j-th molecule varies by $\delta\boldsymbol{E}^0(j)$, it changes the dipole moment of the same j-th molecule. As a consequence, the modified dipole changes the polarization of a neighboring i-th molecule by $\delta\boldsymbol{\mu}(i)$ through the electrostatic intermolecular coupling. The first-order variation of $\delta\boldsymbol{\mu}(i)$ is given with the variation of $\delta\boldsymbol{E}^0(j)$ by

$$\begin{aligned} \delta\boldsymbol{\mu}(i) &= \sum_{j=1}^{N} \boldsymbol{\alpha}^{\text{eff}}(ij)\,\delta\boldsymbol{E}^0(j) \\ &= \sum_{j=1}^{N} \boldsymbol{\alpha}(i)\,\boldsymbol{g}(ij)\,\delta\boldsymbol{E}^0(j) = \boldsymbol{\alpha}(i)\,\delta\boldsymbol{E}^{\text{loc}}(i). \end{aligned} \tag{5.16}$$

The last expression of Eq. (5.16) is obtained by using the differential relation of Eq. (5.9) with respect to $\boldsymbol{E}^0$,

$$\delta\boldsymbol{E}(i) = \sum_{j} \boldsymbol{g}(ij)\,\delta\boldsymbol{E}^0(j) \equiv \delta\boldsymbol{E}^{\text{loc}}(i),$$

and $\delta \boldsymbol{E}(i)$ is denoted with the superscript "loc" to designate the local field of i-th molecule.[3] The last expression of Eq. (5.16) means that $\delta \boldsymbol{\mu}(i)$ is also represented as the local response of i-th molecule to the change in the local field $\delta \boldsymbol{E}^{\text{loc}}(i)$.

Then we deal with a variation of the external field of light, $\delta \boldsymbol{E}^{\text{ext}}$. This gives rise to the uniform variation of $\boldsymbol{E}^0$, $\delta \boldsymbol{E}^0(1) = \delta \boldsymbol{E}^0(2) = \cdots = \delta \boldsymbol{E}^0(N) = \delta \boldsymbol{E}^{\text{ext}}$, since the external field of light is considered uniform in the microscopic scale in question. Accordingly, Eq. (5.16) indicates $\delta \boldsymbol{\mu}(i)$ to be

$$\delta \boldsymbol{\mu}(i) = \sum_{j=1}^{N} \boldsymbol{\alpha}^{\text{eff}}(ij)\, \delta \boldsymbol{E}^0(j) = \left(\sum_{j=1}^{N} \boldsymbol{\alpha}^{\text{eff}}(ij) \right) \delta \boldsymbol{E}^{\text{ext}}.$$

This formula is written in the component representation using the factor $\boldsymbol{g}(ij)$ or $\boldsymbol{f}(i)$,

$$\begin{aligned}
\delta \mu_p(i) &= \sum_{j}^{N} \sum_{q} \alpha_{pq}^{\text{eff}}(ij)\, \delta E_q^{\text{ext}} = \sum_{j}^{N} \sum_{q,r}^{x\sim z} \alpha_{pr}(i)\, g_{rq}(ij)\, \delta E_q^{\text{ext}} \\
&= \sum_{q,r}^{x\sim z} \alpha_{pr}(i)\, f_{rq}(i)\, \delta E_q^{\text{ext}} = \sum_{r}^{x\sim z} \alpha_{pr}(i)\, \delta E_r^{\text{loc}}(i).
\end{aligned} \tag{5.17}$$

Equation (5.17) includes the variation of the local field $\delta E_r^{\text{loc}}(i)$ given by

$$\delta E_r^{\text{loc}}(i) = \sum_{j} \sum_{q}^{x\sim z} g_{rq}(ij)\, \delta E_q^{\text{ext}} = \sum_{q}^{x\sim z} f_{rq}(i)\, \delta E_q^{\text{ext}}. \tag{5.18}$$

Equation (5.18) clearly indicates that $\boldsymbol{f}(i)$ is the correction factor for the local field at the i-th molecule with respect to the uniform external field. By addition of the uniform external field, the dipole moment of the whole system $\delta \boldsymbol{M}$ changes by

$$\delta \boldsymbol{M} = \sum_{i=1}^{N} \delta \boldsymbol{\mu}(i) = \left(\sum_{i=1}^{N} \sum_{j=1}^{N} \boldsymbol{\alpha}^{\text{eff}}(ij) \right) \delta \boldsymbol{E}^{\text{ext}}.$$

This formula indicates the polarizability of the whole system $\boldsymbol{A} = \partial \boldsymbol{M} / \partial \boldsymbol{E}^{\text{ext}}$, which presents the response of $\boldsymbol{M}$ with respect to the uniform external field $\boldsymbol{E}^{\text{ext}}$, as

$$\boldsymbol{A} = \sum_{i=1}^{N} \sum_{j=1}^{N} \boldsymbol{\alpha}^{\text{eff}}(ij).$$

[3]In the following we employ $\boldsymbol{E}^{\text{loc}}$ with the superscript "loc" for $\boldsymbol{E}$ to distinguish the local field from other kinds of field.

The components of $\boldsymbol{A}$ are written by

$$A_{pq} = \sum_{i,j}^{N} \alpha_{pq}^{\text{eff}}(ij) = \sum_{i,j}^{N} \sum_{r}^{x\sim z} \alpha_{pr}(i) g_{rq}(ij) = \sum_{i}^{N} \sum_{r}^{x\sim z} \alpha_{pr}(i) f_{rq}(i). \quad (5.19)$$

Equation (5.19) shows that $\boldsymbol{A}$ is not simply the sum of $\boldsymbol{\alpha}(i)$, but involves the local field correction factor $\boldsymbol{f}(i)$.

5.2 Local Field Correction for $\chi^{(2)}$

Next, we irradiate the interface system with oscillating electric fields of visible light $\boldsymbol{E}^{\text{ext}}(\omega_1)$ and infrared light $\boldsymbol{E}^{\text{ext}}(\omega_2)$. The superscript "ext" designates the external fields of perturbation.[4] Then the nonlinear susceptibility of interface generates the sum-frequency polarization at the i-th molecule, namely $\boldsymbol{\mu}^{(2),0}(i, \Omega)$. Here we formulate the sum-frequency polarization generated at the interface from the sum-frequency polarization at each molecule.

Let us recall the microscopic derivation of nonlinear polarization from the second-order perturbation on the density matrix in Chap. 3. When we have defined the perturbation Hamiltonian $\hat{H}' = -\hat{\boldsymbol{\mu}} \cdot \boldsymbol{E}^{\text{ext}}(t)$ in Eq. (3.18) in the semiclassical theory, $\hat{\boldsymbol{\mu}}$ designates the dipole *operator* of the whole material system (including the interface) and $\boldsymbol{E}^{\text{ext}}(t)$ is the external electric field as the perturbation. The dipole operator for the whole material system $\hat{\boldsymbol{\mu}}$ is given by the sum of dipole *operators* of constituent molecules $\hat{\boldsymbol{\mu}}(j)$, i.e. $\hat{\boldsymbol{\mu}} = \sum_j \hat{\boldsymbol{\mu}}(j)$. The second-order polarization in Eq. (3.29) has been represented with the dipole operator $\hat{\boldsymbol{\mu}}$ and the second-order density matrix $\rho^{(2)}$ by

$$P_p^{(2),0}(t) = \text{Tr}\left[\hat{\mu}_p\, \rho^{(2)}(t)\right] = \text{Tr}\left[\left(\sum_j^N \hat{\mu}_p(j)\right) \rho^{(2)}(t)\right] = \sum_j^N \mu_p^{(2),0}(j, t). \quad (5.20)$$

We put the superscript 0 in the left-hand side to emphasize that Eq. (5.20) is the simple sum of the bare nonlinear polarizations of the molecules. This relation is converted to the frequency domain by taking the Fourier transformation and extracting the $\exp(-i\Omega t)$ component,

$$P_p^{(2),0}(\Omega) = \sum_j^N \mu_p^{(2),0}(j, \Omega). \quad (5.21)$$

[4] We note the distinction between $\boldsymbol{E}^{\text{ext}}(\omega)$ and the incident field $\boldsymbol{E}_I(\omega)$ in Chap. 2, as detailed in Sect. 5.3.

Since $\boldsymbol{P}^{(2),0}(\Omega)$ in Eq. (5.21) is induced by $\boldsymbol{E}^{\text{ext}}(\omega_1)$ and $\boldsymbol{E}^{\text{ext}}(\omega_2)$, we could define a nonlinear susceptibility $\chi_{pqr}^{(2),0}(\Omega, \omega_1, \omega_2)$ by

$$P_p^{(2),0}(\Omega) = \chi_{pqr}^{(2),0}(\Omega, \omega_1, \omega_2) E_q^{\text{ext}}(\omega_1) E_r^{\text{ext}}(\omega_2). \tag{5.22}$$

As discussed in Chap. 3, the nonlinear susceptibility $\boldsymbol{\chi}^{(2),0}$ consists of the vibrationally resonant and nonresonant parts. The resonant part $\boldsymbol{\chi}^{(2),0,\text{res}}$ is represented with the classical time correlation function between the polarizability $\boldsymbol{A}$ and dipole moment $\boldsymbol{M}$ of the whole system (i.e. A_{pq} in Eq. (5.19) and M_r in Eq. (5.13)),

$$\chi_{pqr}^{(2),0,\text{res}}(\Omega, \omega_1, \omega_2) = \frac{i\omega_2}{k_B T} \int_0^\infty dt \left\langle A_{pq}(t) M_r \right\rangle \exp(i\omega_2 t), \tag{5.23}$$

where

$$A_{pq} = \sum_{j=1}^{N} \sum_{r}^{x\sim z} \alpha_{pr}(j) f_{rq}(j), \qquad M_r = \sum_{k=1}^{N} \sum_{s}^{x\sim z} \mu_s^{0'}(k) f_{sr}(k).$$

We note that the local field correction factors $f_{rq}(i)$ and $f_{sr}(k)$ in the above equations are associated to the visible and infrared fields, respectively. This role becomes evident by multiplying the respective fields in Eq. (5.22) as follows,

$$A_{pq} E_q^{\text{ext}}(\omega_1) = \sum_{j=1}^{N} \sum_{r}^{x\sim z} \alpha_{pr}(j) f_{rq}(j) E_q^{\text{ext}}(\omega_1) = \sum_{j=1}^{N} \sum_{r}^{x\sim z} \alpha_{pr}(j) E_r^{\text{loc}}(j, \omega_1),$$

$$M_r E_r^{\text{ext}}(\omega_2) = \sum_{k=1}^{N} \sum_{s}^{x\sim z} \mu_s^{0'}(k) f_{sr}(k) E_r^{\text{ext}}(\omega_2) = \sum_{k=1}^{N} \sum_{s}^{x\sim z} \mu_s^{0'}(k) E_s^{\text{loc}}(k, \omega_2).$$

These equations apparently indicate that the $\boldsymbol{f}$ factor changes the external visible/infrared field $\boldsymbol{E}^{\text{ext}}(\omega_{1/2})$ to the local field $\boldsymbol{E}^{\text{loc}}(\omega_{1/2})$, respectively.

The above formulas (5.20), (5.21), (5.22) have treated the bare nonlinear polarizations induced by the second-order perturbation. However, the bare nonlinear polarization $\boldsymbol{\mu}^{(2),0}(i, \Omega)$ interacts each other by the dielectric coupling in the same way as discussed in Sect. 5.1, and thereby modifies itself from $\boldsymbol{\mu}^{(2),0}(i, \Omega)$ to $\boldsymbol{\mu}^{(2)}(i, \Omega)$. This effect of dielectric coupling has to be considered to derive the observed polarization at the sum frequency Ω. The relation between $\boldsymbol{\mu}^{(2),0}(i, \Omega)$ and $\boldsymbol{\mu}^{(2)}(i, \Omega)$ is given after Eq. (5.8) in Sect. 5.1 by

$$\mu_p^{(2)}(i, \Omega) = \mu_p^{(2),0}(i, \Omega) - \sum_{q}^{x\sim z} \alpha_{pq}(i) \sum_{j(\neq i)}^{N} \sum_{r}^{x\sim z} T_{qr}(ij) \mu_r^{(2)}(j, \Omega),$$

or

$$\begin{aligned}\boldsymbol{\mu}^{(2)}(\Omega) &= \boldsymbol{\mu}^{(2),0}(\Omega) - \boldsymbol{\alpha T \mu}^{(2)}(\Omega) \\ &= [\mathbf{1} + \boldsymbol{\alpha T}]^{-1}\boldsymbol{\mu}^{(2),0}(\Omega) = \boldsymbol{g}^T\boldsymbol{\mu}^{(2),0}(\Omega)\end{aligned}$$

in the matrix form, where the external field $\boldsymbol{E}^0(\Omega)$ at the sum frequency Ω in Eq. (5.8) is assumed to be absent. Consequently, the sum-frequency polarization of the whole system $P_p^{(2)}(\Omega)$ is

$$P_p^{(2)}(\Omega)=\sum_i^N \mu_p^{(2)}(i,\Omega)=\sum_{i,j}^N\sum_s^{x\sim z} g_{sp}(ji)\mu_s^{(2),0}(j,\Omega)=\sum_j^N\sum_s^{x\sim z} f_{sp}(j)\mu_s^{(2),0}(j,\Omega). \tag{5.24}$$

The second-order polarization including the dielectric coupling is thus represented in the semi-classical manner by

$$P_p^{(2)}(t) = \sum_j^N\sum_s^{x\sim z} f_{sp}(j)\,\mathrm{Tr}\left[\hat{\mu}_s(j)\rho^{(2)}(t)\right]. \tag{5.25}$$

The above argument clarifies the distinction between $\boldsymbol{P}^{(2),0}(\Omega)$ in Eq. (5.21) and $\boldsymbol{P}^{(2)}(\Omega)$ in Eq. (5.24) for describing the sum-frequency polarization. Accordingly, we introduce another form of nonlinear susceptibility $\boldsymbol{\chi}^{(2)}$, instead of $\boldsymbol{\chi}^{(2),0}$, which corresponds to $\boldsymbol{P}^{(2)}(\Omega)$ as

$$P_p^{(2)}(\Omega) = \chi_{pqr}^{(2)}(\Omega,\omega_1,\omega_2)E_q^{\text{ext}}(\omega_1)E_r^{\text{ext}}(\omega_2). \tag{5.26}$$

The difference between $\boldsymbol{\chi}^{(2),0}$ and $\boldsymbol{\chi}^{(2)}$ comes from the difference in the definition of the sum-frequency polarization, $\boldsymbol{P}^{(2),0}(\Omega)$ and $\boldsymbol{P}^{(2)}(\Omega)$. The observed sum-frequency polarization and the corresponding nonlinear susceptibility are represented by the latter quantities, $\boldsymbol{P}^{(2)}(\Omega)$ and $\boldsymbol{\chi}^{(2)}$, in Eq. (5.26).

The vibrationally resonant part of $\boldsymbol{\chi}^{(2)}$ is therefore expressed as

$$\chi_{pqr}^{(2),\text{res}}(\Omega,\omega_1,\omega_2) = \frac{i\omega_2}{k_B T}\int_0^\infty dt\left\langle A_{\text{eff},pq}(t)M_r\right\rangle\exp(i\omega_2 t), \tag{5.27}$$

where

$$\begin{aligned}A_{\text{eff},pq} &= \sum_{j=1}^N\sum_{r,s}^{x\sim z} f_{sp}(j)\alpha_{sr}(j)f_{rq}(j) \equiv \sum_{j=1}^N \alpha_{\text{eff}}(j), \\ M_r &= \sum_{k=1}^N\sum_s^{x\sim z}\mu_s^{0'}(k)f_{sr}(k) \equiv \sum_{k=1}^N \mu_{\text{eff}}(k).\end{aligned} \tag{5.28}$$

The three local field correction factors, $f_{sp}(j)$, $f_{rq}(j)$ and $f_{sr}(k)$, in Eq. (5.28) are associated to the sum-frequency Ω, visible ω_1, and infrared ω_2 fields, respectively. The computational analysis of SFG spectra by MD simulation employs the expression of Eq. (5.27).

To summarize the above computational procedure of $\boldsymbol{A}_{\text{eff}}$ and $\boldsymbol{M}$, the calculation of $\boldsymbol{\chi}^{(2)}$ by MD simulation requires the following properties of constituent molecules at each time:

(i) dipole-dipole coupling tensor $\boldsymbol{T}(ij)$,
(ii) external field $\boldsymbol{E}^0(j)$,
(iii) polarizability tensor $\boldsymbol{\alpha}(j)$,
(iv) permanent dipole moment vector $\boldsymbol{\mu}^0(j)$.

Among these properties, (i) and (ii) are readily obtained from the instantaneous molecular configuration $\{\boldsymbol{r}(j)\}$. Conventional force fields for MD simulation allow for calculating $\boldsymbol{E}^0(j)$ at each time step, as it is a necessary quantity to evaluate electrostatic forces. On the other hand, usual force fields of MD do not provide (iii) and (iv). We have to obtain the instantaneous values of $\boldsymbol{\alpha}(j)$ and $\boldsymbol{\mu}^0(j)$ at each time step in addition to the force field calculations. The modeling of $\boldsymbol{\alpha}(j)$ and $\boldsymbol{\mu}^0(j)$ will be discussed in the next Chap. 6.

5.3 Interfacial Dielectric Constant ε'

The preceding argument on the local field is quite relevant to the microscopic understanding of the interfacial dielectric constant ε' mentioned in Chap. 2. As we have summarized in Sect. 2.3 the factors to determine the SFG and SHG spectra, the factors related to interface properties include the nonlinear susceptibility $\boldsymbol{\chi}^{(2)}$ and the dielectric constant of the interface ε'. The preceding Sect. 5.2 provided microscopic formulation of $\boldsymbol{\chi}^{(2)}$ on the basis of the molecular theory of local field. In relation to the above argument, we discuss here the interfacial dielectric constant ε' from the same molecular viewpoint. The microscopic local field is connected to the concept of ε' in the phenomenological three-layer model.

External field vs. local field Let us consider the optical geometry of SFG measurement in Fig. 2.1. We notice that Fig. 2.1 adopts the three-layer model and assumes the infinite thin layer of interface at $z \approx 0$. However, here we discuss the structure and properties in the interface layer. In the following discussion the medium α is assumed to be vacuum for simplicity, and hence $\varepsilon^\alpha = 1$, though extension to other situations is straightforward.

For microscopic treatment of electric fields at the interface, we distinguish the incident field $\boldsymbol{E}_I$, external field $\boldsymbol{E}^{\text{ext}}$, and local field $\boldsymbol{E}^{\text{loc}}$. Suppose a plane wave $\boldsymbol{E}_I^\alpha(\omega_f)$ ($f = 1$ or 2) in Eq. (2.11) is incident from the medium α (vacuum) to the interface, the interface region near $z \approx 0$ feels the *external* field of the frequency ω_f, $\boldsymbol{E}^{\text{ext}}(\omega_f)$. This is a superposition of the incident and reflected radiation fields, and thus related to the incident field through the optical factor $\boldsymbol{L}_I(\omega_f)$,

$$\boldsymbol{E}^{\text{ext}}(\omega_f) = \boldsymbol{L}_I(\omega_f)\boldsymbol{E}_I^{\alpha}(\omega_f) \qquad (\text{for } f = 1 \text{ or } 2), \tag{5.29}$$

where

$$\boldsymbol{L}_I(\omega_f) = \begin{pmatrix} \dfrac{2q^\beta}{\varepsilon^\beta q^\alpha + q^\beta} & 0 & 0 \\ 0 & \dfrac{2q^\alpha}{q^\alpha + q^\beta} & 0 \\ 0 & 0 & \dfrac{2\varepsilon^\beta q^\alpha}{\varepsilon^\beta q^\alpha + q^\beta} \end{pmatrix}. \tag{5.30}$$

The notations included in Eq. (5.30) are common with those of the Fresnel factor $\boldsymbol{F}^{i\to j}$ in Eq. (2.18). $\boldsymbol{F}^{i\to j}$ for the three-layer model coincides with $\boldsymbol{L}_I$ by setting $\varepsilon^\alpha = \varepsilon' = 1$. $\boldsymbol{E}^{\text{ext}}(\omega_f)$ in Eq. (5.29) omits its z dependence, since we treat the field near the interface ($z \approx 0$) in a microscopic dimension much shorter than the light wavelength.

The external field at the interface, $\boldsymbol{E}^{\text{ext}}(\omega_f)$, is *not* identical to the local field that is felt by an individual molecule there, due to the local field correction in Sect. 5.1. The microscopic local field near the interface $\boldsymbol{E}^{\text{loc}}$ is related to $\boldsymbol{E}^{\text{ext}}$ with the local field factor $\boldsymbol{f}(z, \omega_f)$,[5]

$$\boldsymbol{E}^{\text{loc}}(z, \omega_f) = \boldsymbol{f}(z, \omega_f)\boldsymbol{E}^{\text{ext}}(\omega_f) = \boldsymbol{f}(z, \omega_f)\boldsymbol{L}_I(\omega_f)\boldsymbol{E}_I^{\alpha}(\omega_f). \tag{5.31}$$

$\boldsymbol{E}^{\text{loc}}(z, \omega_f)$ and $\boldsymbol{f}(z, \omega_f)$ in Eq. (5.31) depend on the z coordinate in the microscopic dimension ($z \approx 0$) due to the inhomogeneous structure of the interface, in contrast to $\boldsymbol{E}^{\text{ext}}(\omega_f)$. $\boldsymbol{f}(z, \omega_f)$ is a diagonal tensor for an azimuthally isotropic interface, where $f_x = f_y$ and f_z is distinct from the other two, for symmetry reasons.

The above formulation of the local field involves both the optical factor $\boldsymbol{L}$ and the local field factor $\boldsymbol{f}$, which have distinct physical origins of interactions. $\boldsymbol{L}$ is associated to the radiation field emitted by oscillating dipoles, while $\boldsymbol{f}$ is to the near field or the dipole-dipole coupling. The $\boldsymbol{f}(z, \omega_f)$ in Eq. (5.31) is essentially equivalent to the microscopic quantity $\boldsymbol{f}(i)$ of Eq. (5.14) in Sect. 5.1. The latter is defined at an instantaneous molecular configuration, while the former is obtained by statistically averaging the molecular configurations of the interface.

Local field and dielectric constant We deal with the z dependence of $\boldsymbol{f}(z, \omega_f)$ in relation to the dielectric constant. First, consider the bulk region ($z \ll 0$) of uniform medium with a dielectric constant ε. Then the classical Lorentz model [3] illustrated in Fig. 5.1a, b leads to

$$f_x = f_y = \frac{\varepsilon + 2}{3} \quad \text{and} \quad f_z = \frac{\varepsilon + 2}{3\varepsilon}. \tag{5.32}$$

[5]Note that Ref. [12] denotes the local field factor by $\boldsymbol{s}$.

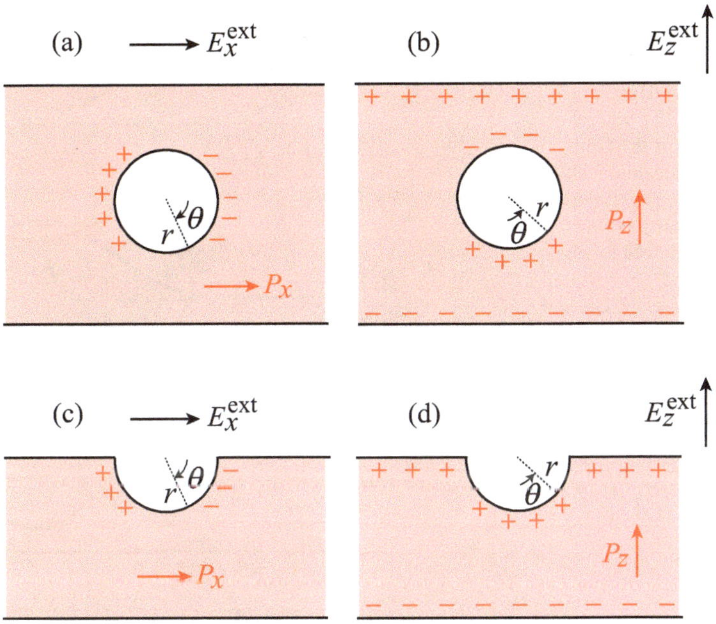

Fig. 5.1 Lorentz models in the bulk (**a**, **b**) and at the interface (**c**, **d**). (**a**, **c**) External electric field E_x^{ext} parallel to the interface, (**b**, **d**) E_z^{ext} perpendicular to the interface

We notice that ε corresponds to the ratio of the local field factors, $\varepsilon = f_x/f_z$. The dielectric constant acts as a screening factor of the electric field normal to a surface by induced surface charges. In a similar manner, Zhuang et al. [16] applied the Lorentz model to describe the dielectric property at the interface, and proposed

$$\varepsilon' = \frac{\varepsilon(\varepsilon+5)}{2(2\varepsilon+1)}. \tag{5.33}$$

[Problem 5.1] Let us derive ε' in Eq. (5.33) on the basis of the Lorentz model illustrated in Fig. 5.1a–d.

(a, b) Bulk First, suppose a spherical cavity of radius r embedded in the bulk of a slab with a dielectric constant ε, and put the system in an electric field E_x^{ext} or E_z^{ext}. Show that the local fields inside the cavity in the two cases are

$$E_x^{loc} = f_x E_x^{ext} = \frac{\varepsilon+2}{3} E_x^{ext} \quad \text{and} \quad E_z^{loc} = f_z E_z^{ext} = \frac{\varepsilon+2}{3\varepsilon} E_z^{ext}. \tag{5.32}$$

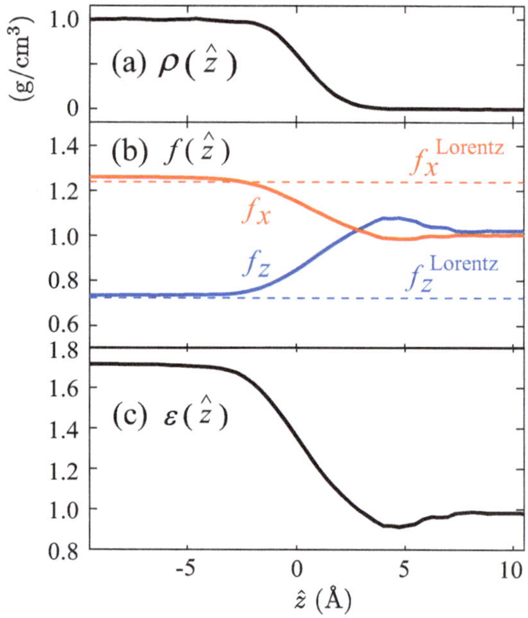

Fig. 5.2 MD results of dielectric properties at air/water interface as a function of the depth coordinate $\hat{z}$ [12], where $\hat{z} = 0$ stands for the Gibbs dividing surface, and $\hat{z} > 0$ ($\hat{z} < 0$) for the gas (liquid) phase. (**a**) density profile $\rho(\hat{z})$. (**b**) local field factors $f_x(\hat{z})$ (red) and $f_z(\hat{z})$ (blue) at optical frequency. Dashed horizontal lines of f_x^{Lorentz} and f_z^{Lorentz} show the results of Lorentz model in Eq. (5.32). (**c**) dielectric constant $\varepsilon(\hat{z}) = f_x(\hat{z})/f_z(\hat{z})$. (Reprinted with permission from Ref. [12]. Copyright 2011, American Institute of Physics)

(c, d) Surface Next, suppose a hemisphere cavity at the interface. Show that the local fields E_x^{loc} and E_z^{loc} at the center of the hemisphere in the two cases are

$$E_x^{\text{loc}} = f_x^{\text{surf}} E_x^{\text{ext}} = \frac{\varepsilon + 5}{6} E_x^{\text{ext}} \quad \text{and} \quad E_z^{\text{loc}} = f_z^{\text{surf}} E_z^{\text{ext}} = \frac{2\varepsilon + 1}{3\varepsilon} E_z^{\text{ext}}. \tag{5.34}$$

Take the ratio of the local field factors to derive $\varepsilon' = f_x^{\text{surf}}/f_z^{\text{surf}}$.

We should note, however, that the quantitative applicability of Eq. (5.33) is not clear because of the limitation of the dielectric continuum model in a molecular scale. Alternatively, the molecular theory of local field in Sect. 5.1 allows us to directly calculate $\boldsymbol{f}(z, \omega)$ near the surface by MD simulation [12]. Figure 5.2 shows an example of the MD results of $\boldsymbol{f}(z, \omega)$ at optical frequency as a function of z for air/water interface. Panel (b) shows that $\boldsymbol{f} \to 1$ at $z \gg 0$ (gas phase), while it approaches the values of the Lorentz model in Eq. (5.32) at $z \ll 0$ (liquid phase). The transition of $\boldsymbol{f}$ takes place within a nanometer of the interface region. The ratio $\varepsilon(z) = f_x(z)/f_z(z)$ is plotted as a function of z in Panel (c). The value varies from $\varepsilon = 1$ (gas) to 1.72 (liquid), and its transition behavior is similar with that of the density profile $\rho(z)$ in Panel (a).

Microscopic $\boldsymbol{\varepsilon'}$ The interfacial dielectric constant ε' is a phenomenological parameter used in the three-layer model in Chap. 2. Essentially it accounts for the local field effect on the nonlinear polarization at the interface for the SFG or SHG spectra. Now the local field at the interface is formulated in the microscopic level, ε' can be represented as well.

As we discussed in Sect. 5.2, the bare second-order polarization induced at the interface, $\boldsymbol{P}^{(2),0}(\Omega)$ in Eq. (5.21), is influenced by the local field and modified to $\boldsymbol{P}^{(2)}(\Omega)$ in Eq. (5.24). The two polarization are given by

$$P_p^{(2),0}(\Omega) = \sum_j^N \mu_p^{(2),0}(j,\Omega), \tag{5.21}$$

$$P_p^{(2)}(\Omega) = \sum_j^N \sum_s^{x\sim z} f_{sp}(j)\, \mu_s^{(2),0}(j,\Omega), \tag{5.24}$$

where $p = x \sim z$ and j denotes the constituent molecules. Equations (5.21) and (5.24) express the difference between $\boldsymbol{P}^{(2),0}$ and $\boldsymbol{P}^{(2)}$ with the local field factors of molecules $f_{sp}(j)$ (or $\boldsymbol{f}(j)^T$). The ratio between $\boldsymbol{P}^{(2),0}$ and $\boldsymbol{P}^{(2)}$ is thus represented by

$$f_p^{\text{surf}} = \frac{P_p^{(2)}(\Omega)}{P_p^{(2),0}(\Omega)} = \frac{\displaystyle\sum_j^N \sum_s^{x\sim z} f_{sp}(j)\, \mu_s^{(2),0}(j,\Omega)}{\displaystyle\sum_j^N \mu_p^{(2),0}(j,\Omega)}. \tag{5.35}$$

The ratio f_p^{surf} accounts for the local field correction for the nonlinear polarization at the interface. Therefore, it allows for determining the effective local field factor of the interface $\boldsymbol{f}^{\text{surf}}$ for the SHG and SFG spectroscopy in the microscopic level. $\boldsymbol{f}^{\text{surf}}$ should be a diagonal tensor,

$$\boldsymbol{f}^{\text{surf}} = \begin{pmatrix} f_x^{\text{surf}} & & \\ & f_y^{\text{surf}} & \\ & & f_z^{\text{surf}} \end{pmatrix},$$

for an azimuthally isotropic interface with $f_x^{\text{surf}} = f_y^{\text{surf}}$ for symmetry reasons, and the diagonal elements are presented by Eq. (5.35). The interfacial dielectric constant ε' is accordingly defined as $\varepsilon' = f_x^{\text{surf}}/f_z^{\text{surf}}$, using $\boldsymbol{f}^{\text{surf}}$ in Eq. (5.35) instead of Eq. (5.34).

Equation (5.35) shows that the local field factor of the interface $\boldsymbol{f}^{\text{surf}}$ is the average of the local field factor of j-th molecule $\boldsymbol{f}(j)$ with a weight of its nonlinear polarization $\boldsymbol{\mu}^{(2),0}(j)$. This definition of $\boldsymbol{f}^{\text{surf}}$ provides a microscopic basis to the phenomenological parameter ε'.

We also note that the $\boldsymbol{\chi}^{(2)}$ formula of Eqs. (5.27) and (5.28) in Sect. 5.2 incorporates the effect of local field through molecular interactions. Therefore, the computation of $\boldsymbol{\chi}^{(2)}$ by Eq. (5.27) or (5.28) obviates explicit use of the phenomenological parameter ε' to evaluate the nonlinear polarization at interface.

5.4 Solutions to Problems

5.4.1 Interfacial Dielectric Constant

[Problem 5.1] Let us derive ε' in Eq. (5.33) on the basis of the Lorentz model illustrated in Fig. 5.1a–d.

(a, b) Bulk First, suppose a spherical cavity of radius r embedded in the bulk of a slab with a dielectric constant ε, and put the system in an electric field E_x^{ext} or E_z^{ext}. Show that the local fields inside the cavity in the two cases are

$$E_x^{\text{loc}} = f_x E_x^{\text{ext}} = \frac{\varepsilon + 2}{3} E_x^{\text{ext}} \quad \text{and} \quad E_z^{\text{loc}} = f_z E_z^{\text{ext}} = \frac{\varepsilon + 2}{3\varepsilon} E_z^{\text{ext}}. \tag{5.32}$$

(c, d) Surface Next, suppose a hemisphere cavity at the interface. Show that the local fields E_x^{loc} and E_z^{loc} at the center of the hemisphere in the two cases are

$$E_x^{\text{loc}} = f_x^{\text{surf}} E_x^{\text{ext}} = \frac{\varepsilon + 5}{6} E_x^{\text{ext}} \quad \text{and} \quad E_z^{\text{loc}} = f_z^{\text{surf}} E_z^{\text{ext}} = \frac{2\varepsilon + 1}{3\varepsilon} E_z^{\text{ext}}. \tag{5.34}$$

Take the ratio of the local field factors to derive $\varepsilon' = f_x^{\text{surf}} / f_z^{\text{surf}}$.

Here we discuss the relation between the local field and the dielectric constant. The relation is distinct in the bulk and at the interface, and the difference could be used to define the dielectric constant at the interface. The local field inside the cavity is evaluated by the Lorentz model [3].

(1) Bulk In case (a),

(a)

$$\begin{aligned} E_x^{\text{loc}} &= E_x^{\text{ext}} + \int_{\text{sphere}} d\sigma \frac{P_x \cos\theta}{r^2} \cos\theta = E_x^{\text{ext}} + \frac{4\pi}{3} P_x \\ &= \frac{\varepsilon + 2}{3} E_x^{\text{ext}} = f_x E_x^{\text{ext}}, \end{aligned} \tag{5.36}$$

where σ denotes the surface of the cavity, and the polarization $P_x = \dfrac{\varepsilon - 1}{4\pi} E_x^{\text{ext}}$. In case (b),

(b)

$$E_z^{\text{loc}} = E_z^{\text{ext}} + \int_{\text{sphere}} d\sigma \frac{P_z \cos\theta}{r^2} \cos\theta - 4\pi P_z = E_z^{\text{ext}} - \frac{8\pi}{3} P_z$$

$$= \frac{\varepsilon + 2}{3\varepsilon} E_z^{\text{ext}} = f_z E_z^{\text{ext}}, \tag{5.37}$$

where $P_z = \frac{\varepsilon - 1}{4\pi}(E_z^{\text{ext}} - 4\pi P_z) = \frac{\varepsilon - 1}{4\pi\varepsilon} E_z^{\text{ext}}$. We note that the induced charge at the surface of the slab is involved in case (b). By comparing Eqs. (5.36) and (5.37), the dielectric constant ε corresponds to the ratio of the two local fields, $\varepsilon = f_x/f_z$.

(2) Surface Next, we adopt the similar argument on the local field at the interface using a hemisphere cavity in Fig. 5.1c, d. In case (c),

$$\begin{aligned}
\text{(c)} \qquad E_x^{\text{loc}} &= E_x^{\text{ext}} + \int_{\text{hemisphere}} d\sigma \frac{P_x \cos\theta}{r^2} \cos\theta \\
&= E_x^{\text{ext}} + \int_0^{\pi} d\phi \int_0^{\pi} \sin\theta \, d\theta \, r^2 \frac{P_x \cos^2\theta}{r^2} \\
&= E_x^{\text{ext}} + \frac{2\pi}{3} P_x = \frac{\varepsilon + 5}{6} E_x^{\text{ext}} = f_x^{\text{surf}} E_x^{\text{ext}}.
\end{aligned} \tag{5.38}$$

In case (d),

$$\begin{aligned}
\text{(d)} \qquad E_z^{\text{loc}} &= E_z^{\text{ext}} + \int_{\text{hemisphere}} d\sigma \frac{P_z \cos\theta}{r^2} \cos\theta - 2\pi P_z \\
&= E_z^{\text{ext}} + \int_0^{2\pi} d\phi \int_0^{\pi/2} \sin\theta \, d\theta \, r^2 \frac{P_z \cos^2\theta}{r^2} - 2\pi P_z \\
&= E_z^{\text{ext}} - \frac{4\pi}{3} P_z = \frac{2\varepsilon + 1}{3\varepsilon} E_z^{\text{ext}} = f_z^{\text{surf}} E_z^{\text{ext}}.
\end{aligned} \tag{5.39}$$

Therefore, the dielectric constant of the interface ε' is

$$\varepsilon' = f_x^{\text{surf}} / f_z^{\text{surf}} = \left(\frac{\varepsilon + 5}{6}\right) \Big/ \left(\frac{2\varepsilon + 1}{3\varepsilon}\right) = \frac{\varepsilon(\varepsilon + 5)}{2(2\varepsilon + 1)}. \tag{5.33}$$

Bibliography

1. Applequist J, Carl JR, Fung KK (1972) Atom dipole interaction model for molecular polarizability. Application to polyatomic molecules and determination of atom polarizabilities. J Am Chem Soc 94:2952–2960
2. Burnham CJ, Xantheas SS (2002) Development of transferable interaction models for water. III. Reparametrization of an all-atom polarizable rigid model (TTM2-R) from first principles. J Chem Phys 116:1500–1510
3. Frohlich H (1960) Theory of dielectrics, 2nd edn. Clarendon Press, Oxford
4. Gray CG, Gubbins KE (1984) Theory of molecular fluids: fundamentals, vol 1. Clarendon, Oxford

5. Lamoureux G, Roux B (2003) Modeling induced polarization with classical Drude oscillators: theory and molecular dynamics simulation algorithm. J Chem Phys 119:3025–3039
6. Morita A, Hynes JT (2002) A theoretical analysis of the sum frequency generation spectrum of the water surface II. Time-dependent approach. J Phys Chem B 106:673–685
7. Morita A, Ishiyama T (2008) Recent progress in theoretical analysis of vibrational sum frequency generation spectroscopy. Phys Chem Chem Phys 10:5801–5816
8. Morita A, Kato S (1997) Ab initio molecular orbital theory on intramolecular charge polarization: effect of hydrogen abstraction on the charge sensitivity of aromatic and nonaromatic species. J Am Chem Soc 119:4021–4032
9. Mukamel S (1995) Principles of nonlinear optical spectroscopy. Oxford University Press, New York
10. Rick SW, Stuart SJ, Berne BJ (1994) Dynamical fluctuating charge force fields: application to liquid water. J Chem Phys 101:6141–6156
11. Rullmann JAC, van Duijnen PT (1988) A polarizable water model for calculation of hydration energies. Mol Phys 63:451–475
12. Shiratori K, Morita A (2011) Molecular theory on dielectric constant at interfaces: a molecular dynamics study of the water/vapor interface. J Chem Phys 134:234705
13. Sprik M, Klein ML (1988) A polarizable model for water using distributed charge sites. J Chem Phys 89:7556–7560
14. Straatsma TP, McCammon JA (1990) Molecular dynamics simulations with interaction potentials including polarization. Development of a noniterative method and application to water. Mol Sim 5:181–192
15. Yu H, van Gunsteren WF (2004) Charge-on-spring polarizable water models revisited: from water clusters to liquid water to ice. J Chem Phys 121:9549–9564
16. Zhuang X, Miranda PB, Kim D, Shen YR (1999) Mapping molecular orientation and conformation at interfaces by surface nonlinear optics. Phys Rev B 59:12632–12640

Chapter 6
Charge Response Kernel for Electronic Polarization

Abstract The charge response kernel (CRK) is a non-empirical and universal method to represent polarization of molecules. It is particularly powerful to the MD calculations of SFG spectroscopy. This chapter explains the CRK theory and its application to polarizable MD simulation. The CRK is rigorously formulated as a second-order derivative of electronic energy on the basis of electronic structure theory, either the ab initio molecular orbital or density functional theory. The calculated CRK offers a general method of fluctuation charges to MD simulation, and also represents the dipole, polarizability and second-order nonlinear susceptibility on the same footing with equivalent accuracy to the underlying ab initio calculations.

Keywords Charge response kernel (CRK) · Electrostatic potential (ESP) charge · Coupled perturbed Hartree-Fock (CPHF) · Polarizable force field

This chapter deals with molecular modeling methods of $\boldsymbol{A}_{\text{eff}}$ and $\boldsymbol{M}$ in Eq. (5.28) for actual interface systems. To construct $\boldsymbol{A}_{\text{eff}}$ and $\boldsymbol{M}$ from molecular simulation, we need the molecular polarizability tensor $\boldsymbol{\alpha}(j)$ and dipole moment vector $\boldsymbol{\mu}(j)$. These molecular properties should be defined at instantaneous configuration, as they vary with molecular rotation and vibration. Since conventional force fields of MD do not offer these properties, we need to evaluate them at each time step in addition to the force field calculations. The modeling of $\boldsymbol{\alpha}(j)$ and $\boldsymbol{\mu}(j)$ is the heart of the SFG calculation by MD simulation, and the reliability of the results critically depends on the accuracy of this modeling. One straightforward method to describe them would be to carry out direct electronic structure calculation of these quantities at each time step of MD simulation like the ab initio MD simulation [4, 12, 14]. It could obviate the modeling in principle, whereas the computational cost is substantially larger than that of the usual ab initio MD simulation, since computation of polarizability is more demanding than that of energy or force.

Here we present a general solution for the modeling of $\boldsymbol{\alpha}(j)$ and $\boldsymbol{\mu}(j)$ on the basis of the charge response kernel (CRK) theory [3, 7, 9]. This theory provides a universal method to describe molecular polarization from non-empirical

A. Morita, *Theory of Sum Frequency Generation Spectroscopy*,
Lecture Notes in Chemistry 97, https://doi.org/10.1007/978-981-13-1607-4_6

calculations of ab initio molecular orbital (MO) or density functional theory (DFT). This is particularly useful to SFG calculations, as it reconciles the ab initio accuracy in the above modeling and the computational efficiency of classical MD simulation.

6.1 Charge Response Kernel (CRK)

The charge response kernel (CRK) is introduced to represent electronic polarization in a site representation of molecular model. As illustrated in the site model in Fig. 6.1, CRK is defined with the following derivative quantity [7],

$$K_{ab} = \left(\frac{\partial Q_a}{\partial V_b} \right), \tag{6.1}$$

where Q_a is the partial charge at the site a, and V_b is the electrostatic potential at the site b. The locations of the sites are arbitrary in principle, though they are usually set at nuclear positions. Equation (6.1) expresses the intramolecular redistribution of electron density induced by the external electrostatic potential. When the sites a and b are different, K_{ab} represents non-local response of electron density.

The CRK theory defines Eq. (6.1) non-empirically on the basis of ab initio MO or DFT. The partial charge Q_a is defined as the electrostatic potential (ESP) charge [2, 6], which is derived from quantum chemical calculation of electron density so as to reproduce the surrounding electrostatic potential by the least square fitting. The ESP charge is suitable to represent the intermolecular interactions, and is widely used for molecular simulations. We employ the ESP charge in the following, though other definitions of Q_a are also feasible as long as Q_a is uniquely determined from the electron density.

The CRK theory offers a straightforward route of calculating K_{ab} in Eq. (6.1) by extending the definition of Q_a. The calculation method of K_{ab} is analogous to that of Hessian or polarizability, as these quantities are commonly considered as response quantities to external perturbation (nuclear position, electric field, etc.), or second-order derivatives of the electronic energy.

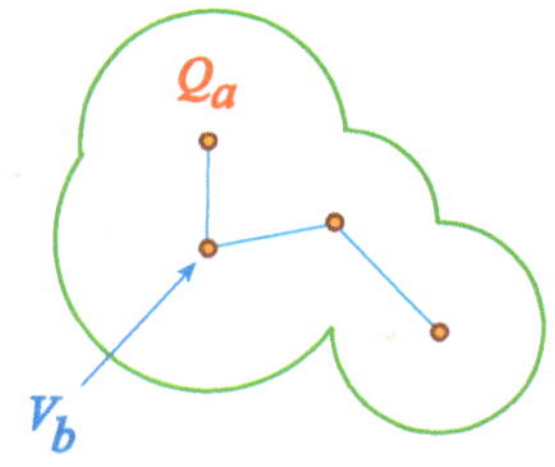

Fig. 6.1 Interaction site model of a molecule. Partial charge at site a and electrostatic potential at site b are illustrated

6.2 Electronic Structure Theory of CRK

The CRK model is defined as a second-order derivative quantity in an arbitrary electronic structure theory of ab initio MO or DFT. In the present section, it is formulated with the closed-shell Hartree-Fock wavefunction for an example. Application to other levels of theory, including DFT [3], is straightforward once the perturbation Hamiltonian is given in the following.

Hamiltonian with perturbation The whole Hamiltonian of the electronic structure with the perturbation is given by

$$\hat{H} = \hat{H}_0 + \hat{H}', \tag{6.2}$$

where $\hat{H}_0$ is the conventional (non-relativistic) electronic Hamiltonian for an isolated molecule, consisting of the kinetic energy, nuclear attraction and electron repulsion,

$$\begin{aligned}\hat{H}_0 &= \sum_{i=1}^{N_e}\left(-\frac{\hbar^2}{2m_e}\nabla_i^2 - \sum_c^{\text{nuc}}\frac{Z_c e^2}{|\boldsymbol{r}_i - \boldsymbol{R}_N(c)|}\right) + \sum_i^{N_e}\sum_{>j}\frac{e^2}{|\boldsymbol{r}_i - \boldsymbol{r}_j|} \\ &= \sum_{i=1}^{N_e}\hat{h}_0(i) + \sum_i^{N_e}\sum_{>j}\frac{e^2}{|\boldsymbol{r}_i - \boldsymbol{r}_j|},\end{aligned} \tag{6.3}$$

where N_e is the number of electrons in the molecule, m_e is the electron mass, and $\boldsymbol{r}_i$, $\boldsymbol{r}_j$ are the coordinates of electrons. Z_c and $\boldsymbol{R}_N(c)$ are the charge and coordinate of the c-th nucleus, respectively. Equation (6.3) consists of the sum of the one-electron operators,

$$\hat{h}_0(i) = -\frac{\hbar^2}{2m_e}\nabla_i^2 - \sum_c^{\text{nuc}}\frac{Z_c e^2}{|\boldsymbol{r}_i - \boldsymbol{R}_N(c)|}, \tag{6.4}$$

and the electron-electron repulsion terms, $\sum_i\sum_{>j}\frac{e^2}{|\boldsymbol{r}_i - \boldsymbol{r}_j|}$.

$\hat{H}'$ in Eq. (6.2) is the perturbation Hamiltonian by the external electrostatic potential in the site representation, i.e.

$$\hat{H}' = \sum_a^{\text{site}}\hat{Q}_a V_a, \tag{6.5}$$

where V_a is the electrostatic potential at the site a, and $\hat{Q}_a$ is the ESP partial charge operator at the site a [7]. V_a is treated as the parameter for the external perturbation on the electronic structure. The partial charge Q_a is given with the expected value at an electronic (many-body) wavefunction Ψ by

$$Q_a = \left\langle \Psi | \hat{Q}_a | \Psi \right\rangle .$$

The partial charge Q_a is related to the derivative of the total energy $\mathcal{E}$ by the Hellmann-Feynman theorem,

$$\frac{\partial \mathcal{E}}{\partial V_a} = \frac{\partial}{\partial V_a} \left\langle \Psi | \hat{H} | \Psi \right\rangle = \left\langle \Psi | \frac{\partial \hat{H}}{\partial V_a} | \Psi \right\rangle = \left\langle \Psi | \hat{Q}_a | \Psi \right\rangle = Q_a. \tag{6.6}$$

This relation (6.6) is valid when the wavefunction Ψ satisfies the variational principle. The CRK is consequently given by the second-order derivative of the total energy $\mathcal{E}$,

$$K_{ab} = \frac{\partial Q_a}{\partial V_b} = \frac{\partial^2 \mathcal{E}}{\partial V_a \partial V_b}. \tag{6.7}$$

The above Eq. (6.7) obviously indicates that K_{ab} is a symmetric matrix, $K_{ab} = K_{ba}$.

ESP charge Then we define the ESP charge Q_a and its operator $\hat{Q}_a$. The partial charge Q_a is determined from electron density and position of nuclei, and the electron density is given as a function of one-electron spatial coordinate. Accordingly, its operator $\hat{Q}_a$ is presented with the one-electron operator for population distribution $\hat{n}_a(i)$ and a function of the nuclear position by

$$\hat{Q}_a = -e \sum_{i=1}^{N_e} \hat{n}_a(i) + Q_a^{\mathrm{nuc}}. \tag{6.8}$$

The first term of Eq. (6.8) is a sum of one-electron operators $\hat{n}_a(i)$ for equivalent N_e electrons, and represents the occupation number of electrons at each site a. The second term Q_a^{nuc} accounts for the contribution of nuclear charge. Q_a is represented as the expectation value of $\hat{Q}_a$,

$$Q_a = \left\langle \Psi | \hat{Q}_a | \Psi \right\rangle = -e \sum_{p,q}^{\mathrm{AO}} D_{pq} \left\langle p | \hat{n}_a | q \right\rangle + Q_a^{\mathrm{nuc}}, \tag{6.9}$$

where p, q denotes the basis sets (atomic orbitals in conventional quantum chemistry), and D_{pq} is the one-electron density matrix. The last expression of Eq. (6.9) including the density matrix is valid also for DFT, which does not explicitly treat the wavefunction.

The ESP charge Q_a is determined so as to reproduce the surrounding electrostatic potential, as illustrated in Fig. 6.2. First we introduce a set of monitoring points

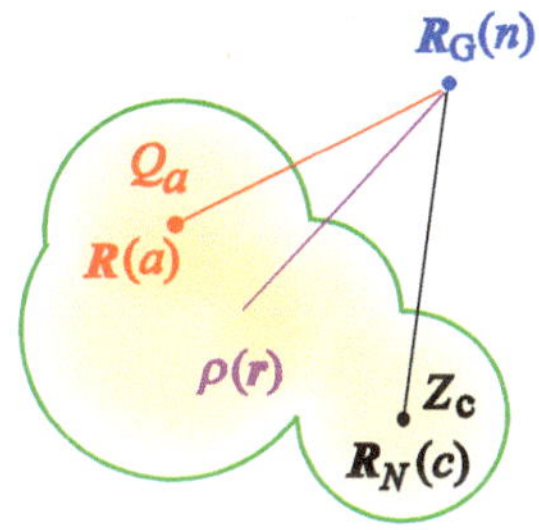

Fig. 6.2 Schematic picture of the electrostatic potential at $\boldsymbol{R}_{\rm G}(n)$

$\{\boldsymbol{R}_{\rm G}(n)\}$ to evaluate the electrostatic potential in outer region of the molecule. The locations of $\{\boldsymbol{R}_{\rm G}(n)\}$ are arbitrary, though they are usually distributed evenly on the van der Waals envelope (green line in Fig. 6.1) or in the outer region. The electrons and nuclei of the molecule generate the electrostatic potential $\phi(\boldsymbol{R}_{\rm G}(n))$ at the point $\boldsymbol{R}_{\rm G}(n)$ by

$$\begin{aligned}\phi(\boldsymbol{R}_{\rm G}(n)) &= -e\int \frac{\rho(\boldsymbol{r})}{|\boldsymbol{r}-\boldsymbol{R}_{\rm G}(n)|}d\boldsymbol{r} + \sum_{c}^{\rm nuc}\frac{Z_c e}{|\boldsymbol{R}_{\rm N}(c)-\boldsymbol{R}_{\rm G}(n)|} \\ &= -e\sum_{p,q}^{\rm AO} D_{pq}\left\langle p\right|\frac{1}{|\boldsymbol{r}-\boldsymbol{R}_{\rm G}(n)|}\left|q\right\rangle + \sum_{c}^{\rm nuc}\frac{Z_c e}{|\boldsymbol{R}_{\rm N}(c)-\boldsymbol{R}_{\rm G}(n)|},\end{aligned} \tag{6.10}$$

where $\boldsymbol{r}$ is the electron coordinate, and $\rho(\boldsymbol{r})$ is the electron density at $\boldsymbol{r}$. The matrix element $\left\langle p\right|\frac{1}{|\boldsymbol{r}-\boldsymbol{R}_{\rm G}(n)|}\left|q\right\rangle$ is equivalent to the conventional one-electron integral for the nuclear attraction in the quantum chemistry, by replacing $\boldsymbol{R}_{\rm G}(n)$ with the nuclear coordinate.

Instead of the quantum chemical formula of Eq. (6.10), the electrostatic potential at $\boldsymbol{R}_{\rm G}(n)$ is alternatively represented using a set of partial charges $\{Q_a\}$ located at $\boldsymbol{R}(a)$. The electrostatic potential is presented with the model of partial charges $\{Q_a\}$ by

$$\phi^{\rm model}(\boldsymbol{R}_{\rm G}(n)) = \sum_{a}^{\rm site}\frac{Q_a}{|\boldsymbol{R}(a)-\boldsymbol{R}_{\rm G}(n)|}. \tag{6.11}$$

(We again note that $\boldsymbol{R}(a)$ is arbitrary in the above formula, though it is often located at the nuclear position, $\boldsymbol{R}_N(c)$.) The set of ESP charges $\{Q_a\}$ are determined so that $\phi^{\rm model}(\boldsymbol{R}_G(n))$ in Eq. (6.11) coincides with $\phi(\boldsymbol{R}_G(n))$ in Eq. (6.10) as much as possible. This procedure of optimization of $\{Q_a\}$ is carried out with the least square fitting that minimizes the following square displacement L,

$$L(\{Q_a\}) = \sum_{n}^{\rm grid}\left\{\phi^{\rm model}(\boldsymbol{R}_{\rm G}(n)) - \phi(\boldsymbol{R}_{\rm G}(n))\right\}^2. \tag{6.12}$$

The least square fitting results in the following definition of $\hat{n}_a$ and Q_a^{nuc} in Eq. (6.9),

$$\hat{n}_a = \sum_b^{\text{site}} (A^{-1})_{ab} \hat{B}_b, \tag{6.13}$$

$$Q_a^{\text{nuc}} = \sum_b^{\text{site}} (A^{-1})_{ab} C_b e, \tag{6.14}$$

where A, B, C are given by

$$A_{ab} = \sum_n^{\text{grid}} \frac{1}{|\boldsymbol{R}(a) - \boldsymbol{R}_{\text{G}}(n)|} \cdot \frac{1}{|\boldsymbol{R}(b) - \boldsymbol{R}_{\text{G}}(n)|}, \tag{6.15}$$

$$\hat{B}_a = \sum_n^{\text{grid}} \frac{1}{|\boldsymbol{r} - \boldsymbol{R}_{\text{G}}(n)|} \cdot \frac{1}{|\boldsymbol{R}(a) - \boldsymbol{R}_{\text{G}}(n)|}, \tag{6.16}$$

$$C_a = \sum_n^{\text{grid}} \sum_c^{\text{nuc}} \frac{Z_c}{|\boldsymbol{R}_{\text{N}}(c) - \boldsymbol{R}_{\text{G}}(n)|} \cdot \frac{1}{|\boldsymbol{R}(a) - \boldsymbol{R}_{\text{G}}(n)|}. \tag{6.17}$$

Equations (6.13) and (6.14) determine the partial charge Q_a and its operator $\hat{Q}_a$ in Eq. (6.9).

[Problem 6.1] Derive $\hat{n}_a$ and Q_a^{nuc} in Eqs. (6.13) and (6.14) by minimizing L in Eq. (6.12).

Note 1. If the locations of the site a coincide with those of nuclei, Q_a^{nuc} in Eq. (6.9) is identical to the nuclear charge at the site in Eq. (6.14).

Note 2. The above procedure of least square fitting is often accompanied with some extra constraints on the total charge or the dipole moment of the molecule, e.g.

$$\sum_a^{\text{site}} Q_a = Q \text{ (constant)},$$

$$\sum_a^{\text{site}} Q_a \boldsymbol{R}(a) = \boldsymbol{\mu} \text{ (constant)},$$

where the values Q and $\boldsymbol{\mu}$ in the right hand side are given as the constraint conditions. Such constraints can be readily implemented in the least square fitting of Problem 6.1 using the Lagrange multipliers [9].

Note 3. The ESP charges tend to be problematic to determine for buried sites of molecules [1]. This problem is resolved with introducing a damping parameter in the least square fitting procedure, as described in Ref. [9]. This modification effectively removes the instability of the definition, while the following procedure of CRK calculations unchanged.

Charge Response Kernel Once the perturbation Hamiltonian $\hat{H}'$ is defined above, the derivative of the total energy $\mathcal{E}$ or the wavefunction Ψ by the external parameter V is uniquely determined. Consequently, the CRK K_{ab} is determined using the derivatives,

$$K_{ab} = \frac{\partial^2 \mathcal{E}}{\partial V_a \partial V_b} = \frac{\partial Q_a}{\partial V_b} = \left\langle \frac{\partial \Psi}{\partial V_b} | \hat{Q}_a | \Psi \right\rangle + \left\langle \Psi | \hat{Q}_a | \frac{\partial \Psi}{\partial V_b} \right\rangle. \tag{6.18}$$

In what follows, we outline the derivation of CRK with a closed-shell Hartree-Fock wavefunction Ψ. The details of the derivation is given in Appendix A.1.

The many-electron wavefunction Ψ is given with a normalized Slater determinant [13],

$$\Psi = \frac{1}{\sqrt{N_e!}} \begin{vmatrix} \psi_1(1)\alpha(1) & \psi_1(1)\beta(1) & \cdots & \psi_{Ne/2}(1)\beta(1) \\ \psi_1(2)\alpha(2) & & & \\ \vdots & & \ddots & \vdots \\ \psi_1(N_e)\alpha(N_e) & & \cdots & \psi_{Ne/2}(N_e)\beta(N_e) \end{vmatrix}, \tag{6.19}$$

where $\psi_i\alpha$ and $\Psi_j\beta$ denote the i-th molecular orbital with α and β spin, respectively. Suppose that the i-th molecular orbital (MO) ψ_i is represented by a linear combination of proper basis functions $\{\chi_p\}$,[1]

$$\psi_i = \sum_p^{\mathrm{AO}} C_{pi} \chi_p. \tag{6.20}$$

Then the derivative of the wavefunction Ψ in Eq. (6.18) can be presented with the derivatives of the MO coefficients $\{C_{pi}\}$. We express these derivatives $(\partial C_{pi}/\partial V_b)$ in the form of linear combination of MO coefficients using a matrix U_{ji}^b by

$$\frac{\partial C_{pi}}{\partial V_b} = \sum_j^{\mathrm{MO}} C_{pj} U_{ji}^b. \tag{6.21}$$

Then U_{ji}^b is determined as the solution of the following linear equation.

$$(\varepsilon_l - \varepsilon_i) U_{li}^b + \sum_j^{\mathrm{occ}} \sum_k^{\mathrm{vir}} H_{likj} U_{kj}^b = e \langle \psi_l | \hat{n}_b | \psi_i \rangle, \tag{6.22}$$

[1]In the present section, the suffixes i, j, k, l stand for the MOs, while p, q, r, s for the basis functions.

where the suffix j denotes the occupied MOs and k denotes the virtual (unoccupied) MOs. ε_i is the canonical orbital energy for the i-th orbital ψ_i. H_{likj} is the two-electron integral in the MO representation,

$$H_{likj} = 4\left(\psi_l \psi_i | \psi_k \psi_j\right) - \left(\psi_l \psi_k | \psi_i \psi_j\right) - \left(\psi_l \psi_j | \psi_k \psi_i\right), \tag{6.23}$$

where

$$\begin{aligned}\left(\psi_i \psi_j | \psi_k \psi_l\right) &= \iint \psi_i(\boldsymbol{r})\psi_j(\boldsymbol{r})\frac{e^2}{|\boldsymbol{r}-\boldsymbol{r}'|}\psi_k(\boldsymbol{r}')\psi_l(\boldsymbol{r}')d\boldsymbol{r}d\boldsymbol{r}' \\ &= \sum_{p,q,r,s}^{\mathrm{AO}} C_{pi}C_{qj}C_{rk}C_{sl}\left(\chi_p\chi_q|\chi_r\chi_s\right)\end{aligned}$$

and

$$\left\langle \psi_l | \hat{n}_b | \psi_i \right\rangle = \sum_{p,q}^{\mathrm{AO}} C_{pl}C_{qi}\langle \chi_p | \hat{n}_b | \chi_q \rangle. \tag{6.24}$$

Equation (6.22) is a set of coupled linear equations for U_{ji}^b, called the Coupled-Perturbed Hartree-Fock (CPHF) equation [5, 10]. Using the solution of U_{li}^b, the CRK K_{ab} is calculated as follows,

$$K_{ab} = \frac{\partial Q_a}{\partial V_b} = -e\sum_{p,q}^{\mathrm{AO}} \frac{\partial D_{pq}}{\partial V_b}\left\langle p|\hat{n}_a|q\right\rangle = -4e\sum_i^{\mathrm{occ}}\sum_j^{\mathrm{vir}}\left\langle \psi_j|\hat{n}_a|\psi_i\right\rangle U_{ji}^b. \tag{6.25}$$

The analogous formulations with CPHF are utilized to other response quantities or second-order derivatives of the electronic energy, such as force constant, polarizability, etc. The CRK is a variant of the general formulations of the second-order derivatives. As a consequence, the extension of CRK to other wavefunctions can be carried out in the common framework with other derivative quantities.

[Problem 6.2] Since the CRK K_{ab} is a symmetric matrix, it is diagonalized by a proper unitary matrix $\boldsymbol{P}$,

$$\boldsymbol{P}^T\boldsymbol{K}\boldsymbol{P} = \begin{pmatrix} \lambda_1 & & \\ & \ddots & \\ & & \lambda_{N_s}\end{pmatrix},$$

where N_s is the number of sites in the molecule. Prove that (i) all the eigenvalues λ_a are not positive ($\lambda_a \leq 0$), and (ii) one eigenvalue is necessarily zero.

(*Hint*) Suppose that the zero-th order wavefunction and the total energy for the ground state are Ψ_0 and $\mathcal{E}_0$, respectively, with no external perturbation $V = 0$. Then, by adding the perturbation V, the total energy $\mathcal{E}$ is expanded in a series of perturbation in the following manner,

$$\mathcal{E} = \mathcal{E}_0 + \mathcal{E}^{(1)} + \mathcal{E}^{(2)} + \cdots = \mathcal{E}_0 + \sum_a Q_a V_a + \frac{1}{2} \sum_{a,b} K_{ab} V_a V_b + \cdots, \tag{6.26}$$

where

$$Q_a = \left(\frac{\partial \mathcal{E}}{\partial V_a}\right)_{V=0} \quad \text{and} \quad K_{ab} = \left(\frac{\partial^2 \mathcal{E}}{\partial V_a \partial V_b}\right)_{V=0}.$$

Recall that the second-order perturbation energy $\mathcal{E}^{(2)}$ for the ground state is always negative.

6.3 Polarizable Model with CRK

The CRK represents the electronic polarization using the ESP site charges, and can be readily utilized as a polarizable model. Here we summarize the merits of CRK modeling and its application to the SFG calculations.

Force Fields Most of current molecular simulations for polyatomic molecules adopt force fields based on the interaction site model of molecules, where intermolecular interactions are constructed with site-site interactions. The CRK modeling is also based on the interaction site model, and readily incorporated in the MD simulation. It allows for varying partial charges at the interaction sites, and thereby representing the electronic polarization in MD simulation [8, 11].

Let us suppose a molecule in condensed phase, such as solution or interface. The electrostatic potential V_{bi} at the site b of molecule i is presented by the intermolecular Coulombic interactions from surrounding molecules $j (\neq i)$ in the following form,

$$V_{bi} = \sum_{j(\neq i)}^{\text{molecule}} \sum_{a}^{\text{site}} \frac{Q_{aj}}{|\boldsymbol{R}(bi) - \boldsymbol{R}(aj)|}, \tag{6.27}$$

where Q_{aj} and $\boldsymbol{R}(aj)$ are the partial charge and coordinates of the site a of molecule j. The partial charge Q_{ai} of molecule i under the electrostatic potential V_{bi} is represented with the CRK by

$$Q_{ai} = Q_a^0 + \sum_{b}^{\text{site}} K_{ab} V_{bi}, \tag{6.28}$$

where Q_a^0 is the partial charge at the site a in the isolated condition free from the external potential. Equation (6.28) shows that the site partial charge Q_{ai} varies in response to the external potential up to the first order.[2] Equations (6.27) and (6.28) provide a self-consistent scheme to determine V_{bi} and Q_{ai} simultaneously. For an actual condensed system, the site coordinates $\boldsymbol{R}(aj)$ changes with molecular motions, and accordingly Eqs. (6.27) and (6.28) are solved to obtain the instantaneous values of V_{bi} and Q_{ai} at each time t during the MD simulation. The effect of electronic polarization is thereby implemented through fluctuation of partial charges during the time evolution.

In the CRK model, the whole electrostatic potential energy of the condensed system is given by

$$U = \sum_i \sum_{>j} \sum_a \sum_b \frac{Q_{ai} Q_{bj}}{|\boldsymbol{R}(ai) - \boldsymbol{R}(bj)|} - \frac{1}{2} \sum_i \sum_a \sum_b K_{ab} V_{ai} V_{bi}. \tag{6.29}$$

The first term of the right hand side stands for the intermolecular site-site Coulomb interactions, and the second term stands for the reorganization energy, as detailed in Appendix A.2. The second term accounts for destabilization energy due to the deformation of electronic polarization. Note that this term is always positive, since the CRK K_{ab} is a non-positive definite matrix (see Problem 6.2). The force acting on the site a of molecule i, $\boldsymbol{F}(ai)$, is derived from Eq. (6.29) by

$$\boldsymbol{F}(ai) = -\frac{\partial U}{\partial \boldsymbol{R}(ai)} = \sum_{j(\neq i)} \sum_b \frac{Q_{ai} Q_{bj} \left(\boldsymbol{R}(ai) - \boldsymbol{R}(bj)\right)}{|\boldsymbol{R}(ai) - \boldsymbol{R}(bj)|^3}. \tag{6.30}$$

We notice that Eq. (6.30) does not include the derivatives $\partial Q/\partial \boldsymbol{R}$ or $\partial V/\partial \boldsymbol{R}$. The expression of $\boldsymbol{F}(ai)$ in Eq. (6.30) apparently coincide with the expression obtained by differentiating the first term of U in Eq. (6.29) simply by $\boldsymbol{R}(ai)$, as if the site charges Q_{ai} and Q_{bj} were fixed. This feature greatly simplifies the calculation of Eq. (6.30), by virtue of the self-consistent conditions of Eqs. (6.27) and (6.28). Appendix A.2 further discuss this feature.

[Problem 6.3] By differentiating Eq. (6.29) with respect to the coordinate $\boldsymbol{R}(ai)$, derive the formula of the force $\boldsymbol{F}(ai)$ in Eq. (6.30). We assume that Q_a^0 and K_{ab} are invariant under the molecular vibration for simplicity.

(*Hint*) Make use of the self-consistent relations of Q_{ai} and V_{ai} in Eqs. (6.27) and (6.28).

---

[2] Equation (6.28) treats a single species for simplicity. Extension to multi-species systems is straightforward, and shown in Eq. (6.34) in Sect. 6.4.

Non-empirical Modeling One of the most significant advantages of the CRK modeling as a polarizable model is that the parameters of polarization are fully determined by ab initio molecular orbital or DFT calculations, based on the common scheme to the ESP charge. Once we employ a method of ESP charge calculation, the extension to the CRK is unique and straightforward with no empirical parameter. Accuracy of the CRK model is entirely conformed to the electronic structure calculation, and it is possible to systematically improve the accuracy by improving the electronic structure calculation.

Fully non-empirical character of the CRK model is quite advantageous to general applications. For example, unstable species or radicals can be readily treated with common accuracy to stable molecules [7, 8], though these unstable species are not suitable to empirical modeling due to scarcity of available experimental properties. It is also straightforward to obtain the conformational dependence of CRK for a vibrating molecule, by performing the CRK calculations for a deformed molecule from its equilibrium conformation. Accurate description of the conformational dependence is a key requisite for MD calculations of vibrational Raman or SFG spectra.

Dipole and Polarizability Next we describe the instantaneous dipole and polarizability of a molecule using the CRK model. We treat an arbitrary vibrating and rotating molecule, and assume that the instantaneous coordinate of its site a is $\boldsymbol{R}(a)$ in the space-fixed coordinates. Then the permanent dipole vector $\boldsymbol{\mu}^0$ and the polarizability tensor $\boldsymbol{\alpha}$ are represented in the same space-fixed coordinates by

$$\mu_p^0 = \sum_a^{\text{site}} Q_a^0 R_p(a), \tag{6.31}$$

$$\alpha_{pq} = -\sum_{a,b}^{\text{site}} K_{ab} R_p(a) R_q(b), \tag{6.32}$$

where the suffixes p, q denote $x \sim z$ in the space-fixed coordinates. In the expressions of Eqs. (6.31) and (6.32), it is noteworthy that Q_a^0 and K_{ab} are invariant with respect to the molecular rotation and thus considered as scalar properties. These scalar characters are advantageous to describe the vector and tensor elements of $\boldsymbol{\mu}$ and $\boldsymbol{\alpha}$ in an arbitrary coordinate system. Once can determine the values of Q_a^0 and K_{ab} as a function of internal coordinates, the vector/tensor elements of μ_p^0/α_{pq} are readily expressed with these values and the coordinates of sites $\{R_p(a)\}$.

[Problem 6.4] Explain the expression of the polarizability α_{pq} in Eq. (6.32). Recall that the polarizability is the derivative of dipole moment with respect to spatially uniform electric field.

In summary, the CRK model offers a general scheme to describe the polarizable force field as well as instantaneous dipole and polarizability in a unified manner. Therefore, it is particularly suitable to calculate the vibrational SFG spectroscopy by MD simulation.

6.4 $\chi^{(2)}$ Formula with CRK Model

Then we present the $\chi^{(2)}$ formula based on the CRK model. We have derived $\chi^{(2)}$ in Sect. 5.2 with the time correlation function of $\boldsymbol{A}_{\text{eff}}$ and $\boldsymbol{M}$ in Eq. (5.27). The purpose here is to express $\boldsymbol{A}_{\text{eff}}$ and $\boldsymbol{M}$ using the CRK model. The following derivation with the CRK model provides an alternative but physically equivalent treatment of the local field effect in Sect. 5.

As we discussed in Sect. 6.3, the partial charge Q_{ai} and electrostatic potential V_{bi} are determined in the following self-consistent equations,

$$V_{bi} = -\boldsymbol{R}(bi) \cdot \boldsymbol{E}_0 + \sum_{j(\neq i)}^{\text{molecule}} \sum_{a}^{\text{site}} \frac{Q_{aj}}{|\boldsymbol{R}(bi) - \boldsymbol{R}(aj)|}, \tag{6.33}$$

$$Q_{ai} = Q_{ai}^0 + \sum_{b}^{\text{site}} K_{abi} V_{bi}. \tag{6.34}$$

The above Eqs. (6.33) and (6.34) are essentially same as Eqs. (6.27) and (6.28), but involve slight extensions. Equation (6.33) temporarily includes an uniform external field $\boldsymbol{E}_0$. Equation (6.34) can describe multi-species systems, since the notations of Q_{ai}^0 and K_{abi} including the suffix of molecule i allows for distinguishing different species among the molecules. The coupled equations of (6.33) and (6.34) lead to the following equations,

$$\sum_j \sum_c^{\text{site}} \left(\delta_{ac}\delta_{ij} - \sum_b^{\text{site}} \frac{K_{bcj}}{|\boldsymbol{R}(ai) - \boldsymbol{R}(bj)|} \right) V_{cj} = -\boldsymbol{R}(ai) \cdot \boldsymbol{E}_0 + \sum_{j(\neq i)} \sum_b^{\text{site}} \frac{Q_{bj}^0}{|\boldsymbol{R}(ai) - \boldsymbol{R}(bj)|}, \tag{6.35}$$

$$\sum_j \sum_c^{\text{site}} \left(\delta_{ac}\delta_{ij} - \sum_b^{\text{site}} \frac{K_{abi}}{|\boldsymbol{R}(bi) - \boldsymbol{R}(cj)|} \right) Q_{cj} = Q_{ai}^0 - \sum_b^{\text{site}} K_{abi} (\boldsymbol{R}(bi) \cdot \boldsymbol{E}_0). \tag{6.36}$$

To solve these equations, we introduce an auxiliary matrix $\boldsymbol{G}$ (or $\boldsymbol{G}^T$) to be

$$[G]_{ai,cj} = \delta_{ac}\delta_{ij} - \sum_b^{\text{site}} \frac{K_{bcj}}{|\boldsymbol{R}(ai) - \boldsymbol{R}(bj)|}, \tag{6.37}$$

$$[G^T]_{ai,cj} = [G]_{cj,ai} = \delta_{ac}\delta_{ij} - \sum_b^{\text{site}} \frac{K_{abi}}{|\boldsymbol{R}(bi) - \boldsymbol{R}(cj)|}, \tag{6.38}$$

and $\boldsymbol{V}^0$,

$$V_{ai}^0 = \sum_{j(\neq i)} \sum_b^{\text{site}} \frac{Q_{bj}^0}{|\boldsymbol{R}(ai) - \boldsymbol{R}(bj)|}. \tag{6.39}$$

Using $\boldsymbol{G}$ and $\boldsymbol{V}^0$, the solutions of Eqs. (6.35) and (6.36) are given as follows:

$$V_{ai} = \sum_j \sum_c^{\text{site}} [G^{-1}]_{ai,cj} \left[V_{cj}^0 - \boldsymbol{R}(cj) \cdot \boldsymbol{E}_0 \right], \tag{6.40}$$

$$\begin{aligned} Q_{ai} &= \sum_j \sum_c^{\text{site}} [G^{T-1}]_{ai,cj} \left[Q_{cj}^0 - \sum_b^{\text{site}} K_{cbj} (\boldsymbol{R}(bj) \cdot \boldsymbol{E}_0) \right] \\ &= \sum_j \sum_c^{\text{site}} [G^{-1}]_{cj,ai} \left[Q_{cj}^0 - \sum_b^{\text{site}} K_{cbj} (\boldsymbol{R}(bj) \cdot \boldsymbol{E}_0) \right]. \end{aligned} \tag{6.41}$$

Now we can present the dipole moment and polarizability of the whole system using the above solutions. The dipole moment is represented by

$$\begin{aligned} \boldsymbol{M} &= \sum_i \sum_a^{\text{site}} Q_{ai} \boldsymbol{R}(ai) \\ &= \sum_i \sum_j \sum_a^{\text{site}} \sum_c^{\text{site}} [G^{-1}]_{cj,ai} Q_{cj}^0 \boldsymbol{R}(ai) \qquad (\boldsymbol{E}_0 = 0). \end{aligned} \tag{6.42}$$

The polarizability tensor of the whole system is given by differentiating $\boldsymbol{M}$ with respect to $\boldsymbol{E}_0$ and then setting to $\boldsymbol{E}_0 = 0$,

$$\begin{aligned} \boldsymbol{A} = \frac{\partial \boldsymbol{M}}{\partial \boldsymbol{E}_0} &= \sum_i \sum_a^{\text{site}} \frac{\partial Q_{ai}}{\partial \boldsymbol{E}_0} \boldsymbol{R}(ai) \\ &= -\sum_i \sum_a^{\text{site}} \sum_b^{\text{site}} K_{ab,i} \boldsymbol{R}(ai) \otimes \left[\sum_j \sum_c^{\text{site}} [G^{-1}]_{bi,cj} \boldsymbol{R}(cj) \right] \qquad (\boldsymbol{E}_0 = 0), \end{aligned} \tag{6.43}$$

where $\otimes$ stands for tensor product. We further consider the local field correction for the sum frequency field in the same manner as in Sect. 5.2, and thereby formulate the effective polarizability $\boldsymbol{A}_{\text{eff}}$ as

$$\boldsymbol{A}_{\text{eff}} = -\sum_i^{\text{site}} \sum_a^{\text{site}} \sum_b K_{abi} \left[\sum_{j'} \sum_{c'}^{\text{site}} [G^{-1}]_{ai,c'j'} \boldsymbol{R}(c'j') \right] \otimes$$

$$\left[\sum_j \sum_c^{\text{site}} [G^{-1}]_{bi,cj} \boldsymbol{R}(cj) \right]. \tag{6.44}$$

Equations (6.42) and (6.44) are the CRK expressions of the dipole and effective polarizability, and are equivalent to Eq. (5.28) in Chap. 5.

6.5 Solutions to Problems

6.5.1 ESP Charge

[Problem 6.1] Derive $\hat{n}_a$ and Q_a^{nuc} in Eqs. (6.13) and (6.14) by minimizing L in Eq. (6.12).

The condition of minimizing L is

$$\frac{\partial L}{\partial Q_a} = 2\sum_n^{\text{grid}} \left(\phi^{\text{model}}(\boldsymbol{R}_G(n)) - \phi(\boldsymbol{R}_G(n)) \right) \frac{\partial \phi^{\text{model}}(\boldsymbol{R}_G(n))}{\partial Q_a} = 0$$

$$= 2\sum_b^{\text{site}} \sum_n^{\text{grid}} \frac{Q_b}{|\boldsymbol{R}(b) - \boldsymbol{R}_G(n)|} \cdot \frac{1}{|\boldsymbol{R}(a) - \boldsymbol{R}_G(n)|}$$

$$- 2\sum_n^{\text{grid}} \left\{ -e \sum_{p,q}^{\text{AO}} D_{pq} \left\langle p \left| \frac{1}{|\boldsymbol{r} - \boldsymbol{R}_G(n)|} \right| q \right\rangle + \sum_c^{\text{nuc}} \frac{Z_c e}{|\boldsymbol{R}_N(c) - \boldsymbol{R}_G(n)|} \right\} \cdot \frac{1}{|\boldsymbol{R}(a) - \boldsymbol{R}_G(n)|}.$$

Therefore, the optimized values of $\{Q_a\}$ satisfy the following equation,

$$\sum_b^{\text{site}} A_{ab} Q_b = -e \sum_{p,q}^{\text{AO}} D_{pq} \left\langle p | \hat{B}_a | q \right\rangle + C_a e, \tag{6.45}$$

where

$$A_{ab} = \sum_n^{\text{grid}} \frac{1}{|\boldsymbol{R}(a) - \boldsymbol{R}_G(n)|} \cdot \frac{1}{|\boldsymbol{R}(b) - \boldsymbol{R}_G(n)|}, \tag{6.15}$$

$$\hat{B}_a = \sum_n^{\text{grid}} \frac{1}{|\boldsymbol{r} - \boldsymbol{R}_G(n)|} \cdot \frac{1}{|\boldsymbol{R}(a) - \boldsymbol{R}_G(n)|}, \tag{6.16}$$

$$C_a = \sum_n^{\text{grid}} \sum_c^{\text{nuc}} \frac{Z_c}{|\boldsymbol{R}_N(c) - \boldsymbol{R}_G(n)|} \cdot \frac{1}{|\boldsymbol{R}(a) - \boldsymbol{R}_G(n)|}. \tag{6.17}$$

The solution of Eq. (6.45) is

$$Q_a = -e \sum_{p,q}^{\text{AO}} D_{pq} \left\langle p \middle| \sum_b^{\text{site}} (A^{-1})_{ab} \hat{B}_b \middle| q \right\rangle + \sum_b^{\text{site}} (A^{-1})_{ab} C_b e.$$

Comparing this result with Eq. (6.9), we derive the following form of $\hat{n}_a$ and Q_a^{nuc},

$$\hat{n}_a = \sum_b^{\text{site}} (A^{-1})_{ab} \hat{B}_b, \tag{6.13}$$

$$Q_a^{\text{nuc}} = \sum_b^{\text{site}} (A^{-1})_{ab} C_b e. \tag{6.14}$$

6.5.2 *Charge Response Kernel*

[Problem 6.2] Since the CRK K_{ab} is a symmetric matrix, it is diagonalized by a proper unitary matrix $\boldsymbol{P}$,

$$\boldsymbol{P}^T \boldsymbol{K} \boldsymbol{P} = \begin{pmatrix} \lambda_1 & & \\ & \ddots & \\ & & \lambda_{N_s} \end{pmatrix},$$

where N_s is the number of sites in the molecule. Prove that (i) all the eigenvalues λ_a are not positive ($\lambda_a \leq 0$), and (ii) one eigenvalue is necessarily zero.

(*Hint*) Suppose that the zero-th order wavefunction and the total energy for the ground state are Ψ_0 and $\mathcal{E}_0$, respectively, with no external perturbation $V = 0$. Then, by adding the perturbation V, the total energy $\mathcal{E}$ is expanded in a series of perturbation in the following manner,

$$\mathcal{E} = \mathcal{E}_0 + \mathcal{E}^{(1)} + \mathcal{E}^{(2)} + \cdots = \mathcal{E}_0 + \sum_a Q_a V_a + \frac{1}{2} \sum_{a,b} K_{ab} V_a V_b + \cdots, \tag{6.26}$$

where

$$Q_a = \left(\frac{\partial \mathcal{E}}{\partial V_a}\right)_{V=0} \quad \text{and} \quad K_{ab} = \left(\frac{\partial^2 \mathcal{E}}{\partial V_a \partial V_b}\right)_{V=0}.$$

Recall that the second-order perturbation energy $\mathcal{E}^{(2)}$ for the ground state is always negative.

The total electronic energy $\mathcal{E}$ under the external potential V is represented by the perturbation expansion,

$$\begin{aligned}\mathcal{E} &= \mathcal{E}_0 + \mathcal{E}^{(1)} + \mathcal{E}^{(2)} + \cdots \\ &= \mathcal{E}_0 + \sum_a^{\text{site}} Q_a V_a + \frac{1}{2}\sum_{a,b}^{\text{site}} K_{ab} V_a V_b + \cdots .\end{aligned} \tag{6.26}$$

The zero-th order wavefunction and energy for the ground state are denoted to be Ψ_0 and $\mathcal{E}_0$, respectively, with no external potential, $V = 0$. The first- and second-order derivatives of the energy correspond to $Q_a = (\partial\mathcal{E}/\partial V_a)$ and $K_{ab} = (\partial^2\mathcal{E}/\partial V_a \partial V_b)$ in Eqs. (6.6) and (6.7), respectively.

Suppose that the ground state Ψ_0 is not degenerated, the second-order energy $\mathcal{E}^{(2)}$ is generally represented in the following sum-over-state expression,

$$\mathcal{E}^{(2)} = \sum_{m(\neq 0)}^{\text{state}} \frac{\left|\left\langle \Psi_0 | \hat{H}' | \Psi_m \right\rangle\right|^2}{\mathcal{E}_0 - \mathcal{E}_m}, \tag{6.46}$$

where the suffix m denote the electronic states other than the ground state. Ψ_m and $\mathcal{E}_m$ are the m-th eigenstate of $\hat{H}_0$ and its energy, i.e. $\hat{H}_0\Psi_m = \mathcal{E}_m\Psi_m$ ($\mathcal{E}_m > \mathcal{E}_0$). Therefore, $\mathcal{E}^{(2)}$ is always negative or zero,

$$\mathcal{E}^{(2)} = \frac{1}{2}\sum_{a,b} K_{ab} V_a V_b \leq 0,$$

which indicates that K_{ab} is a non-positive definite matrix.

The second-order energy $\mathcal{E}^{(2)}$ of Eq. (6.46) vanishes only when $\left\langle \Psi_0 | \hat{H}' | \Psi_m \right\rangle = 0$ holds for all the states $m(\neq 0)$ in Eq. (6.46). This is realized when the external site potentials V_b are constant, $V_1 = V_2 = \cdots = V_{N_s} = \text{const.}$ (this value is assumed to be V_0.) Then the perturbation Hamiltonian also becomes constant,

$$\hat{H}' = \sum_a \hat{Q}_a V_a = \left(\sum_a \hat{Q}_a\right) V_0 = \text{const.}$$

since the sum of the site charges is constant, $\sum_a \hat{Q}_a = \text{const.}$, due to the charge conservation. (This condition is often imposed explicitly as a constraint in the least square fitting for $\hat{Q}_a$.) Therefore, in such case $\left\langle \Psi_0 | \hat{H}' | \Psi_m \right\rangle = 0$ holds for any $m (\neq 0)$.

The above argument concludes that the CRK K_{ab} has always a zero eigenvalue, with the corresponding eigenvector $\boldsymbol{V}$,

$$\boldsymbol{V} = \begin{pmatrix} 1 \\ 1 \\ \vdots \\ 1 \end{pmatrix}.$$

This fact is readily understandable since uniform change in the site potentials has no influence on the electron distribution. Accordingly

$$\Delta Q_a = \left(\sum_b K_{ab} \right) V_0 = 0$$

is derived, which indicates that the sum of the CRK elements in a row (or a column) is inevitably zero.

6.5.3 *Energy and Force with Polarizable Model*

[Problem 6.3] By differentiating Eq. (6.29) with respect to the coordinate $\boldsymbol{R}(ai)$, derive the formula of the force $\boldsymbol{F}(ai)$ in Eq. (6.30). We assume that Q_a^0 and K_{ab} are invariant under the molecular vibration for simplicity.

(*Hint*) Make use of the self-consistent relations of Q_{ai} and V_{ai} in Eqs. (6.27) and (6.28).

The formula of the force $\boldsymbol{F}(ai)$ is derived by differentiating the potential energy U in Eq. (6.29) by the site coordinate $\boldsymbol{R}(ai)$,

$$\begin{aligned}
\boldsymbol{F}(ai) &= -\frac{\partial U}{\partial \boldsymbol{R}(ai)} \\
&= -\frac{\partial}{\partial \boldsymbol{R}(ai)} \left\{ \sum_i \sum_{>j} \sum_a \sum_b \frac{Q_{ai} Q_{bj}}{|\boldsymbol{R}(ai) - \boldsymbol{R}(bj)|} - \frac{1}{2} \sum_i \sum_a \sum_b K_{ab} V_{ai} V_{bi} \right\}
\end{aligned}$$

$$
\begin{aligned}
&= \sum_{j(\neq i)} \sum_b \frac{Q_{ai} Q_{bj} (\boldsymbol{R}(ai) - \boldsymbol{R}(bj))}{|\boldsymbol{R}(ai) - \boldsymbol{R}(bj)|^3} - \sum_j \sum_{k(\neq j)} \sum_b \sum_c \frac{1}{|\boldsymbol{R}(bj) - \boldsymbol{R}(ck)|} \frac{\partial Q_{bj}}{\partial \boldsymbol{R}(ai)} Q_{ck} \\
&\quad + \sum_j \sum_{b,c} K_{bc} V_{bj} \frac{\partial V_{cj}}{\partial \boldsymbol{R}(ai)} \\
&= \sum_{j(\neq i)} \sum_b \frac{Q_{ai} Q_{bj} (\boldsymbol{R}(ai) - \boldsymbol{R}(bj))}{|\boldsymbol{R}(ai) - \boldsymbol{R}(bj)|^3} - \sum_j \sum_b \left\{ \sum_{k(\neq j)} \sum_c \frac{Q_{ck}}{|\boldsymbol{R}(bj) - \boldsymbol{R}(ck)|} \right\} \frac{\partial Q_{bj}}{\partial \boldsymbol{R}(ai)} \\
&\quad + \sum_j \sum_b \left\{ \sum_c K_{bc} \frac{\partial V_{cj}}{\partial \boldsymbol{R}(ai)} \right\} V_{bj} \qquad (6.47) \\
&= \sum_{j(\neq i)} \sum_b \frac{Q_{ai} Q_{bj} (\boldsymbol{R}(ai) - \boldsymbol{R}(bj))}{|\boldsymbol{R}(ai) - \boldsymbol{R}(bj)|^3} - \sum_j \sum_b V_{bj} \frac{\partial Q_{bj}}{\partial \boldsymbol{R}(ai)} + \sum_j \sum_b \frac{\partial Q_{bj}}{\partial \boldsymbol{R}(ai)} V_{bj} \\
&\qquad (6.48) \\
&= \sum_{j(\neq i)} \sum_b \frac{Q_{ai} Q_{bj} (\boldsymbol{R}(ai) - \boldsymbol{R}(bj))}{|\boldsymbol{R}(ai) - \boldsymbol{R}(bj)|^3}. \qquad (6.30)
\end{aligned}
$$

In the derivation from Eq. (6.47) to (6.48), we employed the self-consistent conditions of Eqs. (6.27) and (6.28).

If Q_{ai} and V_{ai} satisfy the self-consistent conditions, the final expression of the force $\boldsymbol{F}(ai)$ does not include the derivatives $\partial Q/\partial \boldsymbol{R}$, $\partial V/\partial \boldsymbol{R}$. This feature is related to the variational principle of polarization, and is discussed in Appendix A.2.

6.5.4 *Polarizability*

[Problem 6.4] Explain the expression of the polarizability α_{pq} in Eq. (6.32). Recall that the polarizability is the derivative of dipole moment with respect to spatially uniform electric field.

Let us assume that a small, uniform electric field ΔE_q is imposed along the direction q ($= x, y, z$ in the space-fixed coordinate). Then the electrostatic potential at the site b changes by[3]

$$
\Delta V_b = -R_q(b)\, \Delta E_q,
$$

[3]Note that an arbitrary constant of the potential associated to the definition of spatial origin has no influence (see Problem 6.2).

where $R_q(b)$ is the q coordinate of the site b. As a consequence, the partial charge at the site a changes by

$$\Delta Q_a = \sum_b^{\text{site}} K_{ab} \Delta V_b = -\sum_b K_{ab} \Delta E_q R_q(b).$$

The change in partial charges gives rise to the change in the dipole moment along the direction p,

$$\Delta \mu_p = \sum_a^{\text{site}} \Delta Q_a R_p(a) = -\sum_{a,b}^{\text{site}} K_{ab} \Delta E_q R_q(b) R_p(a)$$
$$= \left(-\sum_{a,b}^{\text{site}} K_{ab} R_p(a) R_q(b) \right) \Delta E_q$$

Then the polarizability tensor is defined as the derivative of dipole moment with respect to the electric field in the following form,

$$\alpha_{pq} = \lim_{\Delta E \to 0} \frac{\Delta \mu_p}{\Delta E_q} = -\sum_{a,b}^{\text{site}} K_{ab} R_p(a) R_q(b). \quad (6.32)$$

Appendix

A.1 Derivation of CRK from CPHF Equation

In this section we derive the CPHF equation (6.22) and CRK in Eq. (6.25). The CRK is given as a second-order derivative of the total energy, $K_{ab} = (\partial^2 \mathcal{E} / \partial V_a \partial V_b)$, and here it is formulated in the closed-shell Hartree-Fock theory [7]. Other formulations in the unrestricted Hartree-Fock [7] and Kohn-Sham DFT [3] are found in literature.

A.1.1 Wavefunction Under External Field

We briefly summarize the Hartree-Fock wavefunction under the external potential below. The total energy $\mathcal{E}$ for the wavefunction Ψ in Eq. (6.19) is presented by [13]

$$\mathcal{E} = \left\langle \Psi | \hat{H} | \Psi \right\rangle = \left\langle \Psi | \hat{H}_0 + \hat{H}' | \Psi \right\rangle$$

$$
= \sum_{i=1}^{N_e/2} 2\left\langle\psi_i|\hat{h}|\psi_i\right\rangle + \sum_{i=1}^{N_e/2}\sum_{j=1}^{N_e/2} \{2(\psi_i\psi_i|\psi_j\psi_j) - (\psi_i\psi_j|\psi_i\psi_j)\} + \sum_a^{\text{site}} Q_a^{\text{nuc}} V_a
$$

$$
= \sum_i^{\text{occ}} 2h_{ii} + \sum_{i,j}^{\text{occ}} \{2J_{ij} - K_{ij}\} + \sum_a^{\text{site}} Q_a^{\text{nuc}} V_a. \tag{6.49}
$$

In Eq. (6.49), the one-electron operator $\hat{h}$ includes the perturbation,

$$
\hat{h} = \hat{h}_0 - e\sum_a^{\text{site}} \hat{n}_a V_a \tag{6.50}
$$

where $\hat{h}_0$ is the conventional one-electron operator in Eq. (6.4), and the second term indicates the perturbation by the external electrostatic potential.

The Hartree-Fock theory provides a way to determine the wavefunction Ψ and total energy $\mathcal{E}$ through the variational principle. It determines the MO coefficients C_{pi} of Eq. (6.19) by minimizing the energy $\mathcal{E}$ of Eq. (6.49), under the constraint of the following orthonormal conditions among the MOs,

$$
\left\langle\psi_i|\psi_j\right\rangle = \sum_{p,q}^{\text{AO}} C_{pi}C_{qj}s_{pq} = \delta_{ij}, \quad \text{where} \quad s_{pq} = \langle p|q\rangle\,. \tag{6.51}
$$

This variational procedure derives the following general eigenvector problem, called the Hartree-Fock-Roothaan equation, for the MO coefficients,

$$
\sum_q^{\text{AO}} F_{pq}C_{qi} = \varepsilon_i \sum_q^{\text{AO}} s_{pq}C_{qi}, \tag{6.52}
$$

where ε_i corresponds to the eigenvalue, called the i-th canonical orbital energy. F_{pq} is the Fock matrix given as follows,

$$
F_{pq} = \left\langle p|\hat{h}|q\right\rangle + \frac{1}{2}\sum_{r,s}^{\text{AO}} \{2(pq|rs) - (pr|qs)\} D_{rs}. \tag{6.53}
$$

D_{rs} in Eq. (6.53) is called the density matrix,

$$
D_{rs} = 2\sum_i^{\text{occ}} C_{ri}C_{si}. \tag{6.54}
$$

In the above procedure, we note that the parameter of external perturbation V_a is included in Eq. (6.50), which affects on the MD coefficients (6.52), Fock matrix (6.53) and the total energy $\mathcal{E}$.

A.1.2 Derivative of Wavefunction

We have learned that the MO coefficients determined by Eq. (6.52) are affected by the external perturbation parameter. Therefore, the derivative of MO coefficients with respect to V_b is obtained by differentiating Eq. (6.52),

$$
\begin{aligned}
&\frac{\partial}{\partial V_b}\left[\sum_q F_{pq}C_{qi} - \varepsilon_i \sum_q s_{pq}C_{qi}\right] = 0 \\
&= \sum_q F_{pq}\frac{\partial C_{qi}}{\partial V_b} \\
&- e\sum_q \langle p|\hat{n}_b|q\rangle C_{qi} + \sum_q \sum_{r,s} \sum_j^{\text{occ}} \{2\,(pq|rs) - (pr|qs)\}\left(\frac{\partial C_{rj}}{\partial V_b}C_{sj} + C_{rj}\frac{\partial C_{sj}}{\partial V_b}\right) C_{qi} \\
&- \varepsilon_i \sum_q s_{pq}\frac{\partial C_{qi}}{\partial V_b}.
\end{aligned}
\tag{6.55}
$$

In the right hand side of Eq.(6.55), the differential of the Fock matrix $\partial F_{pq}/\partial V_b$ with Eq. (6.53) is included in the second and third terms.

Then the derivative of the MO coefficients, $\partial C/\partial V_b$, in Eq. (6.55) is expressed with U^b in Eq. (6.21). Consequently, the first, third and fourth terms of the right hand side of Eq. (6.55) become

- (first term) + (fourth term)

$$
\begin{aligned}
\sum_q F_{pq}\frac{\partial C_{qi}}{\partial V_b} - \varepsilon_i \sum_q s_{pq}\frac{\partial C_{qi}}{\partial V_b} &= \sum_j \left(\sum_q F_{pq}C_{qj} - \varepsilon_i \sum_q s_{pq}C_{qj}\right) U^b_{ji} \\
&= \sum_j (\varepsilon_j - \varepsilon_i) \sum_q s_{pq}C_{qj}U^b_{ji}
\end{aligned}
$$

- (third term)

$$
\sum_q \sum_{r,s} \sum_j^{\text{occ}} \{2\,(pq|rs) - (pr|qs)\}\left(\frac{\partial C_{rj}}{\partial V_b}C_{sj} + C_{rj}\frac{\partial C_{sj}}{\partial V_b}\right) C_{qi}
$$

$$= \sum_{q,r,s} \sum_{j}^{\text{occ}} \{4\,(pq|rs) - (pr|qs) - (ps|qr)\} \frac{\partial C_{rj}}{\partial V_b} C_{sj} C_{qi}$$

$$= \sum_{q,r,s} \sum_{j}^{\text{occ}} \sum_{k} \{4\,(pq|rs) - (pr|qs) - (ps|qr)\} C_{rk} U_{kj}^{b} C_{sj} C_{qi}$$

where $(pq|rs) = (\chi_p \chi_q | \chi_r \chi_s)$. We have employed Eq. (6.52) to derive the first+fourth terms, and the symmetry relation of the two-electron integral, $(pq|rs) = (pq|sr)$, to derive the third term. By summarizing the above new expressions, Eq. (6.55) becomes

$$\sum_{j} \left(\varepsilon_j - \varepsilon_i\right) \sum_{q} s_{pq} C_{qj} U_{ji}^{b} - e \sum_{q} \left\langle p|\hat{n}_b|q\right\rangle C_{qi}$$

$$+ \sum_{q,r,s} \sum_{j}^{\text{occ}} \sum_{k} \{4\,(pq|rs) - (pr|qs) - (ps|qr)\} C_{rk} U_{kj}^{b} C_{sj} C_{qi} = 0.$$

We further convert this equation into the MO representation by operating $\sum_p C_{pl}$ on both sides of the equation,

$$\sum_{j} \left(\varepsilon_j - \varepsilon_i\right) \sum_{p,q} C_{pl} s_{pq} C_{qj} U_{ji}^{b} - e \sum_{p,q} \left\langle p|\hat{n}_b|q\right\rangle C_{pl} C_{qi}$$

$$+ \sum_{p,q,r,s} \sum_{j}^{\text{occ}} \sum_{k} \{4\,(pq|rs) - (pr|qs) - (ps|qr)\} C_{pl} C_{rk} U_{kj}^{b} C_{sj} C_{qi}$$

$$= \sum_{j} \left(\varepsilon_j - \varepsilon_i\right) \delta_{lj} U_{ji}^{b} - e \left\langle \psi_l|\hat{n}_b|\psi_i\right\rangle + \sum_{j}^{\text{occ}} \sum_{k} H_{likj} U_{kj}^{b} = 0. \tag{6.56}$$

H_{likj} has been defined in Eq. (6.23),

$$H_{likj} = 4\left(\psi_l \psi_i | \psi_k \psi_j\right) - \left(\psi_l \psi_k | \psi_i \psi_j\right) - \left(\psi_l \psi_j | \psi_k \psi_i\right), \tag{6.23}$$

$$\left(\psi_i \psi_j | \psi_k \psi_l\right) = \iint \psi_i(\boldsymbol{r}) \psi_j(\boldsymbol{r}) \frac{e^2}{|\boldsymbol{r} - \boldsymbol{r}'|} \psi_k(\boldsymbol{r}') \psi_l(\boldsymbol{r}') d\boldsymbol{r} d\boldsymbol{r}'$$

$$= \sum_{p,q,r,s}^{\text{AO}} C_{pi} C_{qj} C_{rk} C_{sl} \left(\chi_p \chi_q | \chi_r \chi_s\right),$$

and

$$\left\langle \psi_l|\hat{n}_b|\psi_i\right\rangle = \sum_{p,q}^{\text{AO}} C_{pl} C_{qi} \langle \chi_p|\hat{n}_b|\chi_q\rangle. \tag{6.24}$$

The MO coefficients also satisfy the orthonormal condition of Eq. (6.51). Therefore, Eq. (6.51) is differentiated with V_b,

$$\frac{\partial}{\partial V_b}\sum_{p,q}^{\mathrm{AO}} C_{pi}C_{qj}s_{pq} = \sum_{p,q}\left(\frac{\partial C_{pi}}{\partial V_b}C_{qj}s_{pq} + C_{pi}\frac{\partial C_{qj}}{\partial V_b}s_{pq}\right)$$
$$= \sum_{p,q}^{\mathrm{AO}}\sum_{k}^{\mathrm{MO}}\left(C_{pk}U^b_{ki}C_{qj}s_{pq} + C_{pi}C_{qk}U^b_{kj}s_{pq}\right)$$
$$= \sum_{k}\left(\delta_{kj}U^b_{ki} + \delta_{ik}U^b_{kj}\right) = U^b_{ji} + U^b_{ij} = 0,$$

which indicates that U^b_{ij} is an anti-symmetric matrix ($U^b_{ji} = -U^b_{ij}$). This anti-symmetric character restricts the summation over k in Eq. (6.56) to only the virtual MOs, because the third term of Eq. (6.56) becomes

$$\sum_{j}^{\mathrm{occ}}\sum_{k} H_{likj}U^b_{kj} = \sum_{j}^{\mathrm{occ}}\sum_{k}^{\mathrm{occ}} H_{likj}U^b_{kj} + \sum_{j}^{\mathrm{occ}}\sum_{k}^{\mathrm{vir}} H_{likj}U^b_{kj}$$
$$= \frac{1}{2}\sum_{j}^{\mathrm{occ}}\sum_{k}^{\mathrm{occ}}\left(H_{likj}U^b_{kj} + H_{lijk}U^b_{jk}\right) + \sum_{j}^{\mathrm{occ}}\sum_{k}^{\mathrm{vir}} H_{likj}U^b_{kj} = \sum_{j}^{\mathrm{occ}}\sum_{k}^{\mathrm{vir}} H_{likj}U^b_{kj}$$

using the two following relations, $H_{likj} = H_{lijk}$ and $U^b_{kj} = -U^b_{jk}$. Therefore, Eq. (6.56) derives the following CPHF equation,

$$\left(\varepsilon_l - \varepsilon_i\right)U^b_{li} + \sum_{j}^{\mathrm{occ}}\sum_{k}^{\mathrm{vir}} H_{likj}U^b_{kj} = e\left\langle\psi_l|\hat{n}_b|\psi_i\right\rangle. \tag{6.22}$$

The above CPHF equation determines the coefficients U^b_{kj} between occupied MO j and virtual MO k. The coefficients between the occupied and virtual orbitals are sufficient to describe the deformation of the occupied MOs and the total wavefunction Ψ.

The present CPHF equation (6.22) for the CRK is analogous to those for other response quantities that are represented as second-order derivatives of energy, such as Hessian, polarizability, etc. The difference from other response quantities arises in the right hand side of Eq. (6.22), while the left hand side is common.

A.1.3 Formula of CRK

Using the solution of U_{li}^b thus obtained, the CRK K_{ab} is derived as follows,

$$
\begin{aligned}
K_{ab} = \frac{\partial Q_a}{\partial V_b} &= -e \sum_{p,q}^{\text{AO}} \frac{\partial D_{pq}}{\partial V_b} \langle p|\hat{n}_a|q\rangle \\
&= -2e \sum_{p,q}^{\text{AO}} \sum_{i}^{\text{occ}} \left(\frac{\partial C_{pi}}{\partial V_b} C_{qi} + C_{pi} \frac{\partial C_{qi}}{\partial V_b} \right) \langle p|\hat{n}_a|q\rangle \\
&= -2e \sum_{p,q}^{\text{AO}} \sum_{i}^{\text{occ}} \sum_{j} \left(C_{pj} U_{ji}^b C_{qi} + C_{pi} C_{qj} U_{ji}^b \right) \langle p|\hat{n}_a|q\rangle \\
&= -2e \left\{ \sum_{i}^{\text{occ}} \sum_{j}^{\text{occ}} + \sum_{i}^{\text{occ}} \sum_{j}^{\text{vir}} \right\} \left(\langle \psi_j|\hat{n}_a|\psi_i\rangle U_{ji}^b + \langle \psi_i|\hat{n}_a|\psi_j\rangle U_{ji}^b \right) \\
&= -2e \sum_{i}^{\text{occ}} \sum_{j}^{\text{vir}} \left(\langle \psi_j|\hat{n}_a|\psi_i\rangle U_{ji}^b + \langle \psi_i|\hat{n}_a|\psi_j\rangle U_{ji}^b \right) \\
&= -4e \sum_{i}^{\text{occ}} \sum_{j}^{\text{vir}} \langle \psi_j|\hat{n}_a|\psi_i\rangle U_{ji}^b .
\end{aligned}
\tag{6.25}
$$

A.2 Reorganization Energy of Electronic Polarization

For a condensed system of N molecules, we suppose that the intermolecular interactions are represented with the site charges Q_{ai} and the CRK K_{ab}. In the total potential energy U in Eq. (6.29), the second term

$$
U^{\text{reorg}} = -\frac{1}{2} \sum_{i} \sum_{a,b}^{\text{site}} K_{ab} V_{ai} V_{bi} \tag{6.57}
$$

is called the reorganization energy. U^{reorg} represents destabilization energy due to the electronic polarization, as explained below. This reorganization energy has to be positive for the stability of the polarization, and in fact is proven to be positive in Problem 6.2 because K_{ab} is a non-positive definite matrix. When the external perturbation is imposed on a molecule, the electronic polarization is induced so as to stabilize the interaction with the external potential, at the expense of the polarized (distorted) electronic state. The actual polarization is determined in the balance of the above two factors of stabilization and destabilization. The reorganization energy U^{reorg} corresponds to the latter, destabilizing factor associated to the electronic polarization.

A.2.1 Derivation of Reorganization Energy

In this Appendix we derive the above formula of U^{reorg}. The reorganization energy is defined as the reversible work to polarize the molecule i. In the following we treat the polarization of each molecule i separately, with no intermolecular interactions. The reversible work is regarded to consist of the following two quasi-static processes, (i) and (ii).

$$\boxed{\begin{matrix} V=0 \\ Q=Q^0 \end{matrix}} \xrightarrow{\text{(i)}} \boxed{\begin{matrix} V=V \\ Q=Q \end{matrix}} \xrightarrow{\text{(ii)}} \boxed{\begin{matrix} V=0 \\ Q=Q \end{matrix}}$$

(i) First, the external potential at the site a of molecule i is slowly changed from $V - 0$ to V_{ui}. During this process the site charge changes from Q_a^0 to $Q_{ai} - Q_a^0 + \sum_b K_{ab} V_{bi}$.
(ii) Then, the external potential is slowly changed from $V = V_{ai}$ to 0 while the partial charge Q_{ui} is fixed.

As a consequence, the external potential V is the same before and after the processes $(V = 0)$, while the site charge changes from Q_a^0 to Q_{ai}.

Considering that the charge Q and the electrostatic potential V are conjugate quantities in the thermodynamic sense, an infinitesimal change in the electrostatic potential dV_{ai} gives rise to a infinitesimal work (= increment of the internal energy dU), $dU = Q_{ai} \; dV_{ai}$. During the process (i), the intermediate potential is represented by ξV_{ai}, where ξ is a scaling factor ranging from $\xi = 0$ to 1. The overall change in the internal energy through the process (i), $U^{(\mathrm{i})}$, is thus obtained by integrating dU from $\xi = 0$ to 1,

$$U^{(\mathrm{i})} = \sum_i \sum_a \int_0^1 Q_{ai}(\{\xi V_{ai}\}) V_{ai} d\xi = \sum_i \sum_a \int_0^1 \left(Q_a^0 + \sum_b K_{ab} \xi V_{bi} \right) V_{ai} d\xi$$
$$= \sum_i \sum_a Q_a^0 V_{ai} + \frac{1}{2} \sum_i \sum_{a,b} K_{ab} V_{ai} V_{bi}.$$

We note that the partial charge Q_{ai} varies from Q_{ai}^0 to Q_{ai} along with the change in the intermediate potential ξV during the process (i). On the other hand, the change in the internal energy during the process (ii) is obtained by integrating dU from $\xi = 1$ to 0 in an analogous way,

$$U^{(\mathrm{ii})} = \sum_i \sum_a \int_1^0 Q_{ai} V_{ai} d\xi = - \sum_i \sum_a Q_{ai} V_{ai},$$

where the partial charge Q_{ai} is fixed during this process. Therefore, the reorganization energy U^{reorg} is represented with the sum of the above two contributions,

$$
\begin{aligned}
U^{\text{reorg}} &= U^{(\text{i})} + U^{(\text{ii})} \\
&= \sum_i \sum_a Q_a^0 V_{ai} + \frac{1}{2} \sum_i \sum_{a,b} K_{ab} V_{ai} V_{bi} - \sum_i \sum_a Q_{ai} V_{ai} \qquad (6.58) \\
&= -\frac{1}{2} \sum_i \sum_{a,b} K_{ab} V_{ai} V_{bi}. \qquad (6.57)
\end{aligned}
$$

where the last expression is derived using Eq. (6.28).

A.2.2 Variational Principle of Polarization

The above derivation of the reorganization energy in Eq. (6.58) allows us to define the whole electrostatic potential energy of the condensed system to be

$$
\begin{aligned}
\mathcal{U}(Q, V) &= \frac{1}{2} \sum_i \sum_{j(\neq i)} \sum_{a,b} \frac{Q_{ai} Q_{bj}}{|\boldsymbol{R}(ai) - \boldsymbol{R}(bj)|} \\
&\quad + \sum_i \sum_a Q_{ai}^0 V_{ai} + \frac{1}{2} \sum_i \sum_{a,b} K_{ab} V_{ai} V_{bi} - \sum_i \sum_a Q_{ai} V_{ai}. \qquad (6.59)
\end{aligned}
$$

We confirmed that $\mathcal{U}$ in Eq. (6.59) coincides with U in Eq. (6.29) with the help of Eq. (6.28).

Let us suppose that $\mathcal{U}$ in Eq. (6.59) is a function of $\{Q_{ai}\}$ and $\{V_{ai}\}$. Then the conditions to minimize $\mathcal{U}(Q, V)$ are given as follows.

$$
\frac{\partial \mathcal{U}(Q, V)}{\partial Q_{ai}} = \sum_{j(\neq i)} \sum_b \frac{Q_{bj}}{|\boldsymbol{R}(ai) - \boldsymbol{R}(bj)|} - V_{ai} = 0, \qquad (6.60)
$$

$$
\frac{\partial \mathcal{U}(Q, V)}{\partial V_{ai}} = Q_{ai}^0 + \sum_b K_{ab} V_{bi} - Q_{ai} = 0. \qquad (6.61)
$$

Interestingly, these Eqs. (6.60) and (6.61) are identical to the self-consistent conditions of Eqs. (6.27) and (6.28), respectively. This argument indicates that the self-consistent conditions of Q_{ai} and V_{ai} are regarded as variational principle to minimize the whole energy $\mathcal{U}(Q, V)$ in Eq. (6.59). The minimum energy of $\mathcal{U}(Q, V)$ coincides with the actual energy U in Eq. (6.29).

The variational principle clarifies the feature discussed in Problem 6.3 that the force $\boldsymbol{F}_{ai}$ in Eq. (6.30) does not include the derivatives $\partial Q/\partial \boldsymbol{R}$ or $\partial V/\partial \boldsymbol{R}$. This is because the self-consistent conditions (6.60) and (6.61) are satisfied at any time. The derivative of $\mathcal{U}(\boldsymbol{R}, Q, V)$ with respect to the coordinate $\boldsymbol{R}_{ai}$ is formally represented by

$$\boldsymbol{F}_{ai} = -\frac{\partial \mathcal{U}(\boldsymbol{R}, Q, V)}{\partial \boldsymbol{R}_{ai}} - \sum_j \sum_b \left[\frac{\partial \mathcal{U}(\boldsymbol{R}, Q, V)}{\partial Q_{bj}} \frac{\partial Q_{bj}}{\partial \boldsymbol{R}_{ai}} + \frac{\partial \mathcal{U}(\boldsymbol{R}, Q, V)}{\partial V_{bj}} \frac{\partial V_{bj}}{\partial \boldsymbol{R}_{ai}} \right]$$
$$= -\frac{\partial \mathcal{U}}{\partial \boldsymbol{R}_{ai}}.$$

We can readily see that the terms including $\partial Q/\partial \boldsymbol{R}$ or $\partial V/\partial \boldsymbol{R}$ vanish thanks to Eqs. (6.60) and (6.61).

We also note in passing that the above feature is commonly pertinent to the derivative of energies that satisfy the variational principle. In the electronic structure theories, some methods are based on the variational principle, such as Hartree-Fock and MCSCF. Therefore, the first-order derivative of the energies do not involve the derivative of the wavefunction for the same reason. As a consequence, the calculation of forces requires a little additional cost of computation using these methods. Yet the second-order derivative of energies, such as Hessian, polarizability and CRK, requires the derivative of the wavefunction. In such cases the CPHF equation described in Sect. 6.2 is invoked to calculate the derivative of the wavefunction or MO coefficients.

Bibliography

1. Bayly CI, Cieplak P, Cornell WD, Kollman PA (1993) A well-behaved electrostatic potential based method using charge restraints for deriving atomic charges: the RESP model. J Phys Chem 97:10269–10280
2. Cox SR, Williams DE (1981) Representation of the molecular electrostatic potential by a net atomic charge model. J Comp Chem 2:304–323
3. Ishida T, Morita A (2006) Extended treatment of charge response kernel comprising the density functional theory and charge regulation procedures. J Chem Phys 125:074112
4. Ishiyama T, Takahashi H, Morita A (2012) Vibrational spectrum at a water surface: a hybrid quantum mechanics/molecular mechanics molecular dynamics approach. J Phys Condens Matter 24:124107
5. McWeeny R (1992) Methods of molecular quantum mechanics, 2nd edn. Academic, London
6. Momany FA (1978) Determination of partial atomic charges from ab initio molecular electrostatic potentials. Application to formamide, methanol, and formic acid. J Phys Chem 82:592–601
7. Morita A, Kato S (1997) Ab initio molecular orbital theory on intramolecular charge polarization: effect of hydrogen abstraction on the charge sensitivity of aromatic and nonaromatic species. J Am Chem Soc 119:4021–4032
8. Morita A, Kato S (1998) Molecular dynamics simulation with the charge response kernel: diffusion dynamics of pyrazine and pyrazinyl radical in methanol. J Chem Phys 108:6809–6818

9. Morita A, Kato S (2002) The charge response kernel with modified electrostatic potential charge model. J Phys Chem A 106:3909–3916
10. Pulay P (1987) Analytical derivative methods in quantum chemistry. Adv Chem Phys 69:241–286
11. Rick SW, Stuart SJ, Berne BJ (1994) Dynamical fluctuating charge force fields: application to liquid water. J Chem Phys 101:6141–6156
12. Sulpizi M, Salanne M, Sprik M, Gaigeot M (2013) Vibrational sum frequency generation spectroscopy of the water liquid. Vapor interface from density functional theory-based molecular dynamics simulations. J Phys Chem Lett 4:83–87
13. Szabo A, Ostlund NS (1996) Modern quantum chemistry: introduction to advanced electronic structure theory. Dover, New York
14. Wan Q, Galli G (2015) First-principles framework to compute sum-frequency generation vibrational spectra of semiconductors and insulators. Phys Rev Lett 115:246404

Chapter 7
Quadrupole Contributions from Interface and Bulk

Abstract This chapter provides the comprehensive argument on quadrupole contributions in SFG spectroscopy. Though the preceding chapters have dealt with the induced dipole contribution, the accurate theory of SFG and SHG should take account of the induced quadrupole beside the dipole. The former can arise from the bulk region, while the latter stems only from the interface. We clarify three kinds of quadrupole contributions, namely χ^{IQ}, χ^{IQB} and χ^{B}, in addition to the dipole one χ^{ID}. The four terms have different roles in SFG and SHG spectra, as summarized in Sect. 7.5. The following discussion elucidates these quadrupole mechanisms and characters, and derives their microscopic formulas for calculating them.

Keywords Quadrupole · Bulk contribution · χ^{IQB} and χ^{B}

The present chapter provides comprehensive argument on the quadrupole contributions to the SFG spectroscopy. In the preceding chapters, we have assumed that the sum frequency signal originates from the second-order susceptibility $\chi^{(2)}(\Omega, \omega_1, \omega_2)$ at the interface. We have argued in Problem 1.1 that the $\chi^{(2)}$ vanishes in isotropic bulk media, and this property renders the SFG spectroscopy the surface selectivity.

We recall that this argument dealt with the induced electric dipole $\boldsymbol{P}^{(2)}$ in Eq. (1.4), which exclusively arises from a region where the inversion symmetry is broken. However, the sources of the sum frequency signal may include higher-order, electric quadrupole and magnetic dipole contributions in addition to the electric dipolar polarization of Eq. (1.4). These higher-order contributions are not necessarily forbidden in isotropic media for symmetry reason, and thus affect the surface selectivity of SFG spectroscopy. Even if the intrinsic quadrupole signal in the bulk is weaker than the dipole at the surface, the overwhelmingly larger

---

The original version of the chapter has been revised. Footnotes 4 and 6 has been inserted. Several equations on pages 158, 173, 174, 176, 178, 180, 181, 182, 183, 194, 195, 196, have been updated. A correction to this chapter can be found at https://doi.org/10.1007/978-981-13-1607-4_12

A. Morita, *Theory of Sum Frequency Generation Spectroscopy*,
Lecture Notes in Chemistry 97, https://doi.org/10.1007/978-981-13-1607-4_7

volume of the bulk could compensate for the weak quadrupole signal [10]. Rigorous quantitative theory of SFG should take account of these contributions.[1]

The quadrupole contributions were recognized at an early stage of studies on the second-order optical processes [1, 16, 20]. Bloembergen et al. proposed a theory of SHG involving the quadrupole contribution in centrosymmetric media [2, 3], and the theory of SHG has been sophisticated by subsequent studies[6–8, 18, 19]. Here we formulate the quadrupole contributions in SFG as well as in SHG. While the theory of SFG and SHG is mostly common, the SFG spectroscopy has some different sensitivity to the quadrupole contributions from SHG since the two incident fields can propagate with different wavevectors. Once the quadrupole effect on SFG is understood, SHG is regarded as a special case of SFG.

In the present chapter, some arguments given in preceding chapters have to be modified in order to incorporate the quadrupole contributions. First, the three-layer model of SFG in Chap. 2 is generalized to take account of the structure within the interface in Sect. 7.1. Then the second-order polarization is extended to include the quadrupole originating from both interface and bulk in Sect. 7.2. As a consequence, the second-order susceptibility effectively consists of four terms of different characters, namely χ^{ID}, χ^{IQ}, χ^{IQB}, and χ^{B}. Section 7.3 presents the microscopic formulas of those terms in both the energy representation and time-dependent representation, after the discussion of $\chi^{(2)}$ in Chap. 3. Section 7.4 argues the invariance of the quadrupole formulas with respect to the molecular origin, a fundamental requirement for the sound theory. Summary is then given in Sect. 7.5.

7.1 Beyond the Three-Layer Model

We extend the electrodynamic theory of SFG and generalize the three-layer model in Chap. 2 by taking account of depth- (z-) dependent dielectric properties of the interface. The phenomenological three-layer model is not capable of describing internal structure within the interfacial layer. In order to evaluate the quadrupolar polarizations at the surface, we need to account for the structure of electric field gradient over the surface region in a molecular scale.

Notations for Surface SFG Figure 7.1 summarizes the optical geometry and related notations for the SFG measurement. This geometry is identical to that in Fig. 2.1, though the interface region around $z \approx 0$ is not treated as a phenomenological third layer in Fig. 7.1. While the microscopic definition of the origin $z = 0$ is somewhat arbitrary due to finite thickness of the interface, this arbitrariness does not affect the following discussion toward treating the surface and bulk contributions from a unified view. This is because the thickness of the interface is generally much

[1]The general "quadrupole" term incorporates the electric quadrupole and the magnetic dipole contributions, as we argue later.

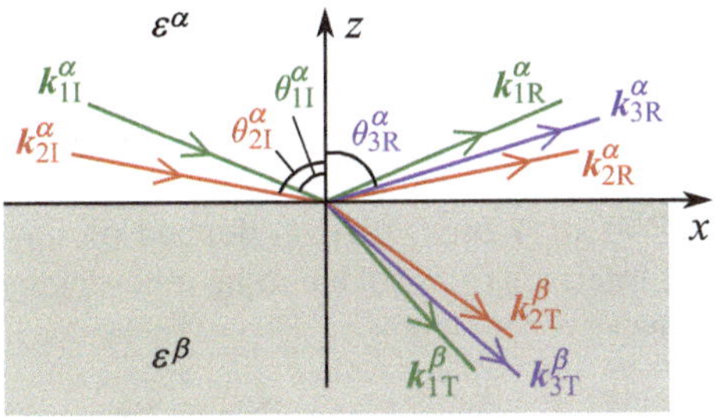

Fig. 7.1 Definition of parameters used for the surface SFG. The subscripts $f = 1, 2, 3$ denote the two incident lights and SFG, respectively. e.g. $\boldsymbol{k}^{\alpha}_{1I} = \boldsymbol{k}^{\alpha}_{I}(\omega_1)$, $\boldsymbol{k}^{\alpha}_{2R} = \boldsymbol{k}^{\alpha}_{R}(\omega_2)$, $\theta^{\alpha}_{3R} = \theta^{\alpha}_{R}(\Omega)$.

shorter than the wavelengths of the light fields. (This condition is usually satisfied in the surface nonlinear spectroscopy.)

Figure 7.1 shows all the wavevectors $\boldsymbol{k}_G$ with the subscripts $G = I, R$ and T, which distinguish the incident, reflected and transmitted fields, respectively. In the following discussion the medium α is assumed to be vacuum for simplicity, and hence $\varepsilon^{\alpha} = 1$, though extension to other situations is straightforward. The absolute values of the wavevectors are obtained by

$$k^i_G(\omega_f) = |\boldsymbol{k}^i_G(\omega_f)| = \frac{\sqrt{\varepsilon^i(\omega_f)}\omega_f}{c},$$

where c is the velocity of light in vacuo, and $i = \alpha$ or β. Due to the continuity of tangential wavevectors at the interface, the x component for each wavevector is invariant under different i and G, e.g. $k^{\alpha}_{I,x}(\omega_f) = k^{\alpha}_{R,x}(\omega_f) = k^{\beta}_{T,x}(\omega_f)$, and hence $k^i_{G,x}(\omega_f)$ is abbreviated to $k_x(\omega_f)$ irrespective of i and G. The tangential components satisfy the relation of momentum conservation,

$$k_x(\Omega) = k_x(\omega_1) + k_x(\omega_2). \tag{7.1}$$

Suppose a plane wave $\boldsymbol{E}^{\alpha}_{I}(\omega_f)$ ($f = 1$ or 2) in Eq. (2.11) is incident from the medium α (vacuum) to the interface. Then the *external* field and *local* field of frequency ω_f are represented near the interface ($z \approx 0$) by

$$\boldsymbol{E}^{\text{ext}}(\boldsymbol{r}, \omega_f, t) = \boldsymbol{E}^{\text{ext}}(\omega_f) \exp\left(ik_x(\omega_f)x - i\omega_f t\right) \tag{7.2}$$

$$\boldsymbol{E}^{\text{loc}}(\boldsymbol{r}, \omega_f, t) = \boldsymbol{E}^{\text{loc}}(z, \omega_f) \exp\left(ik_x(\omega_f)x - i\omega_f t\right), \quad (\text{near } z \approx 0, \quad f = 1 \text{ or } 2) \tag{7.3}$$

respectively. Note that the real fields are given by $\boldsymbol{E}^{\text{ext}}(\boldsymbol{r}, \omega_f, t) + c.c.$ and $\boldsymbol{E}^{\text{ext}}(\boldsymbol{r}, \omega_f, t) + c.c.$ We have discussed in Sect. 5.3 that $\boldsymbol{E}^{\text{ext}}(\omega_f)$ is related to the incident field $\boldsymbol{E}^{\alpha}_{I}(\omega_f)$ with the optical factor $\boldsymbol{L}_I(\omega_f)$ by

$$\boldsymbol{E}^{\text{ext}}(\omega_f) = \boldsymbol{L}_I(\omega_f)\boldsymbol{E}^{\alpha}_{I}(\omega_f), \tag{5.29}$$

and $\boldsymbol{E}^{\mathrm{loc}}(z, \omega_f)$ is represented with the local field factor $\boldsymbol{f}(z, \omega_f)$ by

$$\boldsymbol{E}^{\mathrm{loc}}(z, \omega_f) = \boldsymbol{f}(z, \omega_f)\boldsymbol{E}^{\mathrm{ext}}(\omega_f) = \boldsymbol{f}(z, \omega_f)\boldsymbol{L}_I(\omega_f)\boldsymbol{E}_I^{\alpha}(\omega_f). \tag{5.31}$$

$\boldsymbol{E}^{\mathrm{loc}}(z, \omega_f)$ and $\boldsymbol{f}(z, \omega_f)$ depend on the z coordinate in a microscopic scale near the interface ($z \approx 0$) due to its inhomogeneous environment. $\boldsymbol{f}(z, \omega_f)$ is a diagonal tensor,

$$\boldsymbol{f}(z, \omega_f) = \begin{pmatrix} f_x(z, \omega_f) & 0 & 0 \\ 0 & f_y(z, \omega_f) & 0 \\ 0 & 0 & f_z(z, \omega_f) \end{pmatrix}, \tag{7.4}$$

and $f_x(z, \omega_f) = f_y(z, \omega_f)$ for an azimuthally isotropic interface with $C_{\infty v}$ symmetry.

Z-dependence of nonlinear polarization Next we take account of the z-dependent distribution of interfacial nonlinear polarization. The distribution of the induced second-order polarization at the interface is

$$\boldsymbol{P}_0^{(2)}(\boldsymbol{r}, \Omega, t) = \boldsymbol{P}_0^{(2)}(z, \Omega)\ \exp(ik_x(\Omega)x - i\Omega t), \tag{7.5}$$

where

$$\boldsymbol{P}_0^{(2)}(z, \Omega) = \boldsymbol{\chi}^{\mathrm{D0}}(z, \Omega, \omega_1, \omega_2) : \boldsymbol{E}^{\mathrm{loc}}(z, \omega_1)\boldsymbol{E}^{\mathrm{loc}}(z, \omega_2)$$
$$\left(\text{or } P_{0,p}^{(2)}(z, \Omega) = \sum_{q,r}^{x\sim z} \chi_{pqr}^{\mathrm{D0}}(z, \Omega, \omega_1, \omega_2)E_q^{\mathrm{loc}}(z, \omega_1)E_r^{\mathrm{loc}}(z, \omega_2).\right) \tag{7.6}$$

$\boldsymbol{\chi}^{\mathrm{D0}}(z, \Omega, \omega_1, \omega_2)$ is represented with the molecular hyperpolarizabilities α_l^{D0} in the space-fixed coordinates,

$$\boldsymbol{\chi}^{\mathrm{D0}}(z, \Omega, \omega_1, \omega_2) = \overline{\sum_l^{\text{molecules}} \alpha_l^{\mathrm{D0}}(\Omega, \omega_1, \omega_2)\delta(z - z_l)}, \tag{7.7}$$

with z_l being the z coordinate of the l-th molecule and the over-bar denoting statistical average.[2] $\boldsymbol{\chi}^{\mathrm{D0}}(z, \Omega, \omega_1, \omega_2)$ in Eq. (7.7) is a bare assembly of molecular hyperpolarizabilities, and these molecular hyperpolarizabilities interact with the local fields in Eq. (7.6).

[2] In Eq. (7.6) each factor of $\boldsymbol{\chi}^{\mathrm{D0}}$ and $\boldsymbol{E}^{\mathrm{loc}}$ is statistically averaged before taking the product, whereas in fully microscopic theory their product should be statistically averaged. Therefore, Eq. (7.6) is regarded as an approximated treatment of the fully microscopic theory in Chap. 5, though it is convenient to formulate the z-dependence of polarization. Fully microscopic computation of quadrupolar susceptibilities does not involve this approximation unless they are decomposed along the z.

We have argued in Chap. 5 that the induced nonlinear polarization is also affected by the local field to modify itself. The modified polarization becomes [6–8]

$$\boldsymbol{P}^{(2)}(\boldsymbol{r}, \Omega, t) = \boldsymbol{P}^{(2)}(z, \Omega)\ \exp(ik_x(\Omega)x - i\Omega t), \tag{7.8}$$

where

$$\begin{aligned}\boldsymbol{P}^{(2)}(z, \Omega) &= \boldsymbol{f}(z, \Omega)\boldsymbol{P}_0^{(2)}(z, \Omega)\\ &= \boldsymbol{f}(z, \Omega)\Big[\boldsymbol{\chi}^{\mathrm{D0}}(z, \Omega, \omega_1, \omega_2) : \{\boldsymbol{f}(z, \omega_1)\boldsymbol{L}_I(\omega_1)\boldsymbol{E}_I^{\alpha}(\omega_1)\}\\ &\qquad \{\boldsymbol{f}(z, \omega_2)\boldsymbol{L}_I(\omega_2)\boldsymbol{E}_I^{\alpha}(\omega_2)\}\Big].\end{aligned} \tag{7.9}$$

The last expression is derived using Eqs. (7.6) and (5.31).

The nonlinear polarization $\boldsymbol{P}^{(2)}(z, \Omega)$ in Eq. (7.9) is induced within the interface region (at about $z \approx 0$) whose thickness is much smaller than the wavelength of the fields. In such case, it is represented with a delta function in Eq. (2.12), $\boldsymbol{P}^{(2)}(z, \Omega) \simeq \boldsymbol{P}^S(\Omega)\delta(z)$, in the macroscopic treatment in Chap. 2. Accordingly, $\boldsymbol{P}^S(\Omega)$ is derived by integrating $\boldsymbol{P}^{(2)}(z, \Omega)$ over the interface region,

$$\begin{aligned}\boldsymbol{P}^S(\Omega) &\equiv \int \boldsymbol{P}^{(2)}(z, \Omega)dz \end{aligned} \tag{7.10}$$

$$\begin{aligned}&= \int dz\ \boldsymbol{f}(z, \Omega)\Big[\boldsymbol{\chi}^{\mathrm{D0}}(z, \Omega, \omega_1, \omega_2) : \{\boldsymbol{f}(z, \omega_1)\boldsymbol{L}_I(\omega_1)\hat{\boldsymbol{e}}_I^{\alpha}(\omega_1)\}\\ &\qquad \{\boldsymbol{f}(z, \omega_2)\boldsymbol{L}_I(\omega_2)\hat{\boldsymbol{e}}_I^{\alpha}(\omega_2)\}\Big] \cdot E_I^{\alpha}(\omega_1)E_I^{\alpha}(\omega_2)\\ &= \boldsymbol{\chi}^{\mathrm{ID}}(\Omega, \omega_1, \omega_2) : \{\boldsymbol{L}_I(\omega_1)\hat{\boldsymbol{e}}_I^{\alpha}(\omega_1)\}\ \{\boldsymbol{L}_I(\omega_2)\hat{\boldsymbol{e}}_I^{\alpha}(\omega_2)\}\ E_I^{\alpha}(\omega_1)E_I^{\alpha}(\omega_2).\end{aligned} \tag{7.11}$$

In Eq. (7.11) the incident electric field $\boldsymbol{E}_I^{\alpha}(\omega_f)$ is expressed by $\boldsymbol{E}_I^{\alpha}(\omega_f) = \hat{\boldsymbol{e}}_I^{\alpha}(\omega_f)E_I^{\alpha}(\omega_f)$, where $E_I^{\alpha}(\omega_f)$ and $\hat{\boldsymbol{e}}_I^{\alpha}(\omega_f)$ are the absolute value and unit vector of $\boldsymbol{E}_I^{\alpha}(\omega_f)$, respectively. $\boldsymbol{\chi}^{\mathrm{ID}}$ is introduced as

$$\boldsymbol{\chi}^{\mathrm{ID}}(\Omega, \omega_1, \omega_2) = \int dz\ \boldsymbol{f}(z, \Omega)\left[\boldsymbol{\chi}^{\mathrm{D0}}(z, \Omega, \omega_1, \omega_2) : \boldsymbol{f}(z, \omega_1)\boldsymbol{f}(z, \omega_2)\right]. \tag{7.12}$$

This is an alternative definition of the nonlinear susceptibility that incorporates the local field corrections.

Effective nonlinear susceptibility Then we present the effective second-order susceptibility $\chi_{\mathrm{eff}}^{(2)}$ in Eq. (2.22) of Chap. 2 without resort to the three-layer model.

The effective susceptibility has been introduced in Chap. 2 by Eqs. (2.19) and (2.22) as

$$\boldsymbol{e}(\Omega) \cdot \boldsymbol{P}^S(\Omega) = \chi^{(2)}_{\text{eff}} E^\alpha_I(\omega_1) E^\alpha_I(\omega_2).$$

(Here we have assumed both the incident phases $i1$ and $i2$ to be α in Fig. 7.1.) As shown in Table 2.1, $\boldsymbol{e}(\Omega)$ is given as the result of the Fresnel transformation in the interface, $\boldsymbol{e}(\Omega) = \boldsymbol{F}^{i\to j}(\Omega)\hat{\boldsymbol{e}}^i(\Omega)$, and hence the left side is written by $\boldsymbol{e}(\Omega) \cdot \boldsymbol{P}^S(\Omega) = \hat{\boldsymbol{e}}^i(\Omega) \cdot \boldsymbol{F}^{i\to j}(\Omega)\boldsymbol{P}^S(\Omega)$. The Fresnel factor $\boldsymbol{F}^{i\to j}$ is a symmetric tensor and is given in Eq. (2.18), which involves the interfacial dielectric constant ε'.

In the present chapter, however, we obviate the use of phenomenological parameter ε', and incorporate the effect into the microscopic local field correction f. Accordingly, $\chi^{(2)}_{\text{eff},G}$ is defined here so as to satisfy the following condition,

$$\hat{\boldsymbol{e}}^i_G(\Omega) \cdot \boldsymbol{L}_G(\Omega)\boldsymbol{P}^S(\Omega) = \chi^{(2)}_{\text{eff},G} E^\alpha_I(\omega_1) E^\alpha_I(\omega_2), \tag{7.13}$$

where the subscript G ($= R$ or T) specifies reflected or transmitted SFG signal. In Eq. (7.13) the optical factor $\boldsymbol{L}_G(\Omega)$ is used instead of $\boldsymbol{F}^{i\to j}$, and the local field f at the frequency Ω is incorporated in the definition of $\boldsymbol{P}^S(\Omega)$ (see Eqs. (7.11) and (7.12)). $\chi^{(2)}_{\text{eff},G}$ thus introduced will be expanded to incorporate the quadrupole contributions in the next Sect. 7.2.

By inserting Eq. (7.11) into (7.13), we get the $\chi^{(2)}_{\text{eff},G}$ formula for the electric dipole contribution,

$$\begin{aligned} &\chi^{(2)}_{\text{eff},G}(\Omega, \omega_1, \omega_2) \\ &= \hat{\boldsymbol{e}}^i_G(\Omega) \cdot \boldsymbol{L}_G(\Omega)\left[\boldsymbol{\chi}^{\text{ID}}(\Omega, \omega_1, \omega_2) : \left\{\boldsymbol{L}_I(\omega_1)\hat{\boldsymbol{e}}^\alpha_I(\omega_1)\right\}\left\{\boldsymbol{L}_I(\omega_2)\hat{\boldsymbol{e}}^\alpha_I(\omega_2)\right\}\right]. \end{aligned} \tag{7.14}$$

For an interface of azimuthal $C_{\infty v}$ symmetry, the effective susceptibilities for possible polarization combinations are expressed as follows:

$$\chi^{(2)}_{\text{eff},G,\,\text{SSP}}(\Omega, \omega_1, \omega_2) = L_{G,y}(\Omega)\, L_{I,y}(\omega_1)\, L_{I,z}(\omega_2)\, \sin\theta^\alpha_I(\omega_2)\, \chi^{\text{ID}}_{yyz}(\Omega, \omega_1, \omega_2), \tag{7.15}$$

$$\chi^{(2)}_{\text{eff},G,\,\text{SPS}}(\Omega, \omega_1, \omega_2) = L_{G,y}(\Omega)\, L_{I,z}(\omega_1)\, L_{I,y}(\omega_2)\, \sin\theta^\alpha_I(\omega_1)\, \chi^{\text{ID}}_{yzy}(\Omega, \omega_1, \omega_2), \tag{7.16}$$

$$\chi^{(2)}_{\text{eff},G,\,\text{PSS}}(\Omega, \omega_1, \omega_2) = L_{G,z}(\Omega)\, L_{I,y}(\omega_1)\, L_{I,y}(\omega_2)\, \sin\theta^i_G(\Omega)\, \chi^{\text{ID}}_{zyy}(\Omega, \omega_1, \omega_2), \tag{7.17}$$

$$\begin{aligned} &\chi^{(2)}_{\text{eff},G,\,\text{PPP}}(\Omega, \omega_1, \omega_2) = \\ &- L_{G,x}(\Omega)\, L_{I,x}(\omega_1)\, L_{I,z}(\omega_2)\, \cos\theta^i_G(\Omega)\cos\theta^\alpha_I(\omega_1)\sin\theta^\alpha_I(\omega_2)\, \chi^{\text{ID}}_{xxz}(\Omega, \omega_1, \omega_2) \\ &- L_{G,x}(\Omega)\, L_{I,z}(\omega_1)\, L_{I,x}(\omega_2)\, \cos\theta^i_G(\Omega)\sin\theta^\alpha_I(\omega_1)\cos\theta^\alpha_I(\omega_2)\, \chi^{\text{ID}}_{xzx}(\Omega, \omega_1, \omega_2) \end{aligned}$$

$$+ L_{G,z}(\Omega)\ L_{I,x}(\omega_1)\ L_{I,x}(\omega_2)\ \sin\theta_G^i(\Omega)\cos\theta_I^\alpha(\omega_1)\cos\theta_I^\alpha(\omega_2)\ \chi_{zxx}^{\mathrm{ID}}(\Omega,\omega_1,\omega_2)$$

$$+ L_{G,z}(\Omega)\ L_{I,z}(\omega_1)\ L_{I,z}(\omega_2)\ \sin\theta_G^i(\Omega)\sin\theta_I^\alpha(\omega_1)\sin\theta_I^\alpha(\omega_2)\ \chi_{zzz}^{\mathrm{ID}}(\Omega,\omega_1,\omega_2), \tag{7.18}$$

where the diagonal elements of $\boldsymbol{L}_G(\omega_f)$ are specified with a single subscript of spatial coordinate; i.e. $\boldsymbol{L}_{G,p}(\omega_f) \equiv [\boldsymbol{L}_G(\omega_f)]_{pp}$ $(p = x, y, z)$. The present Eqs. (7.15), (7.16), (7.17), (7.18) correspond to Eqs. (3.49), (3.50), (3.51), (3.52) in Chap. 3 based on the three-layer model.

7.2 Extended Nonlinear Susceptibility

Next we extend the effective susceptibility $\chi_{\mathrm{eff},G}^{(2)}$ in Eq. (7.14) to include the quadrupole, and elucidate the mechanism of surface and bulk contributions beyond the dipole approximation.

7.2.1 Extended Source Polarization

The sum-frequency polarization $\boldsymbol{P}_0^{(2)}(\boldsymbol{r},\Omega,t)$ in Eq. (7.5) has been given by Eq. (7.6). The equivalent formula including the phase factors is

$$\boldsymbol{P}_0^{(2)}(\boldsymbol{r},\Omega,t) = \boldsymbol{\chi}^{\mathrm{D0}}(z,\Omega,\omega_1,\omega_2) : \boldsymbol{E}^{\mathrm{loc}}(\boldsymbol{r},\omega_1,t)\boldsymbol{E}^{\mathrm{loc}}(\boldsymbol{r},\omega_2,t)$$

$$\left(\text{or } P_{0,p}^{(2)}(\boldsymbol{r},\Omega,t) = \sum_{q,r}^{x\sim z} \chi_{pqr}^{\mathrm{D0}}(z,\Omega,\omega_1,\omega_2)E_q^{\mathrm{loc}}(\boldsymbol{r},\omega_1,t)E_r^{\mathrm{loc}}(\boldsymbol{r},\omega_2,t)\right). \tag{7.19}$$

Then, we expand this formula to incorporate the quadrupole contributions as follows,

$$\begin{aligned}
P_{0,p}^{(2)}(\boldsymbol{r},\Omega,t) &= \sum_{q,r}^{x\sim z} \chi_{pqr}^{\mathrm{D0}}(z,\Omega,\omega_1,\omega_2)E_q^{\mathrm{loc}}(\boldsymbol{r},\omega_1,t)E_r^{\mathrm{loc}}(\boldsymbol{r},\omega_2,t)\\
&+ \sum_{q,r,s}^{x-z}\Bigg[\chi_{pqrs}^{\mathrm{D1}}(z,\Omega,\omega_1,\omega_2)\frac{\partial E_q^{\mathrm{loc}}(\boldsymbol{r},\omega_1,t)}{\partial s}E_r^{\mathrm{loc}}(\boldsymbol{r},\omega_2,t)\\
&\quad + \chi_{pqrs}^{\mathrm{D2}}(z,\Omega,\omega_1,\omega_2)E_q^{\mathrm{loc}}(\boldsymbol{r},\omega_1,t)\frac{\partial E_r^{\mathrm{loc}}(\boldsymbol{r},\omega_2,t)}{\partial s}\\
&\quad - \frac{\partial}{\partial s}\left\{\chi_{pqrs}^{\mathrm{Q}}(z,\Omega,\omega_1,\omega_2)E_q^{\mathrm{loc}}(\boldsymbol{r},\omega_1,t)E_r^{\mathrm{loc}}(\boldsymbol{r},\omega_2,t)\right\}\Bigg].
\end{aligned} \tag{7.20}$$

The first line of Eq. (7.20) corresponds to Eq. (7.6), which is inherently surface sensitive. The additional terms involving $\chi^{\mathrm{D1}}_{pqrs}$, $\chi^{\mathrm{D2}}_{pqrs}$ and χ^{Q}_{pqrs} indicate the quadrupole contributions. The second and third lines describe the induced electric dipole in response to the electric field and field gradient. A quantity which couples with the field gradient is of quadrupolar character, as will be evident in Sect. 7.3. The fourth line includes the induced quadrupole,

$$Q^{(2)}_{0,ps}(\boldsymbol{r}, \Omega, t) \equiv \sum_{q,r} \chi^{\mathrm{Q}}_{pqrs}(z, \Omega, \omega_1, \omega_2) E^{\mathrm{loc}}_q(\boldsymbol{r}, \omega_1, t) E^{\mathrm{loc}}_r(\boldsymbol{r}, \omega_2, t), \qquad (7.21)$$

and the gradient of the quadrupole $-\frac{\partial}{\partial s} Q^{(2)}_{0,ps}$ gives rise to a part of the dipole polarization. Note that all the quadrupole terms in Eq. (7.20) are associated to spatial gradient $\partial/\partial s$.

We notice that the quadrupolar susceptibilities, $\boldsymbol{\chi}^{\mathrm{D1}}$, $\boldsymbol{\chi}^{\mathrm{D2}}$ and $\boldsymbol{\chi}^{\mathrm{Q}}$, are non-zero in the bulk region in contrast to the dipolar susceptibility $\boldsymbol{\chi}^{\mathrm{D0}}$ (see Problem 1.1). The contrasting selection rules stem from the fact that $\boldsymbol{\chi}^{\mathrm{D1}}$, $\boldsymbol{\chi}^{\mathrm{D2}}$, $\boldsymbol{\chi}^{\mathrm{Q}}$ are fourth-rank (even-rank) tensors while $\boldsymbol{\chi}^{\mathrm{D0}}$ is third-rank (odd-rank). Material properties of even-rank tensors do not necessarily vanish non-centrosymmetric media for symmetry reason. Therefore, the sum-frequency polarization $\boldsymbol{P}^{(2)}_0(\boldsymbol{r}, \Omega, t)$ in Eq. (7.20) is induced in the bulk region ($z < 0$) in addition to the interface ($z \approx 0$). We accordingly extend the definition of the local field $\boldsymbol{E}^{\mathrm{loc}}(\boldsymbol{r}, \omega_f, t)$ and the total sum-frequency polarization $\boldsymbol{P}^{(2)}(\boldsymbol{r}, \Omega, t)$ in Sect. 7.1 to treat the bulk region as follows.

Therefore, the formula of local field $\boldsymbol{E}^{\mathrm{loc}}(\boldsymbol{r}, \omega_f, t)$ in Eq. (7.3) is extended to be

$$\boldsymbol{E}^{\mathrm{loc}}(\boldsymbol{r}, \omega_f, t) = \boldsymbol{E}^{\mathrm{loc}}(z, \omega_f)\ \exp(i\boldsymbol{k}^{\beta}_T(\omega_f) \cdot \boldsymbol{r} - i\omega_f t), \qquad (7.22)$$

which can describe the local field in both the interface and bulk regions. The nonlinear polarization $\boldsymbol{P}^{(2)}(\boldsymbol{r}, \Omega, t)$ in Eq. (7.8) is analogously extended in the following form,

$$\begin{aligned}
&P^{(2)}_p(\boldsymbol{r}, \Omega, t)\\
&= \sum_{q,r} f_p(z, \Omega) \chi^{\mathrm{D0}}_{pqr}(z, \Omega, \omega_1, \omega_2)\ f_q(z, \omega_1) L_{I,q}(\omega_1) E^{\alpha}_{I,q}(\omega_1) \exp\left(i\boldsymbol{k}^{\beta}_T(\omega_1) \cdot \boldsymbol{r} - i\omega_1 t\right)\\
&\quad \cdot f_r(z, \omega_2) L_{I,r}(\omega_2) E^{\alpha}_{I,r}(\omega_2) \exp\left(i\boldsymbol{k}^{\beta}_T(\omega_2) \cdot \boldsymbol{r} - i\omega_2 t\right)\\
&+ \sum_{q,r,s} f_p(z, \Omega) \chi^{\mathrm{D1}}_{pqrs}(z, \Omega, \omega_1, \omega_2) \frac{\partial}{\partial s} \Big\{ f_q(z, \omega_1) L_{I,q}(\omega_1) E^{\alpha}_{I,q}(\omega_1)\\
&\exp\left(i\boldsymbol{k}^{\beta}_T(\omega_1) \cdot \boldsymbol{r} - i\omega_1 t\right) \Big\} \cdot f_r(z, \omega_2) L_{I,r}(\omega_2) E^{\alpha}_{I,r}(\omega_2) \exp\left(i\boldsymbol{k}^{\beta}_T(\omega_2) \cdot \boldsymbol{r} - i\omega_2 t\right)\\
&+ \sum_{q,r,s} f_p(z, \Omega) \chi^{\mathrm{D2}}_{pqrs}(z, \Omega, \omega_1, \omega_2)\ f_q(z, \omega_1) L_{I,q}(\omega_1) E^{\alpha}_{I,q}(\omega_1) \exp\left(i\boldsymbol{k}^{\beta}_T(\omega_1) \cdot \boldsymbol{r} - i\omega_1 t\right)\\
&\quad \cdot \frac{\partial}{\partial s} \left\{ f_r(z, \omega_2) L_{I,r}(\omega_2) E^{\alpha}_{I,r}(\omega_2) \exp\left(i\boldsymbol{k}^{\beta}_T(\omega_2) \cdot \boldsymbol{r} - i\omega_2 t\right) \right\}
\end{aligned}$$

$$
- \sum_{q,r,s} f_p(z,\Omega)\frac{\partial}{\partial s}\left\{\chi^{\mathrm{Q}}_{pqrs}(z,\Omega,\omega_1,\omega_2)\ f_q(z,\omega_1)L_{I,q}(\omega_1)E^{\alpha}_{I,q}(\omega_1)\right.
$$

$$
\left.\exp\left(i\boldsymbol{k}^{\beta}_{T}(\omega_1)\cdot\boldsymbol{r} - i\omega_1 t\right)\ f_r(z,\omega_2)L_{I,r}(\omega_2)E^{\alpha}_{I,r}(\omega_2)\exp\left(i\boldsymbol{k}^{\beta}_{T}(\omega_2)\cdot\boldsymbol{r} - i\omega_2 t\right)\right\}. \tag{7.23}
$$

This equation (7.23) treats the sum-frequency polarization induced both in the interface ($z \approx 0$) and bulk medium β ($z < 0$) along the common z coordinate. The arbitrariness in the microscopic definition of surface $z = 0$ does not affect the description of the bulk polarization, because the phase of bulk polarization as well as electric fields in the bulk is invariant within the range of arbitrariness (microscopic thickness of the surface region).

7.2.2 Effective Polarization and Susceptibility

Next we discuss the emitted SFG signal in relation to the nonlinear polarization $\boldsymbol{P}^{(2)}(\boldsymbol{r},\Omega,t)$ in Eq. (7.23). The direction G (= R or T) of the emitted SFG signal is determined by the boundary conditions at interface in Eq. (2.13) (see Sect. 2) and is displayed in Fig. 2.1 or 7.1, regardless of whether the quadrupole source polarization is involved. In fact, it is possible to extend the interfacial source polarization $\boldsymbol{P}^{S}(\Omega)$ in Eq. (2.12) so as to involve the quadrupole contributions. Accordingly, we can introduce effective interfacial polarization $\boldsymbol{P}_{\mathrm{eff},G}$ for $\boldsymbol{P}^{S}(\Omega)$ in Eq. (2.12),

$$
\boldsymbol{P}^{S}(\Omega)\exp\left(ik_x(\Omega)x - i\Omega t\right)\delta(z) \longrightarrow \boldsymbol{P}_{\mathrm{eff},G}\exp\left(ik_x(\Omega)x - i\Omega t\right)\delta(z),
$$

so that the radiated SFG field from $\boldsymbol{P}^{(2)}(\boldsymbol{r},\Omega,t)$ in Eq. (7.23) in the direction G coincides with that from $\boldsymbol{P}_{\mathrm{eff},G}$ at the interface (see Fig. 7.2). The suffix G of $\boldsymbol{P}_{\mathrm{eff},G}$ indicates that the effective interfacial polarization $\boldsymbol{P}_{\mathrm{eff},G}$ depends on the direction G. The effective second-order susceptibility $\chi^{(2)}_{\mathrm{eff},G}$ in Eq. (7.13) is analogously extended using $\boldsymbol{P}_{\mathrm{eff},G}$ instead of $\boldsymbol{P}^{S}$ as

$$
\hat{e}^{i}_{G}(\Omega)\cdot \boldsymbol{L}_G(\Omega)\boldsymbol{P}_{\mathrm{eff},G} = \chi^{(2)}_{\mathrm{eff},G}E^{\alpha}_{I}(\omega_1)E^{\alpha}_{I}(\omega_2). \tag{7.24}
$$

The effective polarization $\boldsymbol{P}_{\mathrm{eff},G}$ and susceptibility $\chi^{(2)}_{\mathrm{eff},G}$ thus defined above allow us to represent the SFG signals including both the interface and bulk contributions on the same footing, without apparently modifying the formulas of interfacial SFG.

In what follows, $\boldsymbol{P}_{\mathrm{eff},G}$ including both the dipole and quadrupole terms is derived from $\boldsymbol{P}^{(2)}(\boldsymbol{r},\Omega,t)$ in Eq. (7.23). $\boldsymbol{P}_{\mathrm{eff},G}$ is divided into the contributions from the interface, $\boldsymbol{P}^{\mathrm{I}}$, and the bulk, $\boldsymbol{P}^{\mathrm{B}}_{G}$,

$$
\boldsymbol{P}_{\mathrm{eff},G} = \boldsymbol{P}^{\mathrm{I}} + \boldsymbol{P}^{\mathrm{B}}_{G}. \tag{7.25}
$$

$\boldsymbol{P}^{\mathrm{I}}$ and $\boldsymbol{P}^{\mathrm{B}}_{G}$ are discussed in Sects. 7.2.3 and 7.2.4, respectively. Note that only the latter, bulk contribution $\boldsymbol{P}^{\mathrm{B}}_{G}$ depends on the direction of the emission G.

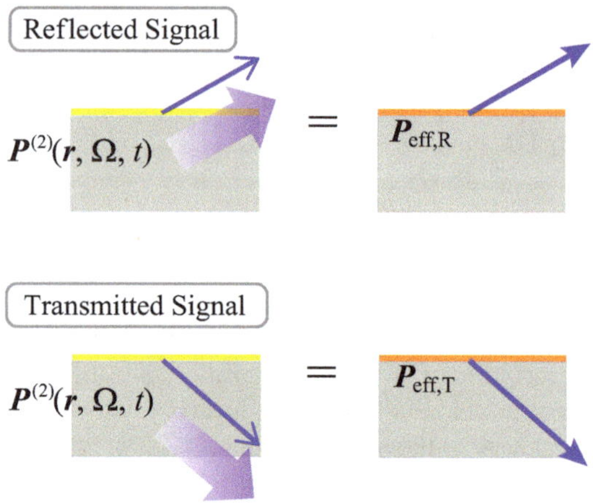

Fig. 7.2 Schematic definition of the effective interfacial polarization $\boldsymbol{P}_{\mathrm{eff},G}$. Left panels illustrate the emission of SFG signal from $\boldsymbol{P}^{(2)}(\boldsymbol{r}, \Omega, t)$ from both interface and bulk. Right panels illustrate the equivalent SFG emission from $\boldsymbol{P}_{\mathrm{eff},G}$ at the interface

To make the following argument clearer, we briefly comment on the qualitatively different mechanisms of the quadrupole contributions to $\boldsymbol{P}^{\mathrm{I}}$ and $\boldsymbol{P}_G^{\mathrm{B}}$. Equation (7.23) shows that all the quadrupolar terms involve the spatial derivative $\partial/\partial s$ ($s = x, y, z$), and the spatial derivative yields two kinds of terms, i.e. the derivative of the local field factor $\boldsymbol{f}(z, \omega_f)$ and of the phase factor $\exp\left(i\boldsymbol{k}_G^i(\omega_f)\cdot\boldsymbol{r} - i\omega_f t\right)$. These two kinds of derivative correspond to distinct mechanisms of SFG. The derivative of the local field factor $\partial\boldsymbol{f}/\partial s$ arises from spatial inhomogeneity of the interface, and thus it is considered as a part of interface polarization $\boldsymbol{P}^{\mathrm{I}}$. On the other hand, the derivative of the phase factor of light fields does not vanish in the bulk region even though the material is homogeneous. It is considered as the bulk polarization $\boldsymbol{P}_G^{\mathrm{B}}$.

7.2.3 Interface Contribution

The interface polarization $\boldsymbol{P}^{\mathrm{I}}$ is derived from Eq. (7.23) by operating the derivative $\partial/\partial s$ on the local field factor $\boldsymbol{f}$ and then by integrating the nonlinear polarization of $\boldsymbol{P}^{(2)}(z, \Omega)$ along z over the interfacial range. We note that the derivative $\partial\boldsymbol{f}/\partial s$ arises along the normal direction ($s = z$). Therefore,

$$
\begin{aligned}
&P_p^{\mathrm{I}} \exp\left(ik_x(\Omega)x - i\Omega x\right)\\
&= \int_{z_b}^{\infty} dz \Bigg[\sum_{q,r} f_p(z,\Omega)\chi_{pqr}^{\mathrm{D0}}(z,\Omega,\omega_1,\omega_2) f_q(z,\omega_1)\, f_r(z,\omega_2)\\
&\quad + \sum_{q,r} f_p(z,\Omega)\chi_{pqrz}^{\mathrm{D1}}(z,\Omega,\omega_1,\omega_2)\frac{\partial f_q(z,\omega_1)}{\partial z} f_r(z,\omega_2)
\end{aligned}
$$

$$
\begin{aligned}
&+\sum_{q,r} f_p(z,\Omega)\chi^{\mathrm{D2}}_{pqrz}(z,\Omega,\omega_1,\omega_2) f_q(z,\omega_1)\frac{\partial f_r(z,\omega_2)}{\partial z} \\
&-\sum_{q,r} f_p(z,\Omega)\frac{\partial}{\partial z}\left\{\chi^{\mathrm{Q}}_{pqrz}(z,\Omega,\omega_1,\omega_2)\ f_q(z,\omega_1)\ f_r(z,\omega_2)\right\}\Bigg] \\
&\cdot L_{I,q}(\omega_1)E^{\alpha}_{I,q}(\omega_1)\exp\left(ik_x(\omega_1)x - i\omega_1 t\right) \\
&\cdot L_{I,r}(\omega_2)E^{\alpha}_{I,r}(\omega_2)\exp\left(ik_x(\omega_2)x - i\omega_2 t\right). \qquad (7.26)
\end{aligned}
$$

In the above integral along the z coordinate, the lower bound z_b is chosen at an arbitrary position sufficiently deep in the medium β so that the integral covers the entire inhomogeneous region of the interface. Consequently, $\boldsymbol{\chi}^{\mathrm{D0}}$ and $\partial \boldsymbol{f}/\partial z$ vanish outside the integral range. In the integral range, the variation in phase factor along z is neglected, $\exp\left(ik^{\beta}_{T,z}(\omega_1)z + ik^{\beta}_{T,z}(\omega_2)z\right) \approx 1$, since the relevant thickness of the interface is much shorter than the light wavelengths. (The derivative with respect to phase factor will be treated separately in the next subsection.)

We further modify the term including $\boldsymbol{\chi}^{\mathrm{Q}}$ in Eq. (7.26) (fifth line) using integration by part as

[fifth line of Eq. (7.26)]:

$$
\begin{aligned}
&-\int_{z_b}^{\infty} dz \sum_{q,r} f_p(z,\Omega)\frac{\partial}{\partial z}\left\{\chi^{\mathrm{Q}}_{pqrz}(z,\Omega,\omega_1,\omega_2) f_q(z,\omega_1)\ f_r(z,\omega_2)\right\} \\
&= \int_{z_b}^{\infty} dz \sum_{q,r} \frac{\partial f_p(z,\Omega)}{\partial z}\left\{\chi^{\mathrm{Q}}_{pqrz}(z,\Omega,\omega_1,\omega_2) f_q(z,\omega_1)\ f_r(z,\omega_2)\right\} \\
&\quad -\sum_{q,r} f_p(z,\Omega)\left\{\chi^{\mathrm{Q}}_{pqrz}(z,\Omega,\omega_1,\omega_2) f_q(z,\omega_1)\ f_r(z,\omega_2)\right\}\Bigg|_{z=z_b}^{z=\infty} \\
&= \int_{z_b}^{\infty} dz \sum_{q,r} \frac{\partial f_p(z,\Omega)}{\partial z}\left\{\chi^{\mathrm{Q}}_{pqrz}(z,\Omega,\omega_1,\omega_2)\ f_q(z,\omega_1)\ f_r(z,\omega_2)\right\} \\
&\quad +\sum_{q,r} f^{\beta}_p(\Omega)\chi^{\mathrm{Q},\beta}_{pqrz}(\Omega,\omega_1,\omega_2) f^{\beta}_q(\omega_1) f^{\beta}_r(\omega_2). \qquad (7.27)
\end{aligned}
$$

In the last expression of the above derivation, the symbols with superscript β denote the quantities in the bulk medium β; for example, $\chi^{\mathrm{Q},\beta}_{pqrz}(\Omega,\omega_1,\omega_2) \equiv \chi^{\mathrm{Q}}_{pqrz}(z_b,\Omega,\omega_1,\omega_2)$ and $f^{\beta}_p(\Omega) \equiv f_p(z_b,\Omega)$, which emerge from the lower bound $z = z_b$. Note that material properties in the interior of bulk medium β have no z-dependence. The upper bound $z = \infty$ correspond to the gas phase, and thus the quantities for the upper bound vanish.

Consequently, $\boldsymbol{P}^{\mathrm{I}}$ in Eq. (7.26) is represented in the following form,

$$P_p^{\mathrm{I}} = \sum_{q,r} \left(\chi_{pqr}^{\mathrm{ID}}(\Omega, \omega_1, \omega_2) + \chi_{pqr}^{\mathrm{IQ}}(\Omega, \omega_1, \omega_2) + \chi_{pqr}^{\mathrm{IQB}}(\Omega, \omega_1, \omega_2) \right)$$
$$\cdot L_{I,q}(\omega_1) L_{I,r}(\omega_2) E_{I,q}^{\alpha}(\omega_1) E_{I,r}^{\alpha}(\omega_2), \tag{7.28}$$

where

$$\chi_{pqr}^{\mathrm{ID}}(\Omega, \omega_1, \omega_2) = \int_{z_b}^{\infty} dz\, \chi_{pqr}^{\mathrm{D0}}(z, \Omega, \omega_1, \omega_2) f_p(z, \Omega) f_q(z, \omega_1) f_r(z, \omega_2), \tag{7.12}$$

$$\chi_{pqr}^{\mathrm{IQ}}(\Omega, \omega_1, \omega_2) = \int_{z_b}^{\infty} dz \left\{ \chi_{pqrz}^{\mathrm{D1}}(z, \Omega, \omega_1, \omega_2) f_p(z, \Omega) \frac{\partial f_q(z, \omega_1)}{\partial z} f_r(z, \omega_2) \right.$$
$$+ \chi_{pqrz}^{\mathrm{D2}}(z, \Omega, \omega_1, \omega_2) f_p(z, \Omega) f_q(z, \omega_1) \frac{\partial f_r(z, \omega_2)}{\partial z}$$
$$\left. + \chi_{pqrz}^{\mathrm{Q}}(z, \Omega, \omega_1, \omega_2) \frac{\partial f_p(z, \Omega)}{\partial z} f_q(z, \omega_1) f_r(z, \omega_2) \right\}, \tag{7.29}$$

$$\chi_{pqr}^{\mathrm{IQB}}(\Omega, \omega_1, \omega_2) = \chi_{pqrz}^{\mathrm{Q},\beta}(\Omega, \omega_1, \omega_2) f_p^{\beta}(\Omega) f_q^{\beta}(\omega_1) f_r^{\beta}(\omega_2) \tag{7.30}$$

The dipole contribution of $\boldsymbol{\chi}^{\mathrm{ID}}$ has been already given in Eq. (7.12). $\boldsymbol{\chi}^{\mathrm{IQ}}$ in Eq. (7.29) indicates the quadrupole contribution associated to the gradient of the local fields at the interface. $\boldsymbol{\chi}^{\mathrm{IQB}}$ in Eq. (7.30) originates from the boundary of integration. We notice that $\chi_{pqr}^{\mathrm{IQB}}$ is entirely determined by bulk properties, and contains no information on the interface [8]. The physical meaning of the $\boldsymbol{\chi}^{\mathrm{IQB}}$ term is further discussed in Appendix A.1.

7.2.4 Bulk Contribution

The remaining portion of the nonlinear polarization in Eq. (7.23) is the terms including the derivative of the phase factors, $\exp\left(i\boldsymbol{k}_G^{\beta}(\omega_f)\cdot\boldsymbol{r} - i\omega_f t\right)$. This portion is denoted with $\boldsymbol{P}^{(2),\mathrm{B}}(\boldsymbol{r}, \Omega, t)$, and is represented including the phase factors by

$$P_p^{(2),\mathrm{B}}(\boldsymbol{r}, \Omega, t)$$
$$= \sum_{q,r,s} f_p^{\beta}(\Omega) \chi_{pqrs}^{\mathrm{D1},\beta}(\Omega, \omega_1, \omega_2)\ f_q^{\beta}(\omega_1) L_{I,q}(\omega_1) E_{I,q}^{\alpha}(\omega_1)$$

$$\left\{\frac{\partial}{\partial s}\exp\left(i\boldsymbol{k}_T^{\beta}(\omega_1)\cdot\boldsymbol{r}-i\omega_1 t\right)\right\} f_r^{\beta}(\omega_2)L_{I,r}(\omega_2)E_{I,r}^{\alpha}(\omega_2)\exp\left(i\boldsymbol{k}_T^{\beta}(\omega_2)\cdot\boldsymbol{r}-i\omega_2 t\right)$$
$$+\sum_{q,r,s} f_p^{\beta}(\Omega)\chi_{pqrs}^{\mathrm{D2},\beta}(\Omega,\omega_1,\omega_2)\; f_q^{\beta}(\omega_1)L_{I,q}(\omega_1)E_{I,q}^{\alpha}(\omega_1)\exp\left(i\boldsymbol{k}_T^{\beta}(\omega_1)\cdot\boldsymbol{r}-i\omega_1 t\right)$$
$$f_r^{\beta}(\omega_2)L_{I,r}(\omega_2)E_{I,r}^{\alpha}(\omega_2)\left\{\frac{\partial}{\partial s}\exp\left(i\boldsymbol{k}_T^{\beta}(\omega_2)\cdot\boldsymbol{r}-i\omega_2 t\right)\right\}$$
$$-\sum_{q,r,s} f_p^{\beta}(\Omega)\chi_{pqrs}^{\mathrm{Q},\beta}(\Omega,\omega_1,\omega_2) f_q^{\beta}(\omega_1)L_{I,q}(\omega_1)E_{I,q}^{\alpha}(\omega_1) f_r^{\beta}(\omega_2)L_{I,r}(\omega_2)E_{I,r}^{\alpha}(\omega_2)$$
$$\cdot\frac{\partial}{\partial s}\left\{\exp\left(i\boldsymbol{k}_T^{\beta}(\omega_1)\cdot\boldsymbol{r}-i\omega_1 t\right)\exp\left(i\boldsymbol{k}_T^{\beta}(\omega_2)\cdot\boldsymbol{r}-i\omega_2 t\right)\right\}. \tag{7.31}$$

$\boldsymbol{P}^{(2),\mathrm{B}}$ is expressed in a similar manner as in Eq. (7.26) by separating the phase factor along x,

$$\boldsymbol{P}^{(2),\mathrm{B}}(\boldsymbol{r},\Omega,t)=\boldsymbol{P}^{\mathrm{B}}(z)\exp\left(ik_x(\Omega)x-i\Omega t\right), \tag{7.32}$$

where $\boldsymbol{P}^{\mathrm{B}}(z)$ is a function of z in the bulk region. By performing the derivative of phase factor by

$$\frac{\partial}{\partial s}\exp\left(i\boldsymbol{k}_G^{\beta}(\omega_f)\cdot\boldsymbol{r}-i\omega_f t\right)=ik_{G,s}^{\beta}\exp\left(i\boldsymbol{k}_G^{\beta}(\omega_f)\cdot\boldsymbol{r}-i\omega_f t\right),$$

the bulk polarization of Eq. (7.31) or (7.32) is written as

$$P_p^{(2),\mathrm{B}}(\boldsymbol{r},\Omega,t)=P_p^{\mathrm{B}}(z)\exp\left(ik_x(\Omega)x-i\Omega t\right)$$
$$=i\sum_{q,r,s} f_p^{\beta}(\Omega)\left\{\chi_{pqrs}^{\mathrm{D1},\beta}(\Omega,\omega_1,\omega_2)k_{T,s}^{\beta}(\omega_1)+\chi_{pqrs}^{\mathrm{D2},\beta}(\Omega,\omega_1,\omega_2)k_{T,s}^{\beta}(\omega_2)\right.$$
$$\left.-\chi_{pqrs}^{\mathrm{Q},\beta}(\Omega,\omega_1,\omega_2)\left(k_{T,s}^{\beta}(\omega_1)+k_{T,s}^{\beta}(\omega_2)\right)\right\}$$
$$\cdot f_q^{\beta}(\omega_1)L_{I,q}(\omega_1)E_{I,q}^{\alpha}(\omega_1)\exp\left(i\boldsymbol{k}_T^{\beta}(\omega_1)\cdot\boldsymbol{r}-i\omega_1 t\right)$$
$$\cdot f_r^{\beta}(\omega_2)L_{I,r}(\omega_2)E_{I,r}^{\alpha}(\omega_2)\exp\left(i\boldsymbol{k}_T^{\beta}(\omega_2)\cdot\boldsymbol{r}-i\omega_2 t\right) \tag{7.33}$$

and thus

$$P_p^{\mathrm{B}}(z)=i\sum_{q,r,s}\left\{\chi_{pqrs}^{\mathrm{D1},\beta}(\Omega,\omega_1,\omega_2)k_{T,s}^{\beta}(\omega_1)+\chi_{pqrs}^{\mathrm{D2},\beta}(\Omega,\omega_1,\omega_2)k_{T,s}^{\beta}(\omega_2)\right.$$
$$\left.-\chi_{pqrs}^{\mathrm{Q},\beta}(\Omega,\omega_1,\omega_2)\left(k_{T,s}^{\beta}(\omega_1)+k_{T,s}^{\beta}(\omega_2)\right)\right\}$$
$$\cdot f_p^{\beta}(\Omega)f_q^{\beta}(\omega_1)f_r^{\beta}(\omega_2)L_{I,q}(\omega_1)L_{I,r}(\omega_2)E_{I,q}^{\alpha}(\omega_1)E_{I,r}^{\alpha}(\omega_2)$$
$$\cdot\exp\left[i\left(k_{T,z}^{\beta}(\omega_1)+k_{T,z}^{\beta}(\omega_2)\right)z\right]. \tag{7.34}$$

Then we derive $\boldsymbol{P}_G^{\mathrm{B}}$ in Eq. (7.25) from $\boldsymbol{P}^{\mathrm{B}}(z)$ in Eq. (7.34) so that $\boldsymbol{P}_G^{\mathrm{B}}$ and $\boldsymbol{P}^{\mathrm{B}}(z)$ should emit equivalent sum-frequency radiation to the direction G, as discussed in Sect. 7.2.2, though $\boldsymbol{P}_G^{\mathrm{B}}$ is supposed to be present at the interface $z = 0$ while $\boldsymbol{P}^{\mathrm{B}}(z)$ is distributed in the bulk β. Figure 7.3 illustrates the geometry for the SFG emission from the bulk polarization $\boldsymbol{P}^{\mathrm{B}}(z)$. Therefore, $\boldsymbol{P}_G^{\mathrm{B}}$ is expressed by integrating the bulk polarization $\boldsymbol{P}^{\mathrm{B}}(z)$ along z with a phase shift $\exp\left(-ik_{G,z}^{\beta}(\Omega)z\right)$ that takes account of the retardation of the radiation [10]. Consequently,

$$P_{G,p}^{\mathrm{B}} = \int_{-\infty}^{0} dz P_p^{\mathrm{B}}(z) \exp\left(-ik_{G,z}^{\beta}(\Omega)z\right) \tag{7.35}$$

$$\begin{aligned} &= i \int_{-\infty}^{0} dz \sum_{q,r,s} \left\{ \chi_{pqrs}^{\mathrm{D1},\beta}(\Omega,\omega_1,\omega_2) k_{T,s}^{\beta}(\omega_1) + \chi_{pqrs}^{\mathrm{D2},\beta}(\Omega,\omega_1,\omega_2) k_{T,s}^{\beta}(\omega_2) \right. \\ &\quad \left. - \chi_{pqrs}^{\mathrm{Q},\beta}(\Omega,\omega_1,\omega_2)(k_{T,s}^{\beta}(\omega_1) + k_{T,s}^{\beta}(\omega_2)) \right\} \\ &\quad \cdot f_p^{\beta}(\Omega) f_q^{\beta}(\omega_1) f_r^{\beta}(\omega_2) L_{I,q}(\omega_1) L_{I,r}(\omega_2) E_{I,q}^{\alpha}(\omega_1) E_{I,r}^{\alpha}(\omega_2) \\ &\quad \cdot \exp\left[i(k_{T,z}^{\beta}(\omega_1) + k_{T,z}^{\beta}(\omega_2) - k_{G,z}^{\beta}(\Omega))z \right], \end{aligned} \tag{7.36}$$

where the integral is taken in the bulk region of the medium β, $-\infty < z < 0$. The upper bound of the integral is set to $z = 0$ with virtually no ambiguity, since the interface region ($z \approx 0$) has negligibly small contribution to Eq. (7.36) in comparison to the bulk.[3] The integral along z in Eq. (7.36) is straightforward;

$$\begin{aligned} & i \int_{-\infty}^{0} dz \exp\left[i(k_{T,z}^{\beta}(\omega_1) + k_{T,z}^{\beta}(\omega_2) - k_{G,z}^{\beta}(\Omega))z \right] \\ &= \frac{1}{k_{T,z}^{\beta}(\omega_1) + k_{T,z}^{\beta}(\omega_2) - k_{G,z}^{\beta}(\Omega)} \equiv l_G, \end{aligned} \tag{7.37}$$

where l_G is the coherence length in the direction G ($= R$ or T). We note that the coherence length l_G is longer in the transmitted direction ($G = T$) than in the reflected direction ($G = R$) [10]. This is because $k_{R,z}^{\beta}(\Omega)$ has the opposite sign to $k_{T,z}^{\beta}(\omega_1) + k_{T,z}^{\beta}(\omega_2)$ while $k_{T,z}^{\beta}(\Omega)$ has the same sign to it (see Fig. 7.3).

Equation (7.36) includes the $\boldsymbol{\chi}^{\mathrm{D1}}$, $\boldsymbol{\chi}^{\mathrm{D2}}$, and $\boldsymbol{\chi}^{\mathrm{Q}}$ tensors in the bulk medium. In an isotropic material, such fourth-rank tensors χ_{pqrs} become greatly simplified. They have following three independent nonvanishing elements in general,

$$\chi_{pqpq} \equiv \chi_1, \quad \chi_{ppqq} \equiv \chi_2, \quad \chi_{pqqp} \equiv \chi_3, \qquad (p \neq q) \tag{7.38}$$

[3]We note in passing that Eq. (7.36) is different from the bulk polarization given in the previous literature [5, 11] (e.g. Eq. (1.2) of Ref. [11]). In Appendix A.3 we detail the difference and recommend the present definition.

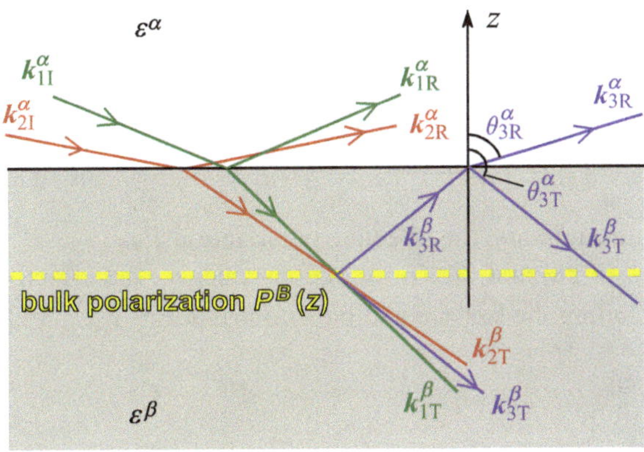

Fig. 7.3 Definitions of parameters for the SFG emission from polarization at z, $\boldsymbol{P}^{\mathrm{B}}(z)$, embedded in the medium β. The subscripts $f = 1, 2, 3$ denote the two incident lights and SFG, respectively, e.g. $\boldsymbol{k}_{1I}^\alpha = \boldsymbol{k}_I^\alpha(\omega_1)$, $\boldsymbol{k}_{2T}^\beta = \boldsymbol{k}_T^\beta(\omega_2)$, $\theta_{3R}^\alpha = \theta_R^\alpha(\Omega)$

and

$$\chi_{pppp} = \chi_1 + \chi_2 + \chi_3. \tag{7.39}$$

[Problem 7.1] Derive Eq. (7.39) from the three independent elements of Eq. (7.38) in the isotropic condition.

Therefore, in isotropic bulk media, Eq. (7.36) is transformed into

$$P_{G,p}^{\mathrm{B}} = \sum_{q,r} \chi_{G,pqr}^{\mathrm{B0}}(\Omega, \omega_1, \omega_2) L_{I,q}(\omega_1) L_{I,r}(\omega_2) E_{I,q}^\alpha(\omega_1) E_{I,r}^\alpha(\omega_2), \tag{7.40}$$

where

$$\chi_{G,pqr}^{\mathrm{B0}}(\Omega, \omega_1, \omega_2) = l_G \Big\{ \zeta_2^{\mathrm{Q1},\beta}(\Omega, \omega_1, \omega_2)\delta_{pq} k_{T,r}^\beta(\omega_1) + \zeta_1^{\mathrm{Q2},\beta}(\Omega, \omega_1, \omega_2)\delta_{pr} k_{T,q}^\beta(\omega_2)$$
$$+ \zeta_3^{\mathrm{Q1},\beta}(\Omega, \omega_1, \omega_2)\delta_{qr} k_{T,p}^\beta(\omega_1) + \zeta_3^{\mathrm{Q2},\beta}(\Omega, \omega_1, \omega_2)\delta_{qr} k_{T,p}^\beta(\omega_2) \Big\} f_p^\beta(\Omega) f_q^\beta(\omega_1) f_r^\beta(\omega_2) \tag{7.41}$$

and the ζ's are expressed using the notations of Eq. (7.38) by

$$\zeta_1^{\mathrm{Q2},\beta}(\Omega, \omega_1, \omega_2) \equiv \chi_1^{\mathrm{D2},\beta}(\Omega, \omega_1, \omega_2) - \chi_1^{\mathrm{Q},\beta}(\Omega, \omega_1, \omega_2), \tag{7.42}$$

$$\zeta_2^{\mathrm{Q1},\beta}(\Omega, \omega_1, \omega_2) \equiv \chi_2^{\mathrm{D1},\beta}(\Omega, \omega_1, \omega_2) - \chi_2^{\mathrm{Q},\beta}(\Omega, \omega_1, \omega_2), \tag{7.43}$$

$$\zeta_3^{\mathrm{Q1},\beta}(\Omega, \omega_1, \omega_2) \equiv \chi_3^{\mathrm{D1},\beta}(\Omega, \omega_1, \omega_2) - \chi_3^{\mathrm{Q},\beta}(\Omega, \omega_1, \omega_2), \tag{7.44}$$

$$\zeta_3^{\mathrm{Q2},\beta}(\Omega, \omega_1, \omega_2) \equiv \chi_3^{\mathrm{D2},\beta}(\Omega, \omega_1, \omega_2) - \chi_3^{\mathrm{Q},\beta}(\Omega, \omega_1, \omega_2). \tag{7.45}$$

We note that Eq. (7.41) is derived using the orthogonal condition between the electric field and the wave vector in the medium β,

$$\sum_p (f_p^\beta(\omega_f) L_{I,p}(\omega_f) E_{I,p}^\alpha(\omega_f)) \cdot k_{T,p}^\beta(\omega_f) = 0 \qquad (f = 1, 2). \tag{7.46}$$

In summary, the effective polarization $\boldsymbol{P}_{\text{eff},G} = \boldsymbol{P}^{\text{I}} + \boldsymbol{P}_G^{\text{B}}$ in Eq. (7.25) consists of Eqs. (7.28) and (7.40). As a consequence, the effective second-order susceptibility including the surface and bulk contribution, $\chi_{\text{eff},G}^{(2)}$ in Eq. (7.24), is expressed as

$$\chi_{\text{eff},G}^{(2)}(\Omega, \omega_1, \omega_2) =$$
$$\hat{\boldsymbol{e}}_G^i(\Omega) \boldsymbol{L}_G(\Omega) \left[\boldsymbol{\chi}_{q0G}^{(2)}(\Omega, \omega_1, \omega_2) : \left\{ \boldsymbol{L}_I(\omega_1) \hat{\boldsymbol{e}}_I^\alpha(\omega_1) \right\} \left\{ \boldsymbol{L}_I(\omega_2) \hat{\boldsymbol{e}}_I^\alpha(\omega_2) \right\} \right], \tag{7.47}$$

where

$$\chi_{q0G,pqr}^{(2)}(\Omega, \omega_1, \omega_2) =$$
$$\chi_{pqr}^{\text{ID}}(\Omega, \omega_1, \omega_2) + \chi_{pqr}^{\text{IQ}}(\Omega, \omega_1, \omega_2) + \chi_{pqr}^{\text{IQB}}(\Omega, \omega_1, \omega_2) + \chi_{G,pqr}^{B0}(\Omega, \omega_1, \omega_2). \tag{7.48}$$

Equation (7.47) is an extended form of Eq. (7.14) to incorporate the dipole and quadrupole contributions. If the quadrupole terms of $\boldsymbol{\chi}^{\text{IQ}}$, $\boldsymbol{\chi}^{\text{IQB}}$ and $\boldsymbol{\chi}_G^{\text{B0}}$ were neglected in Eq. (7.48), Eq. (7.47) would coincide with Eq. (7.14).

7.2.5 *Expression of Bulk Term* χ^{B}

For an interface of azimuthal $C_{\infty v}$ symmetry, the (achiral) SFG signal is detected only in the SSP, SPS, PSS, or PPP combination for symmetry reasons, even though the bulk contribution is taken into account. These polarization combinations are related to specific tensor elements of $\boldsymbol{\chi}^{\text{ID}}$ in Eqs. (7.15), (7.16), (7.17), (7.18), and their relations have been already discussed in Eqs. (3.49), (3.50), (3.51), (3.52) in Chap. 3. However, the bulk contribution in Eq. (7.48) would break the relation of Eqs. (7.15), (7.16), (7.17), (7.18) between the effective susceptibilities and nonvanishing tensor elements. Here we discuss this relation when the bulk contribution is taken into account.

In the original expressions of $\chi_{\text{eff,SSP}}^{(2)}$, $\chi_{\text{eff,SPS}}^{(2)}$, $\chi_{\text{eff,PSS}}^{(2)}$ and $\chi_{\text{eff,PPP}}^{(2)}$ in Eqs. (7.15), (7.16), (7.17), (7.18), the tensor elements of $\boldsymbol{\chi}^{\text{ID}}$ cannot be simply replaced with those of $\boldsymbol{\chi}_{q0G}^{(2)}$ in Eq. (7.48) to incorporate the quadrupole contributions, because several extra tensor elements of $\chi_{G,pqr}^{\text{B0}}$ in Eq. (7.41) are not necessarily zero. Equation (7.41) indicates that non-zero elements of $\chi_{G,pqr}^{\text{B0}}$ are (yyx), (yxy), (xyy),

(zzx), (zxz), (xzz), and (xxx) in addition to the conventional non-zero elements of (yyz), (yzy), (zyy), (xxz), (xzx), (zxx), and (zzz). As a consequence, the effective susceptibility for the SSP combination, for example, would become

$$\chi^{(2)}_{\text{eff},G,\text{SSP}}(\Omega,\omega_1,\omega_2) = \\ L_{G,y}(\Omega) L_{I,y}(\omega_1)\, L_{I,z}(\omega_2)\, \sin\theta_I^{\alpha}(\omega_2)\, \chi^{(2)}_{q0G,yyz}(\Omega,\omega_1,\omega_2) \\ + L_{G,y}(\Omega)\, L_{I,y}(\omega_1)\, L_{I,x}(\omega_2)\, \cos\theta_I^{\alpha}(\omega_2)\, \chi^{(2)}_{q0G,yyx}(\Omega,\omega_1,\omega_2), \tag{7.49}$$

where the second term originates from the extra element $\chi^{\text{B0}}_{G,yyx}$. One might wonder that the (yyx) element in Eq. (7.49) is incompatible with the $C_{\infty v}$ symmetry of the system. This apparent deviation from the symmetry requirement could be understood by noticing that χ^{B0}_G is not an intrinsic property of the medium but depends on the light geometry, since Eq. (7.41) includes the wave vectors of the applied fields.

In what follows, we will resolve this problem by changing the formula of the bulk contribution from χ^{B0}_G to χ^{B}_G so as to preserve the original expressions of Eqs. (7.15), (7.16), (7.17), (7.18). Accordingly, $\chi^{(2)}_{q0G}$ in Eq. (7.49) is replaced with $\chi^{(2)}_{qG}$, which consists of χ^{ID}, χ^{IQ}, χ^{IQB} and χ^{B},

$$\chi^{(2)}_{qG,pqr}(\Omega,\omega_1,\omega_2) = \\ \chi^{\text{ID}}_{pqr}(\Omega,\omega_1,\omega_2) + \chi^{\text{IQ}}_{pqr}(\Omega,\omega_1,\omega_2) + \chi^{\text{IQB}}_{pqr}(\Omega,\omega_1,\omega_2) + \chi^{\text{B}}_{G,pqr}(\Omega,\omega_1,\omega_2), \tag{7.50}$$

using the new definition of χ^{B}_G instead of χ^{B0}_G. $\chi^{(2)}_{qG}$ preserves the form of the effective susceptibility in Eq. (7.15) with obviating the extra (yyx) element in Eq. (7.49), i.e.

$$\chi^{(2)}_{\text{eff},G,\text{SSP}}(\Omega,\omega_1,\omega_2) = L_{G,y}(\Omega)\, L_{I,y}(\omega_1)\, L_{I,z}(\omega_2) \sin\theta_I^{\alpha}(\omega_2)\, \chi^{(2)}_{qG,yyz}(\Omega,\omega_1,\omega_2). \tag{7.51}$$

Such reformulation is possible by considering the relationship between the x and z elements of the Fresnel factors, beam angles, and local field correction factors.

The χ^{B0}_G elements included in Eq. (7.49) are given by Eq. (7.41),

$$\chi^{\text{B0}}_{G,yyz}(\Omega,\omega_1,\omega_2) = l_G\, \zeta_2^{\text{Q1},\beta}(\Omega,\omega_1,\omega_2) f_y^{\beta}(\Omega) f_y^{\beta}(\omega_1) f_z^{\beta}(\omega_2) k^{\beta}_{T,z}(\omega_1), \tag{7.52}$$

$$\chi^{\text{B0}}_{G,yyx}(\Omega,\omega_1,\omega_2) = l_G\, \zeta_2^{\text{Q1},\beta}(\Omega,\omega_1,\omega_2) f_y^{\beta}(\Omega) f_y^{\beta}(\omega_1) f_x^{\beta}(\omega_2) k_x(\omega_1). \tag{7.53}$$

In order that Eq. (7.49) is equivalent to (7.51), these two elements should be converted to

$$\chi^{\mathrm{B}}_{G,yyz}(\Omega,\omega_1,\omega_2)=l_G\zeta_2^{\mathrm{Q1},\beta}(\Omega,\omega_1,\omega_2)\frac{\left(\boldsymbol{k}_T^\beta(\omega_1)\times\boldsymbol{k}_T^\beta(\omega_2)\right)_y}{k_x(\omega_2)}f_y^\beta(\Omega)f_y^\beta(\omega_1)f_z^\beta(\omega_2). \tag{7.54}$$

[Problem 7.2] Derive $\chi^{\mathrm{B}}_{G,yyz}$ in Eq. (7.54) from $\boldsymbol{\chi}^{\mathrm{B0}}_G$ in Eqs. (7.52) and (7.53).

(*Hint*) During the derivation, use the following two relations,

$$\frac{\cos\theta_I^\alpha(\omega_2)}{\sin\theta_I^\alpha(\omega_2)}=\frac{q^\alpha(\omega_2)}{k_x(\omega_2)},\qquad \frac{f_x^\beta}{f_z^\beta}=\varepsilon^\beta,$$

and $L_I(\omega_f)$ in Eq. (5.30).

Analogous reformulation is possible for other polarization combinations. The relevant tensor elements of $\boldsymbol{\chi}^{\mathrm{B}}_G$ in Eq. (7.50) to the four polarization combinations are summarized as follows.

- SSP case:

$$\chi^{\mathrm{B}}_{G,yyz}(\Omega,\omega_1,\omega_2)=$$
$$l_G\zeta_2^{\mathrm{Q1},\beta}(\Omega,\omega_1,\omega_2)\frac{\left(\boldsymbol{k}_T^\beta(\omega_1)\times\boldsymbol{k}_T^\beta(\omega_2)\right)_y}{k_x(\omega_2)}f_y^\beta(\Omega)f_y^\beta(\omega_1)f_z^\beta(\omega_2), \tag{7.54}$$

- SPS case:

$$\chi^{\mathrm{B}}_{G,yzy}(\Omega,\omega_1,\omega_2)=$$
$$l_G\zeta_1^{\mathrm{Q2},\beta}(\Omega,\omega_1,\omega_2)\frac{\left(\boldsymbol{k}_T^\beta(\omega_2)\times\boldsymbol{k}_T^\beta(\omega_1)\right)_y}{k_x(\omega_1)}f_y^\beta(\Omega)f_z^\beta(\omega_1)f_y^\beta(\omega_2), \tag{7.55}$$

- PSS case:

$$\chi^{\mathrm{B}}_{R,zyy}(\Omega,\omega_1,\omega_2)=l_R\frac{1}{k_x(\Omega)}\left\{\zeta_3^{\mathrm{Q1},\beta}(\Omega,\omega_1,\omega_2)\left(\boldsymbol{k}_T^\beta(\omega_1)\times\boldsymbol{k}_R^\beta(\Omega)\right)_y\right.$$
$$\left.+\zeta_3^{\mathrm{Q2},\beta}(\Omega,\omega_1,\omega_2)\left(\boldsymbol{k}_T^\beta(\omega_2)\times\boldsymbol{k}_R^\beta(\Omega)\right)_y\right\}f_z^\beta(\Omega)f_y^\beta(\omega_1)f_y^\beta(\omega_2), \tag{7.56}$$

$$\chi^{\mathrm{B}}_{T,zyy}(\Omega,\omega_1,\omega_2)=$$

$$
l_T \frac{1}{\varepsilon^\alpha(\Omega)k_x(\Omega)} \left\{ \zeta_3^{Q1\beta} \left(\varepsilon^\alpha(\Omega)k^\beta_{T,z}(\omega_1)k_x(\Omega) + \varepsilon^\beta(\Omega)k_x(\omega_1)k^\alpha_{R,z}(\Omega) \right) \right.
$$

$$
\left. + \zeta_3^{Q2\beta} \left(\varepsilon^\alpha(\Omega)k^\beta_{T,z}(\omega_2)k_x(\Omega) + \varepsilon^\beta(\Omega)k_x(\omega_2)k^\alpha_{R,z}(\Omega) \right) \right\} f_z^\beta(\Omega) f_y^\beta(\omega_1) f_y^\beta(\omega_2), \tag{7.57}
$$

- PPP case:

$$
\chi^{\mathrm{B}}_{G,xxz} = \chi^{\mathrm{B}}_{G,yyz}, \quad \chi^{\mathrm{B}}_{G,xzx} = \chi^{\mathrm{B}}_{G,yzy}, \quad \chi^{\mathrm{B}}_{G,zxx} = \chi^{\mathrm{B}}_{G,zyy}, \tag{7.58}
$$

$$
\chi^{\mathrm{B}}_{R,zzz} = l_R \left\{ \zeta_2^{Q1\beta} \frac{\left(\boldsymbol{k}^\beta_T(\omega_1) \times \boldsymbol{k}^\beta_T(\omega_2) \right)_y}{k_x(\omega_2)} + \zeta_1^{Q2\beta} \frac{\left(\boldsymbol{k}^\beta_T(\omega_2) \times \boldsymbol{k}^\beta_T(\omega_1) \right)_y}{k_x(\omega_1)} \right.
$$

$$
\left. + \zeta_3^{Q1\beta} \frac{\left(\boldsymbol{k}^\beta_T(\omega_1) \times \boldsymbol{k}^\beta_R(\Omega) \right)_y}{k_x(\Omega)} + \zeta_3^{Q2\beta} \frac{\left(\boldsymbol{k}^\beta_T(\omega_2) \times \boldsymbol{k}^\beta_R(\Omega) \right)_y}{k_x(\Omega)} \right\} f_z^\beta(\Omega) f_z^\beta(\omega_1) f_z^\beta(\omega_2), \tag{7.59}
$$

$$
\chi^{\mathrm{B}}_{T,zzz} = l_T \left\{ \zeta_2^{Q1\beta} \frac{\left(\boldsymbol{k}^\beta_T(\omega_1) \times \boldsymbol{k}^\beta_T(\omega_2) \right)_y}{k_x(\omega_2)} + \zeta_1^{Q2\beta} \frac{\left(\boldsymbol{k}^\beta_T(\omega_2) \times \boldsymbol{k}^\beta_T(\omega_1) \right)_y}{k_x(\omega_1)} \right.
$$

$$
+ \zeta_3^{Q1\beta} \left(k^\beta_{T,z}(\omega_1) + \frac{\varepsilon^\beta(\Omega)}{\varepsilon^\alpha(\Omega)} \frac{k^\alpha_{R,z}(\Omega)k_x(\omega_1)}{k_x(\Omega)} \right)
$$

$$
\left. + \zeta_3^{Q2\beta} \left(k^\beta_{T,z}(\omega_2) + \frac{\varepsilon^\beta(\Omega)}{\varepsilon^\alpha(\Omega)} \frac{k^\alpha_{R,z}(\Omega)k_x(\omega_2)}{k_x(\Omega)} \right) \right\} f_z^\beta(\Omega) f_z^\beta(\omega_1) f_z^\beta(\omega_2), \tag{7.60}
$$

where $G = T$ or R, and we used $f_x(\omega_f) = f_y(\omega_f)$ to derive Eq. (7.58).

The difference in the definitions for the bulk contribution, χ^{B0}_G and χ^{B}_G, are attributed to different treatment of geometric factors related to the light wavevectors, $\boldsymbol{k}^i_G(\omega_f)$. The latter definition χ^B_G allows us to use the apparently same expressions for the effective susceptibilities in Eqs. (7.15), (7.16), (7.17), (7.18).

7.2.6 Summary of Derivation

To summarize the SSP, SPS, PSS, and PPP measurements of $C_{\infty v}$ systems, the relevant effective susceptibilities incorporating the dipole and quadrupole contributions are given in the following forms,

$$\chi^{(2)}_{\mathrm{eff},G,\mathrm{SSP}}(\Omega,\omega_1,\omega_2) = L_{G,y}(\Omega)\, L_{I,y}(\omega_1)\, L_{I,z}(\omega_2)\, \sin\theta_I^{\alpha}(\omega_2)\, \chi^{(2)}_{q\,G,yyz}(\Omega,\omega_1,\omega_2), \tag{7.61}$$

$$\chi^{(2)}_{\mathrm{eff},G,\mathrm{SPS}}(\Omega,\omega_1,\omega_2) = L_{G,y}(\Omega)\, L_{I,z}(\omega_1)\, L_{I,y}(\omega_2)\, \sin\theta_I^{\alpha}(\omega_1)\, \chi^{(2)}_{q\,G,yzy}(\Omega,\omega_1,\omega_2), \tag{7.62}$$

$$\chi^{(2)}_{\mathrm{eff},G,\mathrm{PSS}}(\Omega,\omega_1,\omega_2) = L_{G,z}(\Omega)\, L_{I,y}(\omega_1)\, L_{I,y}(\omega_2)\, \sin\theta_G^{i}(\Omega)\, \chi^{(2)}_{q\,G,zyy}(\Omega,\omega_1,\omega_2), \tag{7.63}$$

$$\begin{aligned}
&\chi^{(2)}_{\mathrm{eff},G,\mathrm{PPP}}(\Omega,\omega_1,\omega_2) = \\
&- L_{G,x}(\Omega)\, L_{I,x}(\omega_1)\, L_{I,z}(\omega_2)\, \cos\theta_G^{i}(\Omega)\cos\theta_I^{\alpha}(\omega_1)\sin\theta_I^{\alpha}(\omega_2)\, \chi^{(2)}_{q\,G,xxz}(\Omega,\omega_1,\omega_2) \\
&- L_{G,x}(\Omega)\, L_{I,z}(\omega_1)\, L_{I,x}(\omega_2)\, \cos\theta_G^{i}(\Omega)\sin\theta_I^{\alpha}(\omega_1)\cos\theta_I^{\alpha}(\omega_2)\, \chi^{(2)}_{q\,G,xzx}(\Omega,\omega_1,\omega_2) \\
&+ L_{G,z}(\Omega)\, L_{I,x}(\omega_1)\, L_{I,x}(\omega_2)\, \sin\theta_G^{i}(\Omega)\cos\theta_I^{\alpha}(\omega_1)\cos\theta_I^{\alpha}(\omega_2)\, \chi^{(2)}_{q\,G,zxx}(\Omega,\omega_1,\omega_2) \\
&+ L_{G,z}(\Omega)\, L_{I,z}(\omega_1)\, L_{I,z}(\omega_2)\, \sin\theta_G^{i}(\Omega)\sin\theta_I^{\alpha}(\omega_1)\sin\theta_I^{\alpha}(\omega_2)\, \chi^{(2)}_{q\,G,zzz}(\Omega,\omega_1,\omega_2).
\end{aligned} \tag{7.64}$$

Equations (7.61), (7.62), (7.63), (7.64) are straightforward extension of Eqs. (7.15), (7.16), (7.17), (7.18) by substituting $\boldsymbol{\chi}^{\mathrm{ID}}$ with $\boldsymbol{\chi}^{(2)}_{qG}$,

$$\begin{aligned}
&\chi^{(2)}_{q\,G,pqr}(\Omega,\omega_1,\omega_2) = \\
&\chi^{\mathrm{ID}}_{pqr}(\Omega,\omega_1,\omega_2) + \chi^{\mathrm{IQ}}_{pqr}(\Omega,\omega_1,\omega_2) + \chi^{\mathrm{IQB}}_{pqr}(\Omega,\omega_1,\omega_2) + \chi^{\mathrm{B}}_{G,pqr}(\Omega,\omega_1,\omega_2).
\end{aligned} \tag{7.50}$$

The four constituent terms of Eq. (7.50) are given by

$$\chi^{\mathrm{ID}}_{pqr}(\Omega,\omega_1,\omega_2) = \int_{z_b}^{\infty} dz\, \chi^{\mathrm{D0}}_{pqr}(z,\Omega,\omega_1,\omega_2) f_p(z,\Omega) f_q(z,\omega_1) f_r(z,\omega_2), \tag{7.12}$$

$$\begin{aligned}
\chi^{\mathrm{IQ}}_{pqr}(\Omega,\omega_1,\omega_2) = \int_{z_b}^{\infty} dz\, &\left\{ \chi^{\mathrm{D1}}_{pqrz}(z,\Omega,\omega_1,\omega_2) f_p(z,\Omega) \frac{\partial f_q(z,\omega_1)}{\partial z} f_r(z,\omega_2) \right. \\
&+ \chi^{\mathrm{D2}}_{pqrz}(z,\Omega,\omega_1,\omega_2) f_p(z,\Omega) f_q(z,\omega_1) \frac{\partial f_r(z,\omega_2)}{\partial z} \\
&\left. + \chi^{\mathrm{Q}}_{pqrz}(z,\Omega,\omega_1,\omega_2) \frac{\partial f_p(z,\Omega)}{\partial z} f_q(z,\omega_1) f_r(z,\omega_2) \right\},
\end{aligned} \tag{7.29}$$

$$\chi^{\mathrm{IQB}}_{pqr}(\Omega,\omega_1,\omega_2) = \chi^{\mathrm{Q},\beta}_{pqrz}(\Omega,\omega_1,\omega_2) f_p^{\beta}(\Omega) f_q^{\beta}(\omega_1) f_r^{\beta}(\omega_2), \tag{7.30}$$

and $\chi^{\mathrm{B}}_{G,}(\Omega,\omega_1,\omega_2)$ is given in Eqs. (7.54), (7.55), (7.56), (7.57), (7.58), (7.59), (7.60).

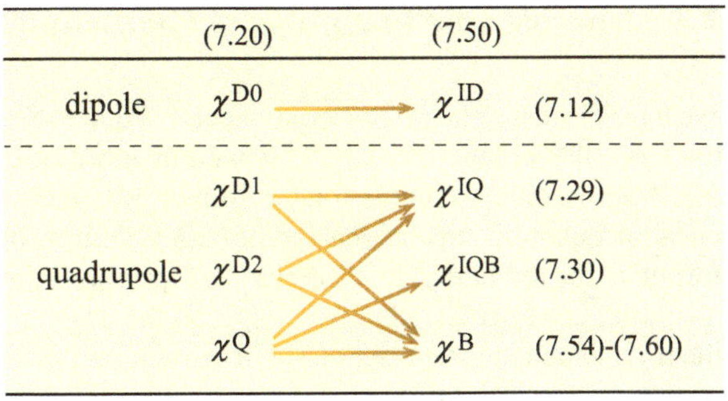

Fig. 7.4 Scheme of derivation for the dipole and quadrupole nonlinear susceptibility terms from ($\boldsymbol{\chi}^{D0}, \boldsymbol{\chi}^{D1}, \boldsymbol{\chi}^{D2}, \boldsymbol{\chi}^{Q}$) in Eq. (7.20) to ($\boldsymbol{\chi}^{ID}, \boldsymbol{\chi}^{IQ}, \boldsymbol{\chi}^{IQB}, \boldsymbol{\chi}^{B}$) in Eq. (7.50)

These four constituent terms include various tensor elements of $\boldsymbol{\chi}^{D0}$, $\boldsymbol{\chi}^{D1}$, $\boldsymbol{\chi}^{D2}$, and $\boldsymbol{\chi}^{Q}$. The nonvanishing and equivalent tensor elements of $\boldsymbol{\chi}^{D0}$, $\boldsymbol{\chi}^{D1}$, $\boldsymbol{\chi}^{D2}$, $\boldsymbol{\chi}^{Q}$ for a system of $C_{\infty v}$ symmetry are summarized as follows [4].

$$\chi^{D0}_{pqr}: \quad (xxz) = (yyz), \quad (xzx) = (yzy), \quad (zxx) = (zyy), \quad (zzz),$$

$$\chi^{D1}_{pqrs}, \ \chi^{D2}_{pqrs}, \ \chi^{Q}_{pqrs}:$$

$$(zzzz), \quad (yzyz) = (xzxz), \quad (yyzz) = (xxzz), \quad (zyyz) = (zxxz),$$

and the $\boldsymbol{\chi}^{Q}$ elements in isotropic bulk β are given on the basis of Eqs. (7.38) and (7.39) by

$$\begin{aligned}
\chi^{Q,\beta}_{pqrs}: (xzxz) = (yzyz) &= \chi_1^{Q,\beta}, \\
(xxzz) = (yyzz) &= \chi_2^{Q,\beta}, \\
(zxxz) = (zyyz) &= \chi_3^{Q,\beta}, \\
(zzzz) &= \chi_1^{Q,\beta} + \chi_2^{Q,\beta} + \chi_3^{Q,\beta}.
\end{aligned}$$

Finally, the extended effective susceptibility beyond the dipole approximation is schematically summarized in Fig. 7.4. The second-order nonlinear optical response to the electric field and its gradient is represented in Eq. (7.20), which includes $\boldsymbol{\chi}^{D0}$, $\boldsymbol{\chi}^{D1}$, $\boldsymbol{\chi}^{D2}$ and $\boldsymbol{\chi}^{Q}$. Those second-order response including both the interface and bulk contributions are represented with the extended susceptibility,

$$\boldsymbol{\chi}^{(2)}_{qG} = \boldsymbol{\chi}^{ID} + \boldsymbol{\chi}^{IQ} + \boldsymbol{\chi}^{IQB} + \boldsymbol{\chi}^{B}_{G}. \tag{7.50}$$

The first term $\boldsymbol{\chi}^{ID}$ originates from the response of electric dipole, while the last three terms $\boldsymbol{\chi}^{IQ} + \boldsymbol{\chi}^{IQB} + \boldsymbol{\chi}^{B}_{G}$ stem from the quadrupole.

The effective susceptibility $\chi^{(2)}_{\mathrm{eff}}$ in Eqs. (3.49), (3.50), (3.51), (3.52) in Chap. 2 is accordingly extended to Eqs. (7.61), (7.62), (7.63), (7.64) in the present chapter. By replacing the original $\boldsymbol{\chi}^{(2)}$ (=$\boldsymbol{\chi}^{ID}$) with $\boldsymbol{\chi}^{(2)}_{qG}$ in Eq. (7.50), the apparently same formula for the effective susceptibility is applied to describe the SFG signal.

7.3 Microscopic Formulas of Quadrupolar Susceptibilities

We have discussed in the preceding Sect. 7.2 that the second-order response includes the term of dipolar origin, χ^{D0}, as well as those of quadrupolar origin, χ^{D1}, χ^{D2}, χ^{Q}. Microscopic formulas for the former χ^{D0} have been given in Chap. 3 on the basis of quantum mechanical perturbation theory. In this section we provide the formulas for the latter, quadrupolar susceptibilities, χ^{D1}, χ^{D2}, and χ^{Q} [14]. These quantities are formulated in two ways, on the basis of the energy representation and the time-dependent representation, by extending the discussion in Chap. 3.

7.3.1 Perturbation Expressions

The quadrupolar susceptibilities, χ^{D1}, χ^{D2} and χ^{Q} are represented with molecular properties in the same way as the dipolar susceptibility in Eq. (7.7),

$$\chi^{D1}(z, \Omega, \omega_1, \omega_2) = \overline{\sum_l^{\text{molecules}} \boldsymbol{\alpha}_l^{D1}(\Omega, \omega_1, \omega_2)\delta(z - z_l)}, \quad (7.65)$$

$$\chi^{D2}(z, \Omega, \omega_1, \omega_2) = \overline{\sum_l^{\text{molecules}} \boldsymbol{\alpha}_l^{D2}(\Omega, \omega_1, \omega_2)\delta(z - z_i)}, \quad (7.66)$$

$$\chi^{Q}(z, \Omega, \omega_1, \omega_2) = \overline{\sum_l^{\text{molecules}} \boldsymbol{\alpha}_l^{Q}(\Omega, \omega_1, \omega_2)\delta(z - z_i)}, \quad (7.67)$$

where $\boldsymbol{\alpha}_l^{D1}(\Omega, \omega_1, \omega_2)$, $\boldsymbol{\alpha}_l^{D2}(\Omega, \omega_1, \omega_2)$, and $\boldsymbol{\alpha}_l^{Q}(\Omega, \omega_1, \omega_2)$ are the quadrupolar hyperpolarizabilities of the l-th molecule represented in the space-fixed coordinates. In the following of this section we focus on the properties of this molecule, and omit the subscript l from the notations.

Extension of perturbation Hamiltonian These properties are associated to the induced electric quadrupole and magnetic dipole in addition to the electric dipole. Therefore, we incorporate these responses in unified formulas. The sum frequency components of the induced dipole $\boldsymbol{\mu}(\Omega)$ and quadrupole $\boldsymbol{q}(\Omega)$ of a molecule are given by

$$\begin{aligned}\mu_p(\Omega) &= \sum_{q,r}^{x-z} \alpha_{pqr}^{D0}(\Omega, \omega_1, \omega_2) E_q(\omega_1) E_r(\omega_2) \\ &+ \sum_{q,r,s} \alpha_{pqrs}^{D1}(\Omega, \omega_1, \omega_2)(\nabla \boldsymbol{E}(\omega_1))_{sq} E_r(\omega_2) \\ &+ \sum_{q,r,s} \alpha_{pqrs}^{D2}(\Omega, \omega_1, \omega_2) E_q(\omega_1)(\nabla \boldsymbol{E}(\omega_2))_{sr}\end{aligned} \quad (7.68)$$

$$+\cdots$$

$$q_{ps}(\Omega) = \sum_{q,r} \alpha^{Q}_{pqrs}(\Omega, \omega_1, \omega_2) E_q(\omega_1) E_r(\omega_2) \tag{7.69}$$

$$+\cdots$$

where $E_p(\omega)$ is the electric field at frequency ω, and

$$(\nabla \boldsymbol{E}(\omega))_{pq} = \partial_p E_q(\omega) = \partial E_q(\omega)/\partial p$$

is the electric field gradient. $\mu_p(\Omega)$ is the dipole moment at the frequency Ω, and $q_{sp}(\Omega)$ is the generalized quadrupole moment that incorporates both the electric quadrupole and the magnetic dipole. It is defined as follows:

$$q_{pq}(\omega) = q^{E}_{pq}(\omega) + \frac{c}{i\omega} \sum_{r}^{x-z} \mu^{M}_{r}(\omega) \varepsilon_{pqr}, \tag{7.70}$$

where $q^{E}_{pq}(\omega)$ is the electric quadrupole moment defined in the Cartesian coordinates. It is given by

$$q^{E}_{pq}(\omega) = \frac{1}{2\pi} \int dt \exp(i\omega t) \left\{ \frac{1}{2} \int d\boldsymbol{r}\; pq\; \rho(\boldsymbol{r}, t) \right\}, \tag{7.71}$$

where $\rho(\boldsymbol{r}, t)$ is the charge density at the position $\boldsymbol{r}$ and time t, and p, q denote $x \sim z$ coordinates of $\boldsymbol{r}$. $\mu^{M}_{r}(\omega)$ is the magnetic dipole moment, and ε_{pqr} is the Levi-Civita permutation symbol (see Appendix A.2). The first term of the electric quadrupole in Eq. (7.70) is symmetric with respect to the exchange of p and q, while the second term of the magnetic dipole is antisymmetric. In the following discussion, the quadrupole moment includes both the electric quadrupole and magnetic dipole unless otherwise noted.[4]

[4]Note that the gradient of the generalized quadrupole moment $-\sum_{q} \partial_q q_{pq}(\omega)$ gives rise to the polarization, as we have mentioned in relation to Eq. (7.21). The current $\boldsymbol{J}$ is related to the multipole moments by

$$\begin{aligned} J_p &= \left[\frac{d}{dt} \left(\boldsymbol{\mu} - \nabla \cdot \boldsymbol{q}^{\mathrm{E}} \right) + c \nabla \times \boldsymbol{\mu}^{\mathrm{M}} + \cdots \right]_p \\ &= -i\omega \left(\mu_p - \sum_q \partial_q q^{\mathrm{E}}_{pq} \right) + c \sum_{q,r} \varepsilon_{pqr} \partial_q \mu^{\mathrm{M}}_r + \cdots \\ &= -i\omega \left(\mu_p - \sum_q \partial_q q_{pq}(\omega) \right) \end{aligned}$$

where we assumed the phase factor of $\exp(-i\omega t)$ (see Eq. (1.3)).

The quadrupolar hyperpolarizabilities $\boldsymbol{\alpha}^{\mathrm{D1}}$, $\boldsymbol{\alpha}^{\mathrm{D2}}$, $\boldsymbol{\alpha}^{\mathrm{Q}}$, are derived by the quantum mechanical perturbation theory with extending the perturbation Hamiltonian. The perturbation Hamiltonian for the light-matter interactions is

$$\hat{H}' = \hat{H}^{\mathrm{int}}(\omega_1) + \hat{H}^{\mathrm{int}}(\omega_2) \tag{7.72}$$

including two incident frequencies ω_f ($f = 1, 2$). Each component of the frequency ω_f involves the electric field $E_p(\omega_f)$, electric field gradient $(\nabla \boldsymbol{E}(\omega_f))_{pq}$ and magnetic field $B_r(\omega_f)$ as

$$\hat{H}^{\mathrm{int}}(\omega_f) = -\sum_p \hat{\mu}_p E_p(\omega_f) - \sum_{p,q} \hat{q}^E_{pq} (\nabla \boldsymbol{E}(\omega_f))_{pq} - \sum_r \hat{\mu}^M_r B_r(\omega_f). \tag{7.73}$$

The first term of Eq. (7.73) denotes the conventional dipole interaction, already treated in Eq. (3.18) of Chap. 3. The second and third terms of Eq. (7.73) stand for the interactions with the electric field gradient and the magnetic field, respectively. Those extra terms give rise to the extended second-order response beyond the dipole approximation. Equation (7.73) is expressed by lumping the second and third terms to be

$$\hat{H}^{\mathrm{int}}(\omega_f) = -\sum_p \hat{\mu}_p E_p(\omega_f) - \sum_{p,q} \hat{q}_{pq}(\omega_f) (\nabla \boldsymbol{E}(\omega_f))_{pq} \tag{7.74}$$

using the generalized quadrupole moment operator,

$$\hat{q}_{pq}(\omega) = \hat{q}^E_{pq} + \frac{c}{i\omega} \sum_r^{x-z} \hat{\mu}^M_r \varepsilon_{pqr}. \tag{7.75}$$

The equivalence of Eqs. (7.73) and (7.74) is shown by

$$\begin{aligned} -\sum_{p,q} \hat{q}_{pq}(\omega_f)(\nabla \boldsymbol{E}(\omega_f))_{pq} &= -\sum_{p,q} \left(\hat{q}^E_{pq} + \frac{c}{i\omega} \sum_r \hat{\mu}^M_r \varepsilon_{pqr} \right) \partial_p E_q(\omega_f) \\ &= -\sum_{p,q} \hat{q}^E_{pq} \partial_p E_q(\omega_f) - \sum_r \hat{\mu}^M_r B_r(\omega_f), \end{aligned}$$

where the following Maxwell equation is invoked,

$$(\nabla \times \boldsymbol{E}(\omega_f))_r = \sum_{p,q} \varepsilon_{rpq} \partial_p E_q(\omega_f) = -\frac{1}{c} \frac{\partial B_r(\omega_f)}{\partial t} = \frac{i\omega_f}{c} B_r(\omega_f).$$

Sum-over-state expressions Then we derive the quadrupolar hyperpolarizabilities by using the perturbation Hamiltonian in Eq. (7.74) and the second-order perturbation theory [14, 15]. This derivation is quite in parallel to that in Sect. 3.2. In the previous section, we have employed the perturbation Hamiltonian of electric dipole interaction, i.e. $\hat{H}^{\mathrm{int}}(\omega_f) = -\sum_p \hat{\mu}_p E_p(\omega_f)$, and derived the second-order hyperpolarizability $\boldsymbol{\alpha}^{\mathrm{D0}}$,

$$\alpha^{\mathrm{D0}}_{pqr}(\Omega, \omega_1, \omega_2) = \frac{1}{\hbar^2} \sum_{g,m,n}^{\mathrm{states}} (\rho_g^{(0)} - \rho_m^{(0)})$$
$$\left[\frac{\langle g|\mu_p|n\rangle\langle n|\mu_r|m\rangle\langle m|\mu_q|g\rangle}{(\omega_1 - \omega_{mg} + i_{mg})(\Omega - \omega_{ng} + i\Gamma_{ng})} - \frac{\langle g|\mu_r|n\rangle\langle n|\mu_p|m\rangle\langle m|\mu_q|g\rangle}{(\omega_1 - \omega_{mg} + i\Gamma_{mg})(\Omega - \omega_{mn} + i\Gamma_{mn})} \right.$$
$$\left. + \frac{\langle g|\mu_p|n\rangle\langle n|\mu_q|m\rangle\langle m|\mu_r|g\rangle}{(\omega_2 - \omega_{mg} + i\Gamma_{mg})(\Omega - \omega_{ng} + i\Gamma_{ng})} - \frac{\langle g|\mu_q|n\rangle\langle n|\mu_p|m\rangle\langle m|\mu_r|g\rangle}{(\omega_2 - \omega_{mg} + i\Gamma_{mg})(\Omega - \omega_{mn} + i\Gamma_{mn})} \right], \tag{7.76}$$

where g, m, n denote eigenstates of the molecule, $\rho_g^{(0)}$ is a diagonal element of the density matrix at thermal equilibrium, and $\Gamma_{mn} = \Gamma_{nm}$ is the damping rate. In this equation and hereafter, the symbol ^ for an operator is omitted in a matrix element. This Eq. (7.76) describes the second-order molecular hyperpolarizability, and is equivalent to Eq. (3.31) in Chap. 3.

Equation (7.76) can be decomposed into the vibrationally resonant ($\boldsymbol{\alpha}^{\mathrm{D0,res}}$) and non-resonant ($\boldsymbol{\alpha}^{\mathrm{D0,nonres}}$) terms, $\boldsymbol{\alpha}^{\mathrm{D0}} = \boldsymbol{\alpha}^{\mathrm{D0,res}} + \boldsymbol{\alpha}^{\mathrm{D0,nonres}}$, where

$$\alpha^{\mathrm{D0,res}}_{pqr}(\Omega, \omega_1, \omega_2) = \frac{1}{\hbar^2} \sum_{g,m,n}^{\mathrm{states}} (\rho_g^{(0)} - \rho_m^{(0)})$$
$$\left[\frac{\langle g|\mu_p|n\rangle\langle n|\mu_q|m\rangle\langle m|\mu_r|g\rangle}{(\omega_2 - \omega_{mg} + i\Gamma_{mg})(\Omega - \omega_{ng} + i\Gamma_{ng})} - \frac{\langle g|\mu_q|n\rangle\langle n|\mu_p|m\rangle\langle m|\mu_r|g\rangle}{(\omega_2 - \omega_{mg} + i\Gamma_{mg})(\Omega - \omega_{mn} + i\Gamma_{mn})} \right], \tag{7.77}$$

$$\alpha^{\mathrm{D0,nonres}}_{pqr}(\Omega, \omega_1, \omega_2) = \frac{1}{\hbar^2} \sum_{g,m,n}^{\mathrm{states}} (\rho_g^{(0)} - \rho_m^{(0)})$$
$$\left[\frac{\langle g|\mu_p|n\rangle\langle n|\mu_r|m\rangle\langle m|\mu_q|g\rangle}{(\omega_1 - \omega_{mg} + i\Gamma_{mg})(\Omega - \omega_{ng} + i\Gamma_{ng})} - \frac{\langle g|\mu_r|n\rangle\langle n|\mu_p|m\rangle\langle m|\mu_q|g\rangle}{(\omega_1 - \omega_{mg} + i\Gamma_{mg})(\Omega - \omega_{mn} + i\Gamma_{mn})} \right]. \tag{7.78}$$

The former includes $(\omega_2 - \omega_{mg} + i\Gamma_{mg})$ in the denominator, indicating possible resonance with the infrared frequency ω_2, whereas the latter does not.

Then we employ the extended perturbation Hamiltonian including electric quadrupole and magnetic dipole interactions in Eq. (7.74) and carry out the perturbation expansion up to the second order after Sect. 3.2. Since the derivation is in parallel with that in Sect. 3.2, we omit here to repeat the derivation procedure. The outcome of the second-order perturbation gives rise to additional hyperpolarizability terms, $\boldsymbol{\alpha}^{\mathrm{D1}}$, $\boldsymbol{\alpha}^{\mathrm{D2}}$ and $\boldsymbol{\alpha}^{\mathrm{Q}}$, due to the extra term $-\hat{\boldsymbol{q}} : \nabla \boldsymbol{E}$ of Eq. (7.74). The sum-over-state expressions of these terms are as follows.

$$\alpha^{\mathrm{D1}}_{pqrs}(\Omega, \omega_1, \omega_2) = \frac{1}{\hbar^2} \sum_{g,m,n}^{\text{states}} (\rho_g^{(0)} - \rho_m^{(0)})$$
$$\left[\frac{\langle g|\mu_p|n\rangle\langle n|\mu_r|m\rangle\langle m|q_{sq}(\omega_1)|g\rangle}{(\omega_1 - \omega_{mg} + i\Gamma_{mg})(\Omega - \omega_{ng} + i\Gamma_{ng})} - \frac{\langle g|\mu_r|n\rangle\langle n|\mu_p|m\rangle\langle m|q_{sq}(\omega_1)|g\rangle}{(\omega_1 - \omega_{mg} + i\Gamma_{mg})(\Omega - \omega_{mn} + i\Gamma_{mn})} \right.$$
$$\left. + \frac{\langle g|\mu_p|n\rangle\langle n|q_{sq}(\omega_1)|m\rangle\langle m|\mu_r|g\rangle}{(\omega_2 - \omega_{mg} + i\Gamma_{mg})(\Omega - \omega_{ng} + i\Gamma_{ng})} - \frac{\langle g|q_{sq}(\omega_1)|n\rangle\langle n|\mu_p|m\rangle\langle m|\mu_r|g\rangle}{(\omega_2 - \omega_{mg} + i\Gamma_{mg})(\Omega - \omega_{mn} + i\Gamma_{mn})} \right], \quad (7.79)$$

$$\alpha^{\mathrm{D2}}_{pqrs}(\Omega, \omega_1, \omega_2) = \frac{1}{\hbar^2} \sum_{g,m,n}^{\text{states}} (\rho_g^{(0)} - \rho_m^{(0)})$$
$$\left[\frac{\langle g|\mu_p|n\rangle\langle n|q_{sr}(\omega_2)|m\rangle\langle m|\mu_q|g\rangle}{(\omega_1 - \omega_{mg} + i\Gamma_{mg})(\Omega - \omega_{ng} + i\Gamma_{ng})} - \frac{\langle g|q_{sr}(\omega_2)|n\rangle\langle n|\mu_p|m\rangle\langle m|\mu_q|g\rangle}{(\omega_1 - \omega_{mg} + i\Gamma_{mg})(\Omega - \omega_{mn} + i\Gamma_{mn})} \right.$$
$$\left. + \frac{\langle g|\mu_p|n\rangle\langle n|\mu_q|m\rangle\langle m|q_{sr}(\omega_2)|g\rangle}{(\omega_2 - \omega_{mg} + i\Gamma_{mg})(\Omega - \omega_{ng} + i\Gamma_{ng})} - \frac{\langle g|\mu_q|n\rangle\langle n|\mu_p|m\rangle\langle m|q_{sr}(\omega_2)|g\rangle}{(\omega_2 - \omega_{mg} + i\Gamma_{mg})(\Omega - \omega_{mn} + i\Gamma_{mn})} \right], \quad (7.80)$$

$$\alpha^{\mathrm{Q}}_{pqrs}(\Omega, \omega_1, \omega_2) = \frac{1}{\hbar^2} \sum_{g,m,n}^{\text{states}} (\rho_g^{(0)} - \rho_m^{(0)})$$
$$\left[\frac{\langle g|q_{ps}(\Omega)|n\rangle\langle n|\mu_r|m\rangle\langle m|\mu_q|g\rangle}{(\omega_1 - \omega_{mg} + i\Gamma_{mg})(\Omega - \omega_{ng} + i\Gamma_{ng})} - \frac{\langle g|\mu_r|n\rangle\langle n|q_{ps}(\Omega)|m\rangle\langle m|\mu_q|g\rangle}{(\omega_1 - \omega_{mg} + i\Gamma_{mg})(\Omega - \omega_{mn} + i\Gamma_{mn})} \right.$$
$$\left. + \frac{\langle g|q_{ps}(\Omega)|n\rangle\langle n|\mu_q|m\rangle\langle m|\mu_r|g\rangle}{(\omega_2 - \omega_{mg} + i\Gamma_{mg})(\Omega - \omega_{ng} + i\Gamma_{ng})} - \frac{\langle g|\mu_q|n\rangle\langle n|q_{ps}(\Omega)|m\rangle\langle m|\mu_r|g\rangle}{(\omega_2 - \omega_{mg} + i\Gamma_{mg})(\Omega - \omega_{mn} + i\Gamma_{mn})} \right]. \quad (7.81)$$

We notice that Eqs. (7.76) and (7.79), (7.80), (7.81) have analogous forms except for the operators μ and q in the matrix elements.

In these equations, the first two terms in the square bracket which include $(\omega_1 - \omega_{mg} + i\Gamma_{mg})$ in the denominator are vibrationally non-resonant terms, while the latter two terms including $(\omega_2 - \omega_{mg} + i\Gamma_{mg})$ are vibrationally resonant terms. For example, $\boldsymbol{\alpha}^{\mathrm{D1}}$ in Eq. (7.79) is decomposed as $\boldsymbol{\alpha}^{\mathrm{D1}} = \boldsymbol{\alpha}^{\mathrm{D1,res}} + \boldsymbol{\alpha}^{\mathrm{D1,nonres}}$, where

$$\alpha^{\mathrm{D1,res}}_{pqrs}(\Omega, \omega_1, \omega_2) = \frac{1}{\hbar^2} \sum_{g,m,n}^{\text{states}} (\rho_g^{(0)} - \rho_m^{(0)})$$
$$\left[+ \frac{\langle g|\mu_p|n\rangle\langle n|q_{sq}(\omega_1)|m\rangle\langle m|\mu_r|g\rangle}{(\omega_2 - \omega_{mg} + i\Gamma_{mg})(\Omega - \omega_{ng} + i\Gamma_{ng})} - \frac{\langle g|q_{sq}(\omega_1)|n\rangle\langle n|\mu_p|m\rangle\langle m|\mu_r|g\rangle}{(\omega_2 - \omega_{mg} + i\Gamma_{mg})(\Omega - \omega_{mn} + i\Gamma_{mn})} \right], \quad (7.82)$$

$$\alpha^{\mathrm{D1,nonres}}_{pqrs}(\Omega,\omega_1,\omega_2) = \frac{1}{\hbar^2}\sum_{g,m,n}^{\text{states}}(\rho_g^{(0)}-\rho_m^{(0)})$$

$$\left[\frac{\langle g|\mu_p|n\rangle\langle n|\mu_r|m\rangle\langle m|q_{sq}(\omega_1)|g\rangle}{(\omega_1-\omega_{mg}+i\Gamma_{mg})(\Omega-\omega_{ng}+i\Gamma_{ng})} - \frac{\langle g|\mu_r|n\rangle\langle n|\mu_p|m\rangle\langle m|q_{sq}(\omega_1)|g\rangle}{(\omega_1-\omega_{mg}+i\Gamma_{mg})(\Omega-\omega_{mn}+i\Gamma_{mn})}\right]. \tag{7.83}$$

$\boldsymbol{\alpha}^{\mathrm{D2}}$ and $\boldsymbol{\alpha}^{\mathrm{Q}}$ in Eqs. (7.80) and (7.81) are decomposed into the vibrationally resonant and nonresonant terms in the analogous way.

Expressions for whole system Finally, we note that Eqs. (7.76), (7.77), (7.78), (7.79), (7.80), (7.81), (7.82), (7.83) include the damping parameters Γ, since these formulas are derived with the phenomenological Liouville equation including Γ (see Chap. 3). Accordingly, the states g, m, n in the above equations could refer to a partial system embedded in bath. Equivalent expressions are possible in principle when we define these states g, m, n for the whole system (partial system + bath). Then the corresponding equations do not include the damping parameters;

$$\alpha^{\mathrm{D0}}_{pqr}(\Omega,\omega_1,\omega_2) = \frac{1}{\hbar^2}\sum_{g,m,n}^{\substack{\text{whole}\\ \text{states}}}(\rho_g^{(0)}-\rho_m^{(0)})$$

$$\left[\frac{\langle g|\mu_p|n\rangle\langle n|\mu_r|m\rangle\langle m|\mu_q|g\rangle}{(\omega_1-\omega_{mg})(\Omega-\omega_{ng})} - \frac{\langle g|\mu_r|n\rangle\langle n|\mu_p|m\rangle\langle m|\mu_q|g\rangle}{(\omega_1-\omega_{mg})(\Omega-\omega_{mn})}\right.$$
$$\left.+\frac{\langle g|\mu_p|n\rangle\langle n|\mu_q|m\rangle\langle m|\mu_r|g\rangle}{(\omega_2-\omega_{mg})(\Omega-\omega_{ng})} - \frac{\langle g|\mu_q|n\rangle\langle n|\mu_p|m\rangle\langle m|\mu_r|g\rangle}{(\omega_2-\omega_{mg})(\Omega-\omega_{mn})}\right], \tag{7.84}$$

$$\alpha^{\mathrm{D1}}_{pqrs}(\Omega,\omega_1,\omega_2) = \frac{1}{\hbar^2}\sum_{g,m,n}^{\substack{\text{whole}\\ \text{states}}}(\rho_g^{(0)}-\rho_m^{(0)})$$

$$\left[\frac{\langle g|\mu_p|n\rangle\langle n|\mu_r|m\rangle\langle m|q_{sq}(\omega_1)|g\rangle}{(\omega_1-\omega_{mg})(\Omega-\omega_{ng})} - \frac{\langle g|\mu_r|n\rangle\langle n|\mu_p|m\rangle\langle m|q_{sq}(\omega_1)|g\rangle}{(\omega_1-\omega_{mg})(\Omega-\omega_{mn})}\right.$$
$$\left.+\frac{\langle g|\mu_p|n\rangle\langle n|q_{sq}(\omega_1)|m\rangle\langle m|\mu_r|g\rangle}{(\omega_2-\omega_{mg})(\Omega-\omega_{ng})} - \frac{\langle g|q_{sq}(\omega_1)|n\rangle\langle n|\mu_p|m\rangle\langle m|\mu_r|g\rangle}{(\omega_2-\omega_{mg})(\Omega-\omega_{mn})}\right], \tag{7.85}$$

$$\alpha^{\mathrm{D2}}_{pqrs}(\Omega,\omega_1,\omega_2) = \frac{1}{\hbar^2}\sum_{g,m,n}^{\substack{\text{whole}\\ \text{states}}}(\rho_g^{(0)}-\rho_m^{(0)})$$

$$\left[\frac{\langle g|\mu_p|n\rangle\langle n|q_{sr}(\omega_2)|m\rangle\langle m|\mu_q|g\rangle}{(\omega_1-\omega_{mg})(\Omega-\omega_{ng})} - \frac{\langle g|q_{sr}(\omega_2)|n\rangle\langle n|\mu_p|m\rangle\langle m|\mu_q|g\rangle}{(\omega_1-\omega_{mg})(\Omega-\omega_{mn})}\right.$$
$$\left.+\frac{\langle g|\mu_p|n\rangle\langle n|\mu_q|m\rangle\langle m|q_{sr}(\omega_2)|g\rangle}{(\omega_2-\omega_{mg})(\Omega-\omega_{ng})} - \frac{\langle g|\mu_q|n\rangle\langle n|\mu_p|m\rangle\langle m|q_{sr}(\omega_2)|g\rangle}{(\omega_2-\omega_{mg})(\Omega-\omega_{mn})}\right], \tag{7.86}$$

$$\alpha^{Q}_{pqrs}(\Omega,\omega_1,\omega_2) = \frac{1}{\hbar^2}\sum_{g,m,n}^{\text{whole states}} (\rho_g^{(0)} - \rho_m^{(0)})$$

$$\left[\frac{\langle g|q_{ps}(\Omega)|n\rangle\langle n|\mu_r|m\rangle\langle m|\mu_q|g\rangle}{(\omega_1-\omega_{mg})(\Omega-\omega_{ng})} - \frac{\langle g|\mu_r|n\rangle\langle n|q_{ps}(\Omega)|m\rangle\langle m|\mu_q|g\rangle}{(\omega_1-\omega_{mg})(\Omega-\omega_{mn})}\right.$$

$$\left.+\frac{\langle g|q_{ps}(\Omega)|n\rangle\langle n|\mu_q|m\rangle\langle m|\mu_r|g\rangle}{(\omega_2-\omega_{mg})(\Omega-\omega_{ng})} - \frac{\langle g|\mu_q|n\rangle\langle n|q_{ps}(\Omega)|m\rangle\langle m|\mu_r|g\rangle}{(\omega_2-\omega_{mg})(\Omega-\omega_{mn})}\right], \tag{7.87}$$

The above alternative expressions are based on the sum over eigenstates for the whole system, and will be utilized in Sect. 7.4.

7.3.2 *Time-Dependent Expressions*

Equations (7.79), (7.80), (7.81) for $\boldsymbol{\alpha}^{D1}$, $\boldsymbol{\alpha}^{D2}$ and $\boldsymbol{\alpha}^{Q}$ can be converted to the equivalent formulas based on the time correlation functions. The following derivation is analogous to that in Sect. 4.3, where $\boldsymbol{\chi}^{(2)}$ (or $\boldsymbol{\alpha}^{D0}$) is given with the time correlation function of the polarizability $\boldsymbol{\alpha}$ and the dipole moment $\boldsymbol{\mu}$.

Before extending the time correlation formula, we define the induced dipole and quadrupole of a system (a molecule or the interface system) in response to an electric field at a frequency ω as

$$\mu_p(\omega) = \sum_{q}^{x-z} \alpha_{pq}(\omega)E_q(\omega) + \sum_{q,r}^{x-z} \beta'_{pqr}(\omega)(\nabla \boldsymbol{E}(\omega))_{qr} + \cdots, \tag{7.88}$$

$$q_{pq}(\omega) = \sum_{r}^{x-z} \beta_{pqr}(\omega)E_r(\omega) + \cdots, \tag{7.89}$$

where $\boldsymbol{\beta}$ and $\boldsymbol{\beta}'$ denote quadrupolar polarizability of the system (a molecule or the interface system). $\boldsymbol{\beta}$ and $\boldsymbol{\beta}'$ are related to

$$\beta_{pqr}(\omega) = \beta'_{rqp}(-\omega)^* = \beta'_{rqp}(\omega). \tag{7.90}$$

$\boldsymbol{\alpha}$, $\boldsymbol{\beta}$ and $\boldsymbol{\beta}'$ in Eqs. (7.88) and (7.89) are represented with the perturbation Hamiltonian $\hat{H}^{\text{int}}(\omega)$ in Eq. (7.74) and the first-order perturbation theory of quantum mechanics. They are given on the basis of eigenstates as

$$\alpha_{pq}(\omega) = -\frac{1}{\hbar}\sum_{g,m}^{\text{states}} (\rho_g^{(0)} - \rho_m^{(0)})\frac{\langle g|\mu_p|m\rangle\langle m|\mu_q|g\rangle}{\omega-\omega_{mg}+i\Gamma_{mg}}$$

$$= \frac{1}{\hbar}\sum_{g,m}^{\text{states}} \rho_g^{(0)}\left[-\frac{\langle g|\mu_p|m\rangle\langle m|\mu_q|g\rangle}{\omega-\omega_{mg}+i\Gamma_{mg}} + \frac{\langle g|\mu_q|m\rangle\langle m|\mu_p|g\rangle}{\omega+\omega_{mg}+i\Gamma_{mg}}\right], \tag{7.91}$$

$$\beta_{pqr}(\omega) = -\frac{1}{\hbar}\sum_{g,m}^{\text{states}}(\rho_g^{(0)} - \rho_m^{(0)})\frac{\langle g|q_{pq}(\omega)|m\rangle\langle m|\mu_r|g\rangle}{\omega - \omega_{mg} + i\Gamma_{mg}}$$
$$= \frac{1}{\hbar}\sum_{g,m}^{\text{states}}\rho_g^{(0)}\left[-\frac{\langle g|q_{pq}(\omega)|m\rangle\langle m|\mu_r|g\rangle}{\omega - \omega_{mg} + i\Gamma_{mg}} + \frac{\langle g|\mu_r|m\rangle\langle m|q_{pq}(\omega)|g\rangle}{\omega + \omega_{mg} + i\Gamma_{mg}}\right], \tag{7.92}$$

$$\beta'_{pqr}(\omega) = -\frac{1}{\hbar}\sum_{g,m}^{\text{states}}(\rho_g^{(0)} - \rho_m^{(0)})\frac{\langle g|\mu_p|m\rangle\langle m|q_{qr}(\omega)|g\rangle}{\omega - \omega_{mg} + i\Gamma_{mg}}$$
$$= \frac{1}{\hbar}\sum_{g,m}^{\text{states}}\rho_g^{(0)}\left[-\frac{\langle g|\mu_p|m\rangle\langle m|q_{qr}(\omega)|g\rangle}{\omega - \omega_{mg} + i\Gamma_{mg}} + \frac{\langle g|q_{qr}(\omega)|m\rangle\langle m|\mu_p|g\rangle}{\omega + \omega_{mg} + i\Gamma_{mg}}\right]. \tag{7.93}$$

Equation (7.91) is equivalent to Eqs. (3.25) and (3.26) in Chap. 3. $\alpha_{pq}(\omega)$ in Eq. (7.91) is known to be a thermal average of the Raman tensor (see Appendix A.3),

$$\alpha_{pq}(\omega) = \mathrm{Tr}[\rho\,\alpha(\omega)],$$

where ρ is the density matrix and $\alpha(\omega)$ denotes the Raman tensor. Its matrix element is represented by

$$\langle g|\alpha_{pq}(\omega)|n\rangle = \frac{1}{\hbar}\sum_{m}^{\text{states}}\left[-\frac{\langle g|\mu_p|m\rangle\langle m|\mu_q|n\rangle}{\omega - \omega_{mg} + i\Gamma_{mg}} + \frac{\langle g|\mu_q|m\rangle\langle m|\mu_p|n\rangle}{\omega + \omega_{mn} + i\Gamma_{mn}}\right], \tag{7.94}$$

which is equivalent to Eq. (3.37). We can define the quadrupolar Raman tensors in an analogous manner by

$$\langle g|\beta_{pqr}(\omega)|n\rangle = \frac{1}{\hbar}\sum_{m}^{\text{states}}\left[-\frac{\langle g|q_{pq}(\omega)|m\rangle\langle m|\mu_r|n\rangle}{\omega - \omega_{mg} + i\Gamma_{mg}} + \frac{\langle g|\mu_r|m\rangle\langle m|q_{pq}(\omega)|n\rangle}{\omega + \omega_{mn} + i\Gamma_{mn}}\right], \tag{7.95}$$

$$\langle g|\beta'_{pqr}(\omega)|n\rangle = \frac{1}{\hbar}\sum_{m}^{\text{states}}\left[-\frac{\langle g|\mu_p|m\rangle\langle m|q_{qr}(\omega)|n\rangle}{\omega - \omega_{mg} + i\Gamma_{mg}} + \frac{\langle g|q_{qr}(\omega)|m\rangle\langle m|\mu_p|n\rangle}{\omega + \omega_{mn} + i\Gamma_{mn}}\right]. \tag{7.96}$$

Using the extended Raman tensors, the vibrationally resonant terms of the dipolar and quadrupolar hyperpolarizabilities are represented as follows.

$$\alpha_{pqr}^{\mathrm{D0,res}}(\Omega, \omega_1, \omega_2) = \frac{1}{\hbar^2}\sum_{g,m,n}^{\text{states}}(\rho_g^{(0)} - \rho_m^{(0)})$$

$$\left[\frac{\langle g|\mu_p|n\rangle\langle n|\mu_q|m\rangle\langle m|\mu_r|g\rangle}{(\omega_2-\omega_{mg}+i\Gamma_{mg})(\Omega-\omega_{ng}+i\Gamma_{ng})}-\frac{\langle g|\mu_q|n\rangle\langle n|\mu_p|m\rangle\langle m|\mu_r|g\rangle}{(\omega_2-\omega_{mg}+i\Gamma_{mg})(\Omega-\omega_{mn}+i\Gamma_{mn})}\right] \tag{7.77}$$

$$= -\frac{1}{\hbar}\sum_{g,m}^{\text{states}}(\rho_g^{(0)}-\rho_m^{(0)})\frac{\langle g|\alpha_{pq}(\Omega)|m\rangle\langle m|\mu_r|g\rangle}{\omega_2-\omega_{mg}+i\Gamma_{mg}}, \tag{7.97}$$

$$\alpha_{pqrs}^{\text{D1,res}}(\Omega,\omega_1,\omega_2)=\frac{1}{\hbar^2}\sum_{g,m,n}^{\text{states}}(\rho_g^{(0)}-\rho_m^{(0)})$$

$$\left[+\frac{\langle g|\mu_p|n\rangle\langle n|q_{sq}(\omega_1)|m\rangle\langle m|\mu_r|g\rangle}{(\omega_2-\omega_{mg}+i\Gamma_{mg})(\Omega-\omega_{ng}+i\Gamma_{ng})}-\frac{\langle g|q_{sq}(\omega_1)|n\rangle\langle n|\mu_p|m\rangle\langle m|\mu_r|g\rangle}{(\omega_2-\omega_{mg}+i\Gamma_{mg})(\Omega-\omega_{mn}+i\Gamma_{mn})}\right] \tag{7.82}$$

$$= -\frac{1}{\hbar}\sum_{g,m}^{\text{states}}(\rho_g^{(0)}-\rho_m^{(0)})\frac{\langle g|\beta'_{psq}(\omega_1)|m\rangle\langle m|\mu_r|g\rangle}{\omega_2-\omega_{mg}+i\Gamma_{mg}}, \tag{7.98}$$

$$\alpha_{pqrs}^{\text{D2,res}}(\Omega,\omega_1,\omega_2)=\frac{1}{\hbar^2}\sum_{g,m,n}^{\text{states}}(\rho_g^{(0)}-\rho_m^{(0)})$$

$$\left[\frac{\langle g|\mu_p|n\rangle\langle n|\mu_q|m\rangle\langle m|q_{sr}(\omega_2)|g\rangle}{(\omega_2-\omega_{mg}+i\Gamma_{mg})(\Omega-\omega_{ng}+i\Gamma_{ng})}-\frac{\langle g|\mu_q|n\rangle\langle n|\mu_p|m\rangle\langle m|q_{sr}(\omega_2)|g\rangle}{(\omega_2-\omega_{mg}+i\Gamma_{mg})(\Omega-\omega_{mn}+i\Gamma_{mn})}\right]$$

$$= -\frac{1}{\hbar}\sum_{g,m}^{\text{states}}(\rho_g^{(0)}-\rho_m^{(0)})\frac{\langle g|\alpha_{pq}(\Omega)|m\rangle\langle m|q_{sr}(\omega_2)|g\rangle}{\omega_2-\omega_{mg}+i\Gamma_{mg}}, \tag{7.99}$$

$$\alpha_{pqrs}^{\text{Q,res}}(\Omega,\omega_1,\omega_2)=\frac{1}{\hbar^2}\sum_{g,m,n}^{\text{states}}(\rho_g^{(0)}-\rho_m^{(0)})$$

$$\left[\frac{\langle g|q_{ps}(\Omega)|n\rangle\langle n|\mu_q|m\rangle\langle m|\mu_r|g\rangle}{(\omega_2-\omega_{mg}+i\Gamma_{mg})(\Omega-\omega_{ng}+i\Gamma_{ng})}-\frac{\langle g|\mu_q|n\rangle\langle n|q_{ps}(\Omega)|m\rangle\langle m|\mu_r|g\rangle}{(\omega_2-\omega_{mg}+i\Gamma_{mg})(\Omega-\omega_{mn}+i\Gamma_{mn})}\right]$$

$$= -\frac{1}{\hbar}\sum_{g,m}^{\text{states}}(\rho_g^{(0)}-\rho_m^{(0)})\frac{\langle g|\beta_{psq}(\Omega)|m\rangle\langle m|\mu_r|g\rangle}{\omega_2-\omega_{mg}+i\Gamma_{mg}}. \tag{7.100}$$

These formulas of Eqs. (7.97), (7.98), (7.99), (7.100) are converted to the time correlation ones, via the same procedure as described in Chap. 3. After the derivation of Eq. (4.17), the equivalent formulas are

$$\alpha_{pqr}^{\mathrm{D0,res}}(\Omega,\omega_1,\omega_2)=\frac{i}{\hbar}\int_0^\infty dt\ \overline{\langle\alpha_{pq}(\Omega,t)\mu_r-\mu_r\alpha_{pq}(\Omega,t)\rangle}\exp(i\omega_2 t),\tag{7.101}$$

$$\alpha_{pqrs}^{\mathrm{D1,res}}(\Omega,\omega_1,\omega_2)=\frac{i}{\hbar}\int_0^\infty dt\ \overline{\langle\beta'_{psq}(\omega_1,t)\mu_r-\mu_r\beta'_{psq}(\omega_1,t)\rangle}\exp(i\omega_2 t),\tag{7.102}$$

$$\alpha_{pqrs}^{\mathrm{D2,res}}(\Omega,\omega_1,\omega_2)=\frac{i}{\hbar}\int_0^\infty dt\ \overline{\langle\alpha_{pq}(\Omega,t)q_{sr}(\omega_2)-q_{sr}(\omega_2)\alpha_{pq}(\Omega,t)\rangle}\exp(i\omega_2 t),\tag{7.103}$$

$$\alpha_{pqrs}^{\mathrm{Q,res}}(\Omega,\omega_1,\omega_2)=\frac{i}{\hbar}\int_0^\infty dt\ \overline{\langle\beta_{psq}(\Omega,t)\mu_r-\mu_r\beta_{psq}(\Omega,t)\rangle}\exp(i\omega_2 t).\tag{7.104}$$

The derivation of Eq. (7.101) has been described in detail in Sect. 4.3.1, and the other three equations are derived in the same way.

So far we have treated molecular properties, $\boldsymbol{\alpha}^{\mathrm{D0}}$, $\boldsymbol{\alpha}^{\mathrm{D1}}$, $\boldsymbol{\alpha}^{\mathrm{D2}}$, and $\boldsymbol{\alpha}^{\mathrm{Q}}$. The same discussion should hold for the interface system if the perturbation Hamiltonian $\hat{H}^{\mathrm{int}}$ is defined for the system. The dipolar and quadrupolar nonlinear susceptibilities $\boldsymbol{\chi}^{\mathrm{D0}}$, $\boldsymbol{\chi}^{\mathrm{D1}}$, $\boldsymbol{\chi}^{\mathrm{D2}}$, and $\boldsymbol{\chi}^{\mathrm{Q}}$ are analogously represented with the time correlation formulas,

$$\chi_{pqr}^{\mathrm{D0,res}}(\Omega,\omega_1,\omega_2)=\frac{i}{\hbar}\int_0^\infty dt\ \overline{\langle\alpha_{pq}(\Omega,t)\mu_r-\mu_r\alpha_{pq}(\Omega,t)\rangle}\exp(i\omega_2 t),\tag{7.105}$$

$$\chi_{pqrs}^{\mathrm{D1,res}}(\Omega,\omega_1,\omega_2)=\frac{i}{\hbar}\int_0^\infty dt\ \overline{\langle\beta'_{psq}(\omega_1,t)\mu_r-\mu_r\beta'_{psq}(\omega_1,t)\rangle}\exp(i\omega_2 t),\tag{7.106}$$

$$\chi_{pqrs}^{\mathrm{D2,res}}(\Omega,\omega_1,\omega_2)=\frac{i}{\hbar}\int_0^\infty dt\ \overline{\langle\alpha_{pq}(\Omega,t)q_{sr}(\omega_2)-q_{sr}(\omega_2)\alpha_{pq}(\Omega,t)\rangle}\exp(i\omega_2 t),\tag{7.107}$$

$$\chi_{pqrs}^{\mathrm{Q,res}}(\Omega,\omega_1,\omega_2)=\frac{i}{\hbar}\int_0^\infty dt\ \overline{\langle\beta_{psq}(\Omega,t)\mu_r-\mu_r\beta_{psq}(\Omega,t)\rangle}\exp(i\omega_2 t).\tag{7.108}$$

Note that μ, α, β and β' in Eqs. (7.105), (7.106), (7.107), (7.108) are defined for the interface system.

The classical forms of the time correlation functions are derived after Sect. 4.3.2. When we suppose electronically nonresonant conditions and treat the vibrational resonance, the classical time correlation formulas are given as

$$\chi_{pqr}^{\mathrm{D0,res}}(\Omega,\omega_1,\omega_2) = \frac{i\omega_2}{k_B T}\int_0^\infty dt\ \langle \alpha_{pq}(t)\mu_r\rangle_{cl}\exp(i\omega_2 t), \tag{7.109}$$

$$\chi_{pqrs}^{\mathrm{D1,res}}(\Omega,\omega_1,\omega_2) = \frac{i\omega_2}{k_B T}\int_0^\infty dt\ \langle \beta'_{psq}(t)\mu_r\rangle_{cl}\exp(i\omega_2 t), \tag{7.110}$$

$$\chi_{pqrs}^{\mathrm{D2,res}}(\Omega,\omega_1,\omega_2) = \frac{i\omega_2}{k_B T}\int_0^\infty dt\ \langle \alpha_{pq}(t)q_{sr}(\omega_2)\rangle_{cl}\exp(i\omega_2 t), \tag{7.111}$$

$$\chi_{pqrs}^{\mathrm{Q,res}}(\Omega,\omega_1,\omega_2) = \frac{i\omega_2}{k_B T}\int_0^\infty dt\ \langle \beta_{psq}(t)\mu_r\rangle_{cl}\exp(i\omega_2 t), \tag{7.112}$$

where the subscript cl of $\langle\ \ \rangle_{cl}$ emphasizes the classical time correlation function, and the frequency dependence of α, β and β' are omitted in the electronically non-resonant conditions. The frequency dependence of $q_{sr}(\omega_2)$ in Eq. (7.111) is still necessary to denote the general quadrupole including electronic quadrupole and magnetic dipole in Eq. (7.70).

7.4 Invariance to Molecular Origin

Here we argue a fundamental issue pertinent to the quadrupole. When we treat the quadrupole besides the dipole, we should pay attention to the origin of these moments. In the multipole expansion in general, all the moments except for the lowest-order nonzero one are dependent on the location of the origin. This problem is relevant to the microscopic expressions of dipolar and quadrupolar susceptibilities in Eqs. (7.7) and (7.65), (7.66), (7.67), $\boldsymbol{\chi}^{\mathrm{D0}}(z,\Omega,\omega_1,\omega_2)$ and $\boldsymbol{\chi}^{\mathrm{F}}(z,\Omega,\omega_1,\omega_2)$ (F = D1, D2, Q), which include the location of l-th molecule, z_l, in their definition. The location of z_l could be assigned at the molecular center of mass, though its definition may not be unique. The ambiguity in the definition of molecular origin was first pointed out by Byrnes et al. [5] In this subsection we argue that the effective nonlinear susceptibility, $\chi_{\mathrm{eff},G}^{(2)}$, is well defined regardless of the definition of z_l. This argument is necessary to construct the present SFG theory on a physically solid ground.

Let us change the definition of molecular origin of l-th molecule from z_l to $z_l + \Delta z_l$ in the space-fixed coordinate. This displacement Δz_l of l-th molecule may be arbitrary for each molecule. Consequently, $\boldsymbol{\chi}^{\mathrm{ID}}(\Omega,\omega_1,\omega_2)$ in Eq. (7.12) is transformed into $\boldsymbol{\chi}^{\mathrm{ID}'}(\Omega,\omega_1,\omega_2)$ as follows,

$$\chi^{\mathrm{ID}'}{}_{pqr}(\Omega,\omega_1,\omega_2)$$
$$= \int_{z_b}^{\infty} dz \left(\overline{\sum_{l}^{\text{molecules}} \alpha_{l,pqr}^{\mathrm{D0}}(\Omega,\omega_1,\omega_2)\delta(z-(z_l+\Delta z_l))}\right) f_p(z,\Omega) f_q(z,\omega_1) f_r(z,\omega_2) \tag{7.113}$$

$$
\begin{aligned}
&= \chi^{\mathrm{ID}}_{pqr}(\Omega,\omega_1,\omega_2) \\
&+ \left(\sum_{l}^{\text{molecules}} \Delta z_l \alpha^{\mathrm{D0}}_{l,pqr}(\Omega,\omega_1,\omega_2)\delta(z_b - z_l) \right) f^{\beta}_p(\Omega) f^{\beta}_q(\omega_1) f^{\beta}_r(\omega_2) \\
&+ \int_{z_b}^{\infty} dz \left(\sum_{l}^{\text{molecules}} \Delta z_l \alpha^{\mathrm{D0}}_{l,pqr}(\Omega,\omega_1,\omega_2)\delta(z - z_l) \right) \frac{\partial}{\partial z} \{ f_p(z,\Omega) f_q(z,\omega_1) f_r(z,\omega_2) \} \\
&+ \cdots,
\end{aligned} \tag{7.114}
$$

where the delta function in Eq. (7.113) is expanded in the Taylor series,

$$
\delta(z - (z_l + \Delta z_l)) = \delta(z - z_l) - \left(\frac{d\delta(z)}{dz} \right)_{z=z_l} \Delta z_l + \cdots,
$$

and we neglect the second and higher order derivatives in Eq. (7.114).[5] We note that the dipolar hyperpolarizability $\alpha^{\mathrm{D0}}_{l,pqr}(\Omega,\omega_1,\omega_2)$ in Eq. (7.113) is invariant under the change of the molecular origin, since it is the lower-order nonzero term. (The total charge is unchanged by imposing the electric fields and thus its derivative is zero.)

Next we consider the transformation of the quadrupolar terms, $\boldsymbol{\chi}^{\mathrm{IQ}}$ and $\boldsymbol{\chi}^{\mathrm{IQB}}$. Unlike the dipolar term of $\boldsymbol{\alpha}^{\mathrm{D0}}_l$, the quadrupolar hyperpolarizabilities α^{F}_l (F=D1, D2, Q) vary with the molecular origin. When the origin of l-th molecule shifts by $\Delta \boldsymbol{r}$, then the electric quadrupole operator $\hat{q}^E_{pq}$ and the magnetic dipole operator $\hat{\mu}^M_r$ are transformed into $\hat{q}^{E\,\prime}_{pq}$ and $\hat{\mu}^{M\prime}_r$, respectively, as follows:

$$
\hat{q}^{E\,\prime}_{pq} = \hat{q}^E_{pq} - \frac{1}{2}\hat{\mu}_p \Delta r_q - \frac{1}{2}\hat{\mu}_q \Delta r_p, \tag{7.115}
$$

$$
\hat{\mu}^{M\prime}_r = \hat{\mu}^M_r - \frac{1}{2c} \sum_{p,q}^{x-z} \varepsilon_{rpq} \Delta r_p \hat{j}_q, \tag{7.116}
$$

where $\hat{j}_q$ is the electric current operator. Therefore, the generalized quadrupole operator $\hat{q}_{pq}(\omega_f)$ in Eq. (7.75) is transformed into $\hat{q}_{pq}{}'(\omega_f)$ as

$$
\begin{aligned}
\hat{q}_{pq}{}'(\omega_f) &= \hat{q}^{E\,\prime}_{pq}(\omega_f) + \frac{c}{i\omega_f} \sum_{r}^{x-z} \hat{\mu}^{M\prime}_r(\omega_f)\varepsilon_{pqr} \\
&= \hat{q}_{pq}(\omega_f) - \frac{1}{2}\hat{\mu}_p \Delta r_q - \frac{1}{2}\hat{\mu}_q \Delta r_p - \frac{1}{2i\omega_f}(\Delta r_p \hat{j}_q - \Delta r_q \hat{j}_p).
\end{aligned} \tag{7.117}
$$

[5] We could assume that Δz_l is infinitesimally small without losing generality, since an arbitrary finite displacement is expressed by assembly of infinitesimally small ones.

By replacing $\hat{q}(\omega_f)$ in Eqs. (7.79), (7.80), (7.81) with the shifted operator $\hat{q}'(\omega_f)$, we obtain the transformed quadrupolar hyperpolarizabilities, $\boldsymbol{\alpha}_l^{F\prime}$. During the derivation of $\boldsymbol{\alpha}_l^{F\prime}$, the matrix element of the current operator is represented using the relation $\hat{j} = d\hat{\mu}/dt$ as

$$\left\langle m|\hat{j}|n\right\rangle = \left\langle m\left|\frac{d\hat{\mu}}{dt}\right|n\right\rangle = \frac{i}{\hbar}(E_m - E_n)\left\langle m|\hat{\mu}|n\right\rangle = i\omega_{mn}\left\langle m|\hat{\mu}|n\right\rangle,$$

where the states m, n denote eigenstates of the whole system. Therefore, the matrix element of $\hat{q}_{pq}{}'(\omega_f)$ in Eq. (7.117) becomes

$$\langle m|\hat{q}_{pq}{}'(\omega_f)|n\rangle = \langle m|\hat{q}_{pq}(\omega_f)|n\rangle$$
$$-\frac{1}{2}\langle m|\hat{\mu}_p|n\rangle\Delta r_q - \frac{1}{2}\langle m|\hat{\mu}_q|n\rangle\Delta r_p - \frac{\omega_{mn}}{2\omega_f}\langle m|\hat{\mu}_q|n\rangle\Delta r_p + \frac{\omega_{mn}}{2\omega_f}\langle m|\hat{\mu}_p|n\rangle\Delta r_q. \tag{7.118}$$

By inserting this transformed matrix element of Eq. (7.118) into Eqs. (7.85), (7.86), (7.87), we find that the second and fifth terms in the right-hand-side of Eq. (7.118) cancel each other whereas the third and fourth terms become equivalent. Therefore, the transformed quadrupolar hyperpolarizabilities $\alpha_l^{\mathrm{F}\prime}$ eventually take the following form,

$$\alpha^{\mathrm{F}}_{l,pqrs}{}'(\Omega, \omega_1, \omega_2) = \alpha^{\mathrm{F}}_{l,pqrs}(\Omega, \omega_1, \omega_2) - \alpha^{\mathrm{D0}}_{l,pqr}(\Omega, \omega_1, \omega_2)\Delta r_{l,s} \tag{7.119}$$
$$(\mathrm{F} = \mathrm{D1}, \mathrm{D2}, \mathrm{Q})$$

[Problem 7.3] Derive Eq. (7.119) for F = D1. Use the α^{D1} expression in Eq. (7.85) and Eq. (7.118).

Consequently, $\chi^{\mathrm{F}}_{pqrz}(z, \Omega, \omega_1, \omega_2)$ in Eq. (7.65), (7.66), (7.67) is transformed into

$$\chi^{\mathrm{F}}_{pqrz}{}'(z, \Omega, \omega_1, \omega_2)$$
$$= \left(\sum_{l}^{\text{molecules}} \left(\alpha^{\mathrm{F}}_{l,pqrz}(\Omega, \omega_1, \omega_2) - \alpha^{\mathrm{D0}}_{l,pqr}(\Omega, \omega_1, \omega_2)\Delta z_l\right)\delta(z - (z_l + \Delta z_l))\right)$$
$$(\mathrm{F} = \mathrm{D1}, \mathrm{D2}, \mathrm{Q}). \tag{7.120}$$

Using Eq. (7.120), the $\chi^{\mathrm{IQ}}(\Omega, \omega_1, \omega_2)$ term in Eq. (7.29) is transformed to

$$\chi^{\mathrm{IQ}\ \prime}_{pqr}(\Omega, \omega_1, \omega_2) = \chi^{\mathrm{IQ}}_{pqr}(\Omega, \omega_1, \omega_2)$$
$$- \int_{z_b}^{\infty} dz \left(\overline{\sum_l^{\mathrm{molecule}} \Delta z_l \alpha^{\mathrm{D0}}_{l,pqr}(\Omega, \omega_1, \omega_2)\delta(z - z_l)} \right) \frac{\partial}{\partial z} \{ f_p(z, \Omega) f_q(z, \omega_1) f_r(z, \omega_2) \}$$
$$+ \cdots, \tag{7.121}$$

where we neglected the terms of higher order derivatives. We also get the transformed expression of $\chi^{\mathrm{IQB}\prime}_{pqr}(\Omega, \omega_1, \omega_2)$ from Eqs. (7.30) and (7.119) as follows,

$$\chi^{\mathrm{IQB}\prime}_{pqr}(\Omega, \omega_1, \omega_2) = \chi^{\mathrm{IQB}}_{pqr}(\Omega, \omega_1, \omega_2)$$
$$- \left(\overline{\sum_l^{\mathrm{molecule}} \Delta z_l \alpha^{\mathrm{D0}}_{l,pqr}(\Omega, \omega_1, \omega_2)\delta(z_b - z_l)} \right) f_p^{\beta}(\Omega) f_q^{\beta}(\omega_1) f_r^{\beta}(\omega_2)$$
$$+ \cdots. \tag{7.122}$$

Equations (7.114), (7.121) and (7.122) indicate that the sum of these terms is invariant against the shift of the molecular origin,

$$\chi^{\mathrm{ID}}_{pqr}(\Omega, \omega_1, \omega_2) + \chi^{\mathrm{IQ}}_{pqr}(\Omega, \omega_1, \omega_2) + \chi^{\mathrm{IQB}}_{pqr}(\Omega, \omega_1, \omega_2)$$
$$= \chi^{\mathrm{ID}\ \prime}_{pqr}(\Omega, \omega_1, \omega_2) + \chi^{\mathrm{IQ}\ \prime}_{pqr}(\Omega, \omega_1, \omega_2) + \chi^{\mathrm{IQB}\prime}_{pqr}(\Omega, \omega_1, \omega_2). \tag{7.123}$$

On the other hand, the bulk contribution χ^{B0} is invariant in itself by the change of the molecular origin. This is evident because the factors $\xi_1^{\mathrm{Q2},\beta}$, $\xi_2^{\mathrm{Q1},\beta}$, $\xi_3^{\mathrm{Q1},\beta}$ and $\xi_3^{\mathrm{Q2},\beta}$ involved in Eq. (7.41) are invariant. These four factors are given in Eqs. (7.42), (7.43), (7.44), (7.45) as the difference of two susceptibility terms, e.g. $\xi_1^{\mathrm{Q2},\beta} = \chi_1^{\mathrm{D2},\beta} - \chi_1^{\mathrm{Q},\beta}$ in Eq. (7.42), and thus the additional terms originating from the shift of molecular origin in Eq. (7.120) cancel each other. The alternative expression of the bulk contribution, χ^{B}, given in Sect. 7.2.4 by Eqs. (7.54) and (7.55), (7.56), (7.57), (7.58), (7.59), (7.60) is also invariant for the same reason.

In conclusion, the effective nonlinear susceptibility $\chi^{(2)}_{\mathrm{eff},G}$ in Eq. (7.47) is well defined irrespective of the choice of the molecular origin. This invariance is requisite in order that $\chi^{(2)}_{\mathrm{eff},G}$ is related to experimental observables. Moreover, the sum of the interface contributions, $\chi^{\mathrm{ID}} + \chi^{\mathrm{IQ}} + \chi^{\mathrm{IQB}}$, and the bulk contribution, χ^{B0} or χ^{B}, are respectively invariant. Consequently, the interface and bulk contributions in the SFG/SHG signals are respectively well defined, though each of three terms consisting of the interface contribution, χ^{ID}, χ^{IQ}, χ^{IQB}, is affected by the definition of the molecular origin.

7.5 Summary

In this section we have presented a unified treatment of the dipole and quadrupole contributions to the second-order optical responses. The concept of effective susceptibility $\chi^{(2)}$ is expanded to incorporate both the surface and bulk contributions. This treatment allows us to quantitatively evaluate both surface and bulk contributions in experimental SFG/SHG spectra, and to calculate them from microscopic expressions of general hyperpolarizabilities.

The source of the SFG/SHG signals consists of four terms, i.e. $\chi^{\rm ID}$, $\chi^{\rm IQ}$, $\chi^{\rm IQB}$ and $\chi^{\rm B}$ in Eq. (7.50). These four terms can be characterized in various ways as follows, which help us understand the physical meanings of these terms.

Dipole vs. Quadrupole $\chi^{\rm ID}$ originates from the induced electric dipole, while the other three terms, $\chi^{\rm IQ}$, $\chi^{\rm IQB}$, and $\chi^{\rm B}$ originate from the quadrupole (electric quadrupole and magnetic dipole).

The conventional SFG/SHG theory within the dipole approximation considers only the $\chi^{\rm ID}$ term. The present theory shows that the conventional theory can be straightforwardly extended by replacing $\chi^{\rm ID}$ with $\chi_q = \chi^{\rm ID} + \chi^{\rm IQ} + \chi^{\rm IQB} + \chi^{\rm B}$ if we properly define the latter three terms.

Interface vs. Bulk $\chi^{\rm ID}$ and $\chi^{\rm IQ}$ reflect properties of interface, while $\chi^{\rm IQB}$ and $\chi^{\rm B}$ are determined solely by bulk properties.

It is worth noting that $\chi^{\rm IQB}$ reflects no interface properties, as seen in Eq. (7.30), since it is derived from the lower bound of the integral over the interface region in Eq. (7.27).

Dependence on Optical Geometry $\chi^{\rm ID}$, $\chi^{\rm IQ}$ and $\chi^{\rm IQB}$ are independent of the optical geometry of measurement as they involve no wavevector $\boldsymbol{k}^i_G(\omega_f)$ or related quantities. On the other hand, $\chi^{\rm B}$ depends on the optical geometry as described in Sects. 7.2.4 and 7.2.5.

This difference allows us to distinguish $\chi^{\rm B}$ from the other terms by changing the optical geometry of measurement.

The characteristics of the four terms, $\chi^{\rm ID}$, $\chi^{\rm IQ}$, $\chi^{\rm IQB}$ and $\chi^{\rm B}$, are schematically summarized in Fig. 7.5.

7.6 Solutions to Problems

7.6.1 *Isotropic Tensor Components*

[Problem 7.1] Derive Eq. (7.39) from the three independent elements of Eq. (7.38) in the isotropic condition.

Characters of Nonlinear Susceptibility Terms

	χ^{ID}	χ^{IQ}	χ^{IQB}	χ^{B}
(1) Origin **Dipole vs. Quadrupole**	**Dipole**	**Quadrupole**		
(2) Properties **Surface vs. Bulk**	**Surface**		**Bulk**	
(3) Separate measurement by experiment	**Inseparable**			**Separable**

Fig. 7.5 Characters of four nonlinear susceptibility terms

There are several ways to prove Eqs. (7.38) and (7.39) in the isotropic condition. The most straightforward way is to calculate the spherical average of arbitrary elements,

$$\chi_{pqrs} = \sum_{p'q'r's'}^{x\sim z} \overline{\mathcal{D}_{pp'}\mathcal{D}_{qq'}\mathcal{D}_{rr'}\mathcal{D}_{ss'}}\,\chi^{\mathrm{mol}}_{p'q'r's'},$$

where $\mathcal{D}$ is the rotation matrix in Sect. 3.3, and the overline denotes the isotropic average. However, here we take a more intuitive way to derive Eq. (7.39) as follows. In this section, suffixes of tensors are presented in square parentheses, e.g. $\chi[xxxx] \equiv \chi_{xxxx}$.

First, we identify nonvanishing elements in Eq. (7.38). In the isotropic condition, a tensor element $\chi[pqrs]$ should vanish in general when the suffixes $pqrs$ include either x, y, or z odd times and others even times, such as $\chi[xxxy] = 0$, $\chi[xyyz] = 0$, etc. (Note that a suffix not included is considered as even (zero) times.) This is evident by operating a proper rotation on the isotropic system. A rotation by 180° along a plane consisting of the different types of suffixes should change the sign, though the material is isotropic. For an example of the $\chi[xyyz]$ element, the 180° rotation along the xy plane changes its sign,

$$\chi[xyyz] \longrightarrow \text{“}\chi\,[(-x)(-y)(-y)z]\text{”} = -\chi[xyyz].$$

This indicates that such tensor elements vanish in the isotropic condition.

Therefore, non-zero elements of a fourth-rank tensor must take either form of $\chi[pqpq]$, $\chi[ppqq]$, or $\chi[pqqp]$, where all the $x \sim z$ coordinates are included even times. (It is not possible that all the $x \sim z$ coordinates appear odd times in a fourth-rank tensor element.) Furthermore, the x, y and z coordinates have to be equivalent in the isotropic condition. Therefore, we obtain three independent elements as

$$\begin{aligned}
\chi[xyxy] &= \chi[xzxz] = \chi[yzyz] = \cdots = \chi[pqpq] \quad (\equiv \chi_1)\\
\chi[xxyy] &= \chi[xxzz] = \chi[yyzz] = \cdots = \chi[ppqq] \quad (\equiv \chi_2)\\
\chi[xyyx] &= \chi[xzzx] = \chi[yzzy] = \cdots = \chi[pqqp] \quad (\equiv \chi_3)
\end{aligned}$$

Next we examine the diagonal element $\chi[pppp]$ ($= \chi[xxxx] = \chi[yyyy] = \chi[zzzz]$). As a simple example, let us rotate the coordinate system by 45° around the z axis,

$$x \to \frac{x+y}{\sqrt{2}}, \quad y \to \frac{-x+y}{\sqrt{2}}, \quad z \to z.$$

Then a tensor element $\chi[xxxx]$ is transformed as a direct product of the coordinates,

$$\begin{aligned}
\chi[xxxx] \to \text{``}\chi\left[\frac{x+y}{\sqrt{2}}\,\frac{x+y}{\sqrt{2}}\,\frac{x+y}{\sqrt{2}}\,\frac{x+y}{\sqrt{2}}\right]\text{''} \\
= \frac{1}{4}\{\chi[xxxx] + \chi[xxxy] + \chi[xxyx] + \chi[xxyy] + \chi[xyxx] + \chi[xyxy] \\
+ \chi[xyyx] + \chi[xyyy] + \chi[yxxx] + \chi[yxxy] + \chi[yxyx] + \chi[yxyy] \\
+ \chi[yyxx] + \chi[yyxy] + \chi[yyyx] + \chi[yyyy]\}.
\end{aligned}$$

Since the transformed element should be invariant to the original element in the isotropic condition, the following relation should hold,

$$\begin{aligned}
\chi[xxxx] = \frac{1}{4}\{\chi[xxxx] + \chi[xxxy] + \chi[xxyx] + \chi[xxyy] + \chi[xyxx] + \chi[xyxy] \\
+ \chi[xyyx] + \chi[xyyy] + \chi[yxxx] + \chi[yxxy] + \chi[yxyx] + \chi[yxyy] \\
+ \chi[yyxx] + \chi[yyxy] + \chi[yyyx] + \chi[yyyy]\} \\
= \frac{1}{4}\{\chi[xxxx] + \chi[xxyy] + \chi[xyxy] + \chi[xyyx] \\
+ \chi[yxxy] + \chi[yxyx] + \chi[yyxx] + \chi[yyyy]\},
\end{aligned}$$

where the shaded elements are null. Therefore,

$$\chi[pppp] = \frac{1}{4}\{2\chi[pppp] + 2\chi_1 + 2\chi_2 + 2\chi_3\},$$

and hence

$$\chi[pppp] = \chi_1 + \chi_2 + \chi_3 \tag{7.39}$$

is obtained.

7.6.2 *Bulk Term* χ^{B}

[Problem 7.2] Derive $\chi^{\mathrm{B}}_{G,yyz}$ in Eq. (7.54) from χ^{B0}_{G} in Eqs. (7.52) and (7.53).

(*Hint*) During the derivation, use the following two relations,

$$\frac{\cos\theta_I^\alpha(\omega_2)}{\sin\theta_I^\alpha(\omega_2)} = \frac{q^\alpha(\omega_2)}{k_x(\omega_2)}, \qquad \frac{f_x^\beta}{f_z^\beta} = \varepsilon^\beta,$$

and $L_I(\omega_f)$ in Eq. (5.30).

The expression of $\chi^{(2)}_{\mathrm{eff,G,SSP}}$ in Eq. (7.49) including χ^{B0} can be expanded using Eq. (7.48) as follows,

$$\begin{aligned}
&\chi^{(2)}_{\mathrm{eff},G,\mathrm{SSP}}(\Omega,\omega_1,\omega_2) = \\
&\qquad L_{G,y}(\Omega)L_{I,y}(\omega_1)\,L_{I,z}(\omega_2)\,\sin\theta_I^\alpha(\omega_2)\,\chi^{(2)}_{q0\,G,yyz}(\Omega,\omega_1,\omega_2) \\
&\quad + L_{G,y}(\Omega)\,L_{I,y}(\omega_1)\,L_{I,x}(\omega_2)\,\cos\theta_I^\alpha(\omega_2)\,\chi^{(2)}_{q0\,G,yyx}(\Omega,\omega_1,\omega_2) \qquad (7.49) \\
&\quad = L_{G,y}(\Omega)L_{I,y}(\omega_1)\,L_{I,z}(\omega_2)\,\sin\theta_I^\alpha(\omega_2) \\
&\qquad \cdot\left\{\chi^{\mathrm{ID}}_{yyz}(\Omega,\omega_1,\omega_2)+\chi^{\mathrm{IQ}}_{yyz}(\Omega,\omega_1,\omega_2)+\chi^{\mathrm{IQB}}_{yyz}(\Omega,\omega_1,\omega_2)+\chi^{\mathrm{B0}}_{yyz}(\Omega,\omega_1,\omega_2)\right\} \\
&\quad + L_{G,y}(\Omega)\,L_{I,y}(\omega_1)\,L_{I,x}(\omega_2)\,\cos\theta_I^\alpha(\omega_2)\,\chi^{\mathrm{B0}}_{G,yyx}(\Omega,\omega_1,\omega_2), \qquad (7.124)
\end{aligned}$$

since the yyx element of $\chi^{(2)}_{q0G}$ involves only the χ^{B0}_G term. Equation (7.124) should be equivalent to Eqs. (7.51) and (7.50). Apparent differences are found in the terms related to χ^{B0} and χ^{B}, by comparing the two expressions. To make the two expressions equivalent, the following relation is required,

$$\begin{aligned}
&L_{G,y}(\Omega)L_{I,y}(\omega_1)\,L_{I,z}(\omega_2)\,\sin\theta_I^\alpha(\omega_2)\,\chi^{\mathrm{B}}_{G,yyz}(\Omega,\omega_1,\omega_2) \\
&= L_{G,y}(\Omega)L_{I,y}(\omega_1)\,L_{I,z}(\omega_2)\,\sin\theta_I^\alpha(\omega_2)\,\chi^{\mathrm{B0}}_{G,yyz}(\Omega,\omega_1,\omega_2) \\
&+ L_{G,y}(\Omega)\,L_{I,y}(\omega_1)\,L_{I,x}(\omega_2)\,\cos\theta_I^\alpha(\omega_2)\,\chi^{\mathrm{B0}}_{G,yyx}(\Omega,\omega_1,\omega_2).
\end{aligned}$$

Therefore,

$$\chi^{\mathrm{B}}_{G,yyz}(\Omega,\omega_1,\omega_2)=\chi^{\mathrm{B0}}_{G,yyz}(\Omega,\omega_1,\omega_2)+\frac{L_{I,x}(\omega_2)}{L_{I,z}(\omega_2)}\frac{\cos\theta_I^\alpha(\omega_2)}{\sin\theta_I^\alpha(\omega_2)}\chi^{\mathrm{B0}}_{G,yyx}(\Omega,\omega_1,\omega_2). \qquad (7.125)$$

By inserting Eqs. (7.52) and (7.53) into Eq. (7.125), the above equation (7.125) becomes

$$
\begin{aligned}
&\chi^{\mathrm{B}}_{G,yyz}(\Omega,\omega_1,\omega_2)\\
&= l_G\,\zeta_2^{\mathrm{Q1},\beta}(\Omega,\omega_1,\omega_2) f_y^{\beta}(\Omega) f_y^{\beta}(\omega_1) f_z^{\beta}(\omega_2) k^{\beta}_{T,z}(\omega_1)\\
&\quad + \frac{L_{I,x}(\omega_2)}{L_{I,z}(\omega_2)}\frac{\cos\theta_I^{\alpha}(\omega_2)}{\sin\theta_I^{\alpha}(\omega_2)}\, l_G\,\zeta_2^{\mathrm{Q1},\beta}(\Omega,\omega_1,\omega_2) f_y^{\beta}(\Omega) f_y^{\beta}(\omega_1) f_x^{\beta}(\omega_2) k_x(\omega_1)\\
&= l_G\,\zeta_2^{\mathrm{Q1},\beta}(\Omega,\omega_1,\omega_2) f_y^{\beta}(\Omega) f_y^{\beta}(\omega_1) f_z^{\beta}(\omega_2)\\
&\quad \times \left\{ k^{\beta}_{T,z}(\omega_1) + \frac{L_{I,x}(\omega_2)}{L_{I,z}(\omega_2)}\frac{\cos\theta_I^{\alpha}(\omega_2)}{\sin\theta_I^{\alpha}(\omega_2)}\frac{f_x^{\beta}(\omega_2)}{f_z^{\beta}(\omega_2)} k_x(\omega_1)\right\}.
\end{aligned}
$$

We further employ Eq. (5.30) and the relations mentioned in this Problem, $\frac{\cos\theta_I^{\alpha}(\omega_2)}{\sin\theta_I^{\alpha}(\omega_2)} = \frac{q^{\alpha}(\omega_2)}{k_x(\omega_2)}, \frac{f_x^{\beta}(\omega_2)}{f_z^{\beta}(\omega_2)} = \varepsilon^{\beta}(\omega_2)$, and obtain the following form,

$$
\begin{aligned}
&\chi^{\mathrm{B}}_{G,yyz}(\Omega,\omega_1,\omega_2)\\
&= l_G\,\zeta_2^{\mathrm{Q1},\beta}(\Omega,\omega_1,\omega_2) f_y^{\beta}(\Omega) f_y^{\beta}(\omega_1) f_z^{\beta}(\omega_2)\\
&\quad \cdot \left\{ k^{\beta}_{T,z}(\omega_1) + \frac{2\varepsilon^{\alpha}(\omega_2) q^{\beta}(\omega_2)}{2\varepsilon^{\alpha}(\omega_2)\varepsilon^{\beta}(\omega_2) q^{\alpha}(\omega_2)}\frac{q^{\alpha}(\omega_2)}{k_x(\omega_2)}\varepsilon^{\beta}(\omega_2) k_x(\omega_1)\right\}\\
&= l_G\,\zeta_2^{\mathrm{Q1},\beta}(\Omega,\omega_1,\omega_2) f_y^{\beta}(\Omega) f_y^{\beta}(\omega_1) f_z^{\beta}(\omega_2) \left\{ k^{\beta}_{T,z}(\omega_1) + \frac{q^{\beta}(\omega_2)}{k_x(\omega_2)} k_x(\omega_1)\right\}\\
&= l_G\,\zeta_2^{\mathrm{Q1},\beta}(\Omega,\omega_1,\omega_2)\frac{k^{\beta}_{T,z}(\omega_1) k_x(\omega_2) - k^{\beta}_{T,z}(\omega_2) k_x(\omega_1)}{k_x(\omega_2)} f_y^{\beta}(\Omega) f_y^{\beta}(\omega_1) f_z^{\beta}(\omega_2),
\end{aligned}
\tag{7.54}
$$

where $q^{\beta}(\omega_2) = \left|k^{\beta}_{T,z}(\omega_2)\right| = -k^{\beta}_{T,z}(\omega_2)$ (see Fig. 7.3).

7.6.3 Transformation of Quadrupolar Susceptibility

[Problem 7.3] Derive Eq. (7.119) for F = D1. Use the α^{D1} expression in Eq. (7.85) and Eq. (7.118).

Equation (7.118) shows that the matrix element $\langle m|\hat{q}_{sq}{}'(\omega)|n\rangle$ consists of five terms,

$$\langle m|\hat{q}_{sq}{}'(\omega)|n\rangle =$$

$$\underbrace{\langle m|\hat{q}_{sq}(\omega)|n\rangle}_{\text{A}} - \underbrace{\frac{1}{2}\langle m|\hat{\mu}_s|n\rangle\Delta r_q}_{\text{B}} - \underbrace{\frac{1}{2}\langle m|\hat{\mu}_q|n\rangle\Delta r_s}_{\text{C}} - \underbrace{\frac{\omega_{mn}}{2\omega}\langle m|\hat{\mu}_q|n\rangle\Delta r_s}_{\text{D}}$$

$$+ \underbrace{\frac{\omega_{mn}}{2\omega}\langle m|\hat{\mu}_s|n\rangle\Delta r_q}_{\text{E}} . \tag{7.118$'$}$$

By substituting q_{sq} in Eq. (7.85) with the above $q_{sq}{}'$, we obtain the transformed quadrupolar susceptibility $\alpha^{\mathrm{D1}\prime}(\Omega, \omega_1, \omega_2)$ as follows,

$$\alpha_{pqrs}^{\mathrm{D1}\ \prime}(\Omega, \omega_1, \omega_2) = \frac{1}{\hbar^2} \sum_{g,m,n}^{\substack{\text{whole}\\\text{states}}} (\rho_g^{(0)} - \rho_m^{(0)})$$

$$\left[\frac{\langle g|\hat{\mu}_p|n\rangle\langle n|\hat{\mu}_r|m\rangle\langle m|\hat{q}_{sq}{}'(\omega_1)|g\rangle}{(\omega_1 - \omega_{mg})(\Omega - \omega_{ng})} - \frac{\langle g|\hat{\mu}_r|n\rangle\langle n|\hat{\mu}_p|m\rangle\langle m|\hat{q}_{sq}{}'(\omega_1)|g\rangle}{(\omega_1 - \omega_{mg})(\Omega - \omega_{mn})} \right.$$

$$\left. + \frac{\langle g|\hat{\mu}_p|n\rangle\langle n|\hat{q}_{sq}{}'(\omega_1)|m\rangle\langle m|\hat{\mu}_r|g\rangle}{(\omega_2 - \omega_{mg})(\Omega - \omega_{ng})} - \frac{\langle g|\hat{q}_{sq}{}'(\omega_1)|n\rangle\langle n|\hat{\mu}_p|m\rangle\langle m|\hat{\mu}_r|g\rangle}{(\omega_2 - \omega_{mg})(\Omega - \omega_{mn})} \right]$$

$$= (\text{A}) - (\text{B}) - (\text{C}) - (\text{D}) + (\text{E}) \tag{7.126}$$

where the terms (A), (B), (C), (D), (E) correspond to the first, second, third, fourth and fifth terms in the right hand side of Eq. (7.118). These terms in Eq. (7.126) are shown to be

$$(\text{A}) = \alpha_{pqrs}^{\mathrm{D1}}(\Omega, \omega_1, \omega_2), \tag{7.127}$$

$$(\text{B}) = \frac{1}{2}\alpha_{psr}^{\mathrm{D0}}(\Omega, \omega_1, \omega_2)\Delta r_q, \tag{7.128}$$

$$(\text{C}) = \frac{1}{2}\alpha_{pqr}^{\mathrm{D0}}(\Omega, \omega_1, \omega_2)\Delta r_s, \tag{7.129}$$

$$(\text{D}) = \frac{1}{2\hbar^2} \sum_{g,m,n} (\rho_g^{(0)} - \rho_m^{(0)})$$

$$\left[\frac{\omega_{mg}}{\omega_1} \frac{\langle g|\hat{\mu}_p|n\rangle\langle n|\hat{\mu}_r|m\rangle\langle m|\hat{\mu}_q|g\rangle}{(\omega_1 - \omega_{mg})(\Omega - \omega_{ng})} - \frac{\omega_{mg}}{\omega_1} \frac{\langle g|\hat{\mu}_r|n\rangle\langle n|\hat{\mu}_p|m\rangle\langle m|\hat{\mu}_q|g\rangle}{(\omega_1 - \omega_{mg})(\Omega - \omega_{mn})} \right.$$

$$\left. + \frac{\omega_{nm}}{\omega_1} \frac{\langle g|\hat{\mu}_p|n\rangle\langle n|\hat{\mu}_q|m\rangle\langle m|\hat{\mu}_r|g\rangle}{(\omega_2 - \omega_{mg})(\Omega - \omega_{ng})} - \frac{\omega_{gn}}{\omega_1} \frac{\langle g|\hat{\mu}_q|n\rangle\langle n|\hat{\mu}_p|m\rangle\langle m|\hat{\mu}_r|g\rangle}{(\omega_2 - \omega_{mg})(\Omega - \omega_{mn})} \right] \Delta r_s, \tag{7.130}$$

$$
\begin{aligned}
(\mathrm{E}) = \frac{1}{2\hbar^2} \sum_{g,m,n} (\rho_g^{(0)} - \rho_m^{(0)}) \\
\Bigg[\frac{\omega_{mg}}{\omega_1} \frac{\langle g|\hat{\mu}_p|n\rangle\langle n|\hat{\mu}_r|m\rangle\langle m|\hat{\mu}_s|g\rangle}{(\omega_1 - \omega_{mg})(\Omega - \omega_{ng})} - \frac{\omega_{mg}}{\omega_1} \frac{\langle g|\hat{\mu}_r|n\rangle\langle n|\hat{\mu}_p|m\rangle\langle m|\hat{\mu}_s|g\rangle}{(\omega_1 - \omega_{mg})(\Omega - \omega_{mn})} \\
+ \frac{\omega_{nm}}{\omega_1} \frac{\langle g|\hat{\mu}_p|n\rangle\langle n|\hat{\mu}_s|m\rangle\langle m|\hat{\mu}_r|g\rangle}{(\omega_2 - \omega_{mg})(\Omega - \omega_{ng})} - \frac{\omega_{gn}}{\omega_1} \frac{\langle g|\hat{\mu}_s|n\rangle\langle n|\hat{\mu}_p|m\rangle\langle m|\hat{\mu}_r|g\rangle}{(\omega_2 - \omega_{mg})(\Omega - \omega_{mn})} \Bigg] \Delta r_q .
\end{aligned} \tag{7.131}
$$

In what follows, we prove that (B) = (E) and (C) = (D).

In the term (D) of Eq. (7.130), we apply the three relations,

$$
\begin{aligned}
&\frac{\omega_{mg}}{\omega_1} \frac{1}{(\omega_1 - \omega_{mg})} = \frac{1}{\omega_1 - \omega_{mg}} - \frac{1}{\omega_1}, \\
&\frac{\omega_{nm}}{\omega_1} \frac{1}{(\omega_2 - \omega_{mg})(\Omega - \omega_{ng})} = \frac{1}{(\omega_2 - \omega_{mg})(\Omega - \omega_{ng})} \\
&\qquad - \frac{1}{\omega_1} \left\{ \frac{1}{\omega_2 - \omega_{mg}} - \frac{1}{\Omega - \omega_{ng}} \right\}, \\
&\frac{\omega_{gn}}{\omega_1} \frac{1}{(\omega_2 - \omega_{mg})(\Omega - \omega_{mn})} = \frac{1}{(\omega_2 - \omega_{mg})(\Omega - \omega_{mn})} \\
&\qquad - \frac{1}{\omega_1} \left\{ \frac{1}{\omega_2 - \omega_{mg}} - \frac{1}{\Omega - \omega_{mn}} \right\},
\end{aligned}
$$

and thereby obtain

$$
\begin{aligned}
(\mathrm{D}) = &\frac{1}{2} \alpha_{pqr}^{\mathrm{D0}}(\Omega, \omega_1, \omega_2) \Delta r_s \\
&+ \frac{1}{2\hbar^2} \frac{\Delta r_s}{\omega_1} \sum_{g,m,n} (\rho_g^{(0)} - \rho_m^{(0)}) \\
&\Bigg[- \frac{\langle g|\hat{\mu}_p|n\rangle\langle n|\hat{\mu}_r|m\rangle\langle m|\hat{\mu}_q|g\rangle}{(\Omega - \omega_{ng})} + \frac{\langle g|\hat{\mu}_r|n\rangle\langle n|\hat{\mu}_p|m\rangle\langle m|\hat{\mu}_q|g\rangle}{(\Omega - \omega_{mn})} \\
&- \frac{\langle g|\hat{\mu}_p|n\rangle\langle n|\hat{\mu}_q|m\rangle\langle m|\hat{\mu}_r|g\rangle}{(\omega_2 - \omega_{mg})} + \frac{\langle g|\hat{\mu}_p|n\rangle\langle n|\hat{\mu}_q|m\rangle\langle m|\hat{\mu}_r|g\rangle}{(\Omega - \omega_{ng})} \\
&+ \frac{\langle g|\hat{\mu}_q|n\rangle\langle n|\hat{\mu}_p|m\rangle\langle m|\hat{\mu}_r|g\rangle}{(\omega_2 - \omega_{mg})} - \frac{\langle g|\hat{\mu}_q|n\rangle\langle n|\hat{\mu}_p|m\rangle\langle m|\hat{\mu}_r|g\rangle}{(\Omega - \omega_{mn})} \Bigg].
\end{aligned}
$$

In the above expression of (D), we can show that six terms in the square bracket cancel each other and vanish. By adopting the relation of completeness, $\sum_m |m\rangle \langle m| = 1$, and the commutation relation, $\hat{\mu}_p\hat{\mu}_q = \hat{\mu}_q\hat{\mu}_p$, we find that some terms are canceled and consequently obtain the following expression of (D),

$$
\begin{aligned}
(\mathrm{D}) = &\frac{1}{2}\alpha^{\mathrm{D0}}_{pqr}(\Omega,\omega_1,\omega_2)\Delta r_s \\
&+\frac{1}{2\hbar^2}\frac{\Delta r_s}{\omega_1}\left[\sum_{g,m,n}\rho^{(0)}_m\frac{\langle g|\hat{\mu}_p|n\rangle\langle n|\hat{\mu}_r|m\rangle\langle m|\hat{\mu}_q|g\rangle}{(\Omega-\omega_{ng})}\right. \\
&+\sum_{g,m,n}\rho^{(0)}_g\frac{\langle g|\hat{\mu}_r|n\rangle\langle n|\hat{\mu}_p|m\rangle\langle m|\hat{\mu}_q|g\rangle}{(\Omega-\omega_{mn})} \\
&-\sum_{g,m,n}\rho^{(0)}_m\frac{\langle g|\hat{\mu}_p|n\rangle\langle n|\hat{\mu}_q|m\rangle\langle m|\hat{\mu}_r|g\rangle}{(\Omega-\omega_{ng})} \\
&\left.-\sum_{g,m,n}\rho^{(0)}_g\frac{\langle g|\hat{\mu}_q|n\rangle\langle n|\hat{\mu}_p|m\rangle\langle m|\hat{\mu}_r|g\rangle}{(\Omega-\omega_{mn})}\right] \\
=&\frac{1}{2}\alpha^{\mathrm{D0}}_{pqr}(\Omega,\omega_1,\omega_2)\Delta r_s = (\mathrm{C})
\end{aligned}
$$

where all the terms in the square bracket cancel each other by properly replacing the suffixes, g, m, n. We could also prove (B) = (E) in the same way as above.

Therefore, we derive Eq. (7.119) for F=D1 from Eq. (7.126) as

$$
\begin{aligned}
\alpha^{\mathrm{D1}\,\prime}_{pqrs}(\Omega,\omega_1,\omega_2) &= (\mathrm{A})-(\mathrm{B})-(\mathrm{C})-(\mathrm{D})+(\mathrm{E}) = (\mathrm{A})-2(\mathrm{C}) \\
&= \alpha^{\mathrm{D1}}_{pqrs}(\Omega,\omega_1,\omega_2)-\alpha^{\mathrm{D0}}_{pqr}(\Omega,\omega_1,\omega_2)\Delta r_s .
\end{aligned}
$$

We can also confirm Eq. (7.119) for F=D2 and Q in the same way.

7.6.4 Levi-Civita Tensor

[Problem 7.4] Derive the following formulas (i)-(vi) using the Levi-Civita tensor. ($\boldsymbol{A}, \boldsymbol{B}, \boldsymbol{C}, \boldsymbol{D}$ refer to vectors, and ϕ to a scalar.)

(i) $(\boldsymbol{A}\times\boldsymbol{B})\cdot(\boldsymbol{C}\times\boldsymbol{D}) = (\boldsymbol{A}\cdot\boldsymbol{C})(\boldsymbol{B}\cdot\boldsymbol{D})-(\boldsymbol{A}\cdot\boldsymbol{D})(\boldsymbol{B}\cdot\boldsymbol{C})$

(ii) $\nabla\times(\nabla\times\boldsymbol{A}) = \nabla(\nabla\cdot\boldsymbol{A})-\nabla^2\boldsymbol{A}$

(iii) $\nabla\cdot(\nabla\times\boldsymbol{A}) = 0$

(iv) $\nabla\times(\nabla\phi) = 0$

(v) $\nabla\cdot(\boldsymbol{A}\times\boldsymbol{B}) = \boldsymbol{B}\cdot(\nabla\times\boldsymbol{A})-\boldsymbol{A}\cdot(\nabla\times\boldsymbol{B})$

(vi) $\nabla\times(\boldsymbol{A}\times\boldsymbol{B}) = \boldsymbol{A}(\nabla\cdot\boldsymbol{B})+(\boldsymbol{B}\cdot\nabla)\boldsymbol{A}-\boldsymbol{B}(\nabla\cdot\boldsymbol{A})-(\boldsymbol{A}\cdot\nabla)\boldsymbol{B}$

The formulas (i)-(vi) are derived as follows. Here we employ the Einstein's convention of contraction.

(i) $$(\boldsymbol{A} \times \boldsymbol{B}) \cdot (\boldsymbol{C} \times \boldsymbol{D}) = \varepsilon_{ijk} A_j B_k \varepsilon_{ilm} C_l D_m$$
$$= \left(\delta_{jl}\delta_{km} - \delta_{jm}\delta_{kl}\right) A_j B_k C_l D_m = A_j B_k C_j D_k - A_j B_k C_k D_j$$
$$= (\boldsymbol{A} \cdot \boldsymbol{C})(\boldsymbol{B} \cdot \boldsymbol{D}) - (\boldsymbol{A} \cdot \boldsymbol{D})(\boldsymbol{B} \cdot \boldsymbol{C})$$

(ii) $$[\nabla \times (\nabla \times \boldsymbol{A})]_i = \varepsilon_{ijk}\partial_j[\nabla \times \boldsymbol{A}]_k = \varepsilon_{ijk}\partial_j\left(\varepsilon_{klm}\partial_l A_m\right)$$
$$= \varepsilon_{ijk}\varepsilon_{lmk}\partial_j\partial_l A_m = \left(\delta_{il}\delta_{jm} - \delta_{im}\delta_{jl}\right)\partial_j\partial_l A_m = \partial_j\partial_i A_j - \partial_j\partial_j A_i$$
$$= \partial_i\left(\nabla \cdot \boldsymbol{A}\right) - \nabla^2 A_i$$

(iii) $$\nabla \cdot (\nabla \times \boldsymbol{A}) = \partial_i\left(\varepsilon_{ijk}\partial_j A_k\right) = \varepsilon_{ijk}\partial_i\partial_j A_k = 0$$
$$\left(\because \varepsilon_{ijk}\partial_i\partial_j A_k = -\varepsilon_{jik}\partial_i\partial_j A_k = 0\right)$$

(iv) $$[\nabla \times (\nabla\phi)]_i = \varepsilon_{ijk}\partial_j(\partial_k\phi) = \varepsilon_{ijk}\partial_j\partial_k\phi = 0$$

(v) $$\nabla \cdot (\boldsymbol{A} \times \boldsymbol{B}) = \partial_i\left(\varepsilon_{ijk}A_j B_k\right)$$
$$= \varepsilon_{ijk}(\partial_i A_j)B_k + \varepsilon_{ijk}A_j(\partial_i B_k) = B_k\varepsilon_{kij}\partial_i A_j - A_j\varepsilon_{jik}\partial_i B_k$$
$$= \boldsymbol{B} \cdot (\nabla \times \boldsymbol{A}) - \boldsymbol{A} \cdot (\nabla \times \boldsymbol{B})$$

(vi) $$[\nabla \times (\boldsymbol{A} \times \boldsymbol{B})]_i = \varepsilon_{ijk}\partial_j\left[\boldsymbol{A} \times \boldsymbol{B}\right]_k = \varepsilon_{ijk}\partial_j(\varepsilon_{klm}A_l B_m)$$
$$= (\delta_{il}\delta_{jm} - \delta_{im}\delta_{jl})\partial_j A_l B_m = \partial_j A_i B_j - \partial_j A_j B_i$$
$$= A_i\left(\nabla \cdot \boldsymbol{B}\right) + (\boldsymbol{B} \cdot \nabla)\,\boldsymbol{A} - B_i\left(\nabla \cdot \boldsymbol{A}\right) - (\boldsymbol{A} \cdot \nabla)\,B_i$$

Appendix

A.1 Physical Meaning of χ^{IQB}

Among the four terms of nonlinear susceptibility summarized in Fig. 7.5, χ^{IQB} may be difficult to understand intuitively. It arises from the integral of nonlinear polarization over the interface region in Sect. 7.2.3, though it is a pure bulk property. The χ^{IQB} term is known to have significant contribution in some SFG spectra [12, 13]. Here we explain its origin and physical meaning in an illustrative manner.[6]

As we discussed in Sect. 7.2.3, the χ^{IQB} term in Eq. (7.30) originates from the integral of quadrupole over the interface. Accordingly, we consider the situation in Fig. 7.6 that the induced quadrupole is distributed over the whole interface system,

[6] χ^{IQB} could be interpreted as the contribution of electric quadrupole and magnetic dipole radiation. Here we present a heuristic explanation.

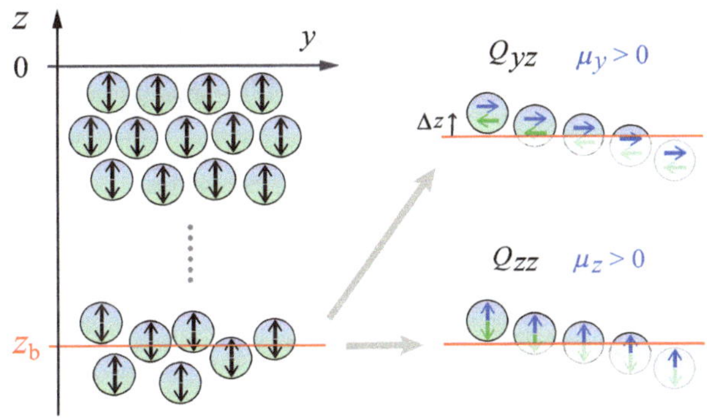

Fig. 7.6 (Left) Schematic picture of the quadrupole (Q_{zz}) distribution in the interface system, where the lower bound of the integral $z = z_b$ is shown by red line. (Right) Divided quadrupole moments (Q_{yz}, Q_{zz}) by the threshold $z = z_b$ bring net dipole moments (μ_y, μ_z) in the integral region $z > z_b$

and evaluate the net polarization of interface $\boldsymbol{P}^I$ in Eq. (7.26) by integrating the dipole polarization over the interface region from $z = z_b$ to ∞. The lower bound $z = z_b$ is arbitrarily chosen in the bulk region so that the integral encompasses the whole interface region. In this integral of polarization, the distributed quadrupole comes into play at the lower bound $z = z_b$. The molecules located across the threshold $z = z_b$ partially contribute to the integral, since the divided quadrupole moments Q_{yz} and Q_{zz} bring net dipole moments μ_y and μ_z, respectively, in the integral of $\boldsymbol{P}^I$. The contribution necessarily arises from the uniformly distributed quadrupole, irrespective of the location of the threshold $z = z_b$. We note that this mechanism is essentially common to the role of quadrupole on the surface potential [22].

The integral of quadrupole contributions could be understood without resorting to the arbitrary threshold of z_b. This mechanism of $\boldsymbol{\chi}^{\mathrm{IQB}}$ is related to the infinite summation of oscillating terms. A quadrupole is regarded as a pair of antiparallel dipoles, as illustrated in Fig. 7.6. Thus the sum of all quadrupole contributions becomes equivalent to the infinite summation of a pair of antiparallel dipoles, $(\mu - \mu) + (\mu - \mu) + (\mu - \mu) + \cdots$, where each pair $(\mu - \mu)$ corresponds to a quadrupole moment. This infinite summation could be defined on the basis of Abel summability [9],

$$(\mu - \mu) + (\mu - \mu) + \cdots = \sum_{n=0}^{\infty} \mu \, (-1)^n = \lim_{x \to -1+0} \frac{\mu}{1-x} = \frac{\mu}{2},$$

which yields a net dipole contribution. The above definition of this infinite sum can be obtained by using a proper convergence factor. Such oscillating sum elucidates the $\boldsymbol{\chi}^{\mathrm{IQB}}$ contribution to the net dipole $\boldsymbol{P}^I$. We will encounter the analogous mechanism in the NaOH aqueous solution surface in Sect. 9.3.

A.2 Levi-Civita Antisymmetric Tensor

The Levi-Civita antisymmetric tensor ε_{ijk} is defined as follows.

$$\varepsilon_{ijk} = \begin{cases} 1 & (ijk = xyz,\ yzx,\ zxy)\,, \\ -1 & (ijk = yxz,\ zyx,\ xzy)\,, \\ 0 & (\text{otherwise})\,. \end{cases} \tag{7.132}$$

This symbol is quite convenient when manipulating various formulas in the vector analysis.

Using the Levi-Civita symbols thus defined, the vector product is represented in the following form,

$$[\boldsymbol{B} \times \boldsymbol{C}]_i = \varepsilon_{ijk} B_j C_k.$$

In this section of Appendix we employ the Einstein's convention of contraction, and omit the summation symbol $\sum$ over i, j, or k.[7] The rotation of a vector is represented in a similar way, $[\nabla \times \boldsymbol{C}]_i = \varepsilon_{ijk}\partial_j C_k$. The scalar triple product is given by $\boldsymbol{A} \cdot (\boldsymbol{B} \times \boldsymbol{C}) = \varepsilon_{ijk} A_i B_j C_k$.

The Levi-Civita tensor satisfies the following contraction formulas,

$$\varepsilon_{ijk}\varepsilon_{pqk} = \delta_{ip}\delta_{jq} - \delta_{iq}\delta_{jp}, \tag{7.133}$$

$$\varepsilon_{ijk}\varepsilon_{pjk} = 2\delta_{ip}, \tag{7.134}$$

$$\varepsilon_{ijk}\varepsilon_{ijk} = 6. \tag{7.135}$$

Proof ε_{ijk} in Eq. (7.132) is expressed using a set of orthonormal vectors $\{\boldsymbol{e}_x, \boldsymbol{e}_y, \boldsymbol{e}_z\}$ as

$$\varepsilon_{ijk} = \begin{vmatrix} (\boldsymbol{e}_x \cdot \boldsymbol{e}_i) & (\boldsymbol{e}_x \cdot \boldsymbol{e}_j) & (\boldsymbol{e}_x \cdot \boldsymbol{e}_k) \\ (\boldsymbol{e}_y \cdot \boldsymbol{e}_i) & (\boldsymbol{e}_y \cdot \boldsymbol{e}_j) & (\boldsymbol{e}_y \cdot \boldsymbol{e}_k) \\ (\boldsymbol{e}_z \cdot \boldsymbol{e}_i) & (\boldsymbol{e}_z \cdot \boldsymbol{e}_j) & (\boldsymbol{e}_z \cdot \boldsymbol{e}_k) \end{vmatrix} \equiv |\, \boldsymbol{U}(xyz, ijk)\,|\,. \tag{7.136}$$

Therefore,

$$\begin{aligned} \varepsilon_{ijk}\varepsilon_{pqr} &= |\, \boldsymbol{U}(xyz, ijk)\,| \cdot |\, \boldsymbol{U}(xyz, pqr)\,| = \left|\, \boldsymbol{U}(xyz, ijk)^T \,\right| \cdot |\, \boldsymbol{U}(xyz, pqr)\,| \\ &= |\, \boldsymbol{U}(ijk, pqr)\,| \\ &= \begin{vmatrix} \delta_{ip} & \delta_{iq} & \delta_{ir} \\ \delta_{jp} & \delta_{jq} & \delta_{jr} \\ \delta_{kp} & \delta_{kq} & \delta_{kr} \end{vmatrix} \\ &= \delta_{ir}(\delta_{jp}\delta_{kq} - \delta_{jq}\delta_{kp}) - \delta_{jr}(\delta_{ip}\delta_{kq} - \delta_{iq}\delta_{kp}) + \delta_{kr}(\delta_{ip}\delta_{jq} - \delta_{iq}\delta_{jp}). \end{aligned}$$

[7]Therefore, $\varepsilon_{ijk} B_j C_k \equiv \sum_{j,k} \varepsilon_{ijk} B_j C_k$.

Equations (7.133), (7.134), (7.135) are proved by taking the contraction.

$$\begin{aligned}
\varepsilon_{ijk}\varepsilon_{pqk} &= \delta_{ik}(\delta_{jp}\delta_{kq} - \delta_{jq}\delta_{kp}) - \delta_{jk}(\delta_{ip}\delta_{kq} - \delta_{iq}\delta_{kp}) + \delta_{kk}(\delta_{ip}\delta_{jq} - \delta_{iq}\delta_{jp}) \\
&= \delta_{iq}\delta_{jp} - \delta_{ip}\delta_{jq} - \delta_{jq}\delta_{ip} + \delta_{jp}\delta_{iq} + 3(\delta_{ip}\delta_{jq} - \delta_{iq}\delta_{jp}) \\
&= \delta_{ip}\delta_{jq} - \delta_{iq}\delta_{jp}, \qquad (7.133) \\
\varepsilon_{ijk}\varepsilon_{pjk} &= \delta_{ip}\delta_{jj} - \delta_{ij}\delta_{jp} = 3\delta_{ip} - \delta ip = 2\delta_{ip}, \qquad (7.134) \\
\varepsilon_{ijk}\varepsilon_{ijk} &= 2\delta_{ii} = 6. \qquad (7.135)
\end{aligned}$$

□

These Eqs. (7.133), (7.134), (7.135), particularly Eq. (7.133), are extensively utilized in manipulating vector formulas. For example, the vector triple product is rearranged as follows,

$$\begin{aligned}
&[\boldsymbol{A} \times (\boldsymbol{B} \times \boldsymbol{C})]_i = \varepsilon_{ijk} A_j [B \times C]_k - \varepsilon_{ijk} A_j (\varepsilon_{klm} B_l C_m) \\
&= \varepsilon_{ijk}\varepsilon_{lmk} A_j B_l C_m = (\delta_{il}\delta_{jm} - \delta_{im}\delta_{jl}) A_j B_l C_m = A_j B_i C_j - A_j B_j C_i \\
&- B_i(\boldsymbol{A} \cdot \boldsymbol{C}) - C_i(\boldsymbol{A} \cdot \boldsymbol{B}),
\end{aligned}$$

where we have employed the permutation relation, $\varepsilon_{klm} = \varepsilon_{lmk}$, and Eq. (7.133) in the above derivation.

[Problem 7.4] Derive the following formulas (i)–(vi) using the Levi-Civita tensor (see Sect. 7.6.4). ($\boldsymbol{A}, \boldsymbol{B}, \boldsymbol{C}, \boldsymbol{D}$ refer to vectors, and ϕ to a scalar.)

(i) $(\boldsymbol{A} \times \boldsymbol{B}) \cdot (\boldsymbol{C} \times \boldsymbol{D}) = (\boldsymbol{A} \cdot \boldsymbol{C})(\boldsymbol{B} \cdot \boldsymbol{D}) - (\boldsymbol{A} \cdot \boldsymbol{D})(\boldsymbol{B} \cdot \boldsymbol{C})$

(ii) $\nabla \times (\nabla \times \boldsymbol{A}) = \nabla(\nabla \cdot \boldsymbol{A}) - \nabla^2 \boldsymbol{A}$

(iii) $\nabla \cdot (\nabla \times \boldsymbol{A}) = 0$

(iv) $\nabla \times (\nabla \phi) = 0$

(v) $\nabla \cdot (\boldsymbol{A} \times \boldsymbol{B}) = \boldsymbol{B} \cdot (\nabla \times \boldsymbol{A}) - \boldsymbol{A} \cdot (\nabla \times \boldsymbol{B})$

(vi) $\nabla \times (\boldsymbol{A} \times \boldsymbol{B}) = \boldsymbol{A}(\nabla \cdot \boldsymbol{B}) + (\boldsymbol{B} \cdot \nabla)\boldsymbol{A} - \boldsymbol{B}(\nabla \cdot \boldsymbol{A}) - (\boldsymbol{A} \cdot \nabla)\boldsymbol{B}$

The Levi-Civita tensor is invariant under rotation of the coordinates,

$$\mathcal{D}_{ip}\mathcal{D}_{jq}\mathcal{D}_{kr}\,\varepsilon_{pqr} = \varepsilon_{ijk}, \qquad (7.137)$$

where $\mathcal{D}$ is the rotational matrix in Eq. (3.43). This feature will be utilized in Chap. 8.

Proof Consider a rotational matrix $\mathcal{D}$ that converts a set of orthonormal vectors $\{\boldsymbol{e}_x, \boldsymbol{e}_y, \boldsymbol{e}_z\}$ to $\{\boldsymbol{e}_{x'}, \boldsymbol{e}_{y'}, \boldsymbol{e}_{z'}\}$,

$$\begin{pmatrix} \boldsymbol{e}_{x'} \\ \boldsymbol{e}_{y'} \\ \boldsymbol{e}_{z'} \end{pmatrix} = \mathcal{D} \begin{pmatrix} \boldsymbol{e}_x \\ \boldsymbol{e}_y \\ \boldsymbol{e}_z \end{pmatrix}.$$

Then the following relation is proved using the unitary character of $\mathcal{D}$ ($|\mathcal{D}| = 1$) and Eq. (7.136),

$$\varepsilon_{ijk} = |\ \boldsymbol{U}(xyz, ijk)\ | = \left| \mathcal{D}^T\ \boldsymbol{U}(x'y'z', ijk) \right| = \left|\ \boldsymbol{U}(x'y'z', ijk)\ \right|.$$

Suppose that pqr denote x', y', z' while ijk denote x, y, z, the following equation is derived.

$$\begin{aligned} \mathcal{D}_{ip}\mathcal{D}_{jq}\mathcal{D}_{kr}\varepsilon_{pqr} = \mathcal{D}_{ip}\mathcal{D}_{jq}\mathcal{D}_{kr}\ |\ \boldsymbol{U}(x'y'z', pqr)\ | = |\ \boldsymbol{U}(x'y'z', ijk)\ | \\ = |\ \boldsymbol{U}(xyz, ijk)\ | = \varepsilon_{ijk}. \end{aligned} \tag{7.137}$$

□

A.3 Definition of Bulk Polarization

In Chap. 7 the bulk polarization $\boldsymbol{P}_G^{\mathrm{B}}$ is defined by Eq. (7.36),

$$\begin{aligned} P_{G,p}^{\mathrm{B}} = i \int_{-\infty}^{0} dz \sum_{q,r,s} \Big\{ &\chi_{pqrs}^{\mathrm{D1},\beta}(\Omega, \omega_1, \omega_2) k_{T,s}^{\beta}(\omega_1) + \chi_{pqrs}^{\mathrm{D2},\beta}(\Omega, \omega_1, \omega_2) k_{T,s}^{\beta}(\omega_2) \\ &- \chi_{pqrs}^{\mathrm{Q},\beta}(\Omega, \omega_1, \omega_2) \left(k_{T,s}^{\beta}(\omega_1) + k_{T,s}^{\beta}(\omega_2) \right) \Big\} \\ &\cdot f_p^{\beta}(\Omega) f_q^{\beta}(\omega_1) f_r^{\beta}(\omega_2) L_{I,q}(\omega_1) L_{I,r}(\omega_2) E_{I,q}^{\alpha}(\omega_1) E_{I,r}^{\alpha}(\omega_2) \\ &\cdot \exp\left[i \left(k_{T,z}^{\beta}(\omega_1) + k_{T,z}^{\beta}(\omega_2) - k_{G,z}^{\beta}(\Omega) \right) z \right], \end{aligned} \tag{7.36}$$

which is consistent to Refs. [21] and [17], while a slightly different expression $\boldsymbol{P}_G^{\mathrm{B}\prime}$ is found in Refs. [11] and [5]. In this Appendix, we clarify the difference between the two expressions.

Some other literature [5, 11] employs a different expression, $\boldsymbol{P}_G^{\mathrm{B}\prime}$, for the bulk polarization,

$$P^{\mathrm{B}}_{G,p}{}' = i \int_{-\infty}^{0} dz \sum_{q,r,s} \left\{ \chi^{\mathrm{D1},\beta}_{pqrs}(\Omega, \omega_1, \omega_2) k^{\beta}_{T,s}(\omega_1) + \chi^{\mathrm{D2},\beta}_{pqrs}(\Omega, \omega_1, \omega_2) k^{\beta}_{T,s}(\omega_2) \right.$$
$$\left. - \chi^{\mathrm{Q},\beta}_{pqrs}(\Omega, \omega_1, \omega_2) k^{\beta}_{G,s}(\Omega) \right\}$$
$$\cdot f^{\beta}_{p}(\Omega) f^{\beta}_{q}(\omega_1) f^{\beta}_{r}(\omega_2) L_{I,q}(\omega_1) L_{I,r}(\omega_2) E^{\alpha}_{I,q}(\omega_1) E^{\alpha}_{I,r}(\omega_2)$$
$$\cdot \exp\left[i \left(k^{\beta}_{T,z}(\omega_1) + k^{\beta}_{T,z}(\omega_2) - k^{\beta}_{G,z}(\Omega) \right) z \right]. \tag{7.138}$$

We notice a slight but significant difference; the third term of the integrand in Eq. (7.36) is $-\chi^{Q,\beta}_{pqrs}(\Omega, \omega_1, \omega_2) \left(k^{\beta}_{T,s}(\omega_1) + k^{\beta}_{T,s}(\omega_2) \right)$, while that in Eq. (7.138) is $-\chi^{Q,\beta}_{pqrs}(\Omega, \omega_1, \omega_2) k^{\beta}_{G,s}(\Omega)$.

We could interpret that $\boldsymbol{P}^{\mathrm{B}\prime}_{G}$ in Eq. (7.138) involves both the bulk polarization $\boldsymbol{P}^{\mathrm{B}}_{G}$ and a part of interfacial polarization attributed to $\boldsymbol{\chi}^{\mathrm{IQB}}$. Equation (7.28) shows that the part of interfacial polarization $\boldsymbol{P}^{\mathrm{I}}$ attributed to $\boldsymbol{\chi}^{\mathrm{IQB}}$ is

$$\begin{aligned} P^{\mathrm{IQB}}_{p} &= \sum_{q,r} \chi^{\mathrm{IQB}}_{pqr}(\Omega, \omega_1, \omega_2) L_{I,q}(\omega_1) L_{I,r}(\omega_2) E^{\alpha}_{I,q}(\omega_1) E^{\alpha}_{I,r}(\omega_2) \\ &= \sum_{q,r} \chi^{\mathrm{Q},\beta}_{pqrz}(\Omega, \omega_1, \omega_2) f^{\beta}_{p}(\Omega) f^{\beta}_{q}(\omega_1) f^{\beta}_{r}(\omega_2) L_{I,q}(\omega_1) L_{I,r}(\omega_2) E^{\alpha}_{I,q}(\omega_1) E^{\alpha}_{I,r}(\omega_2) \\ &= i \int_{-\infty}^{0} dz \sum_{q,r,s} \delta_{sz} \left(k^{\beta}_{T,s}(\omega_1) + k^{\beta}_{T,s}(\omega_2) - k^{\beta}_{G,s}(\Omega) \right) \chi^{Q,\beta}_{pqrz}(\Omega, \omega_1, \omega_2) \\ &\quad \cdot f^{\beta}_{p}(\Omega) f^{\beta}_{q}(\omega_1) f^{\beta}_{r}(\omega_2) L_{I,q}(\omega_1) L_{I,r}(\omega_2) E^{\alpha}_{I,q}(\omega_1) E^{\alpha}_{I,r}(\omega_2) \\ &\quad \cdot \exp\left[i \left(k^{\beta}_{T,z}(\omega_1) + k^{\beta}_{T,z}(\omega_2) - k^{\beta}_{G,z}(\Omega) \right) z \right], \end{aligned} \tag{7.139}$$

where the form of $\boldsymbol{\chi}^{\mathrm{IQB}}$ in Eq. (7.30) has been adopted in the above formula. Therefore, we can readily see the following relation,

$$P^{\mathrm{B}}_{G,p} + P^{\mathrm{IQB}}_{p} = P^{\mathrm{B}}_{G,p}{}'. \tag{7.140}$$

Although either definition for the bulk polarization could be used in principle, we recommend $\boldsymbol{P}^{\mathrm{B}}_{G}$ in Eq. (7.36) to describe the bulk polarization for the following reasons. First, $\boldsymbol{P}^{\mathrm{B}}_{G}$ is a well defined quantity with respect to the choice of origin as described in Sect. 7.4, and can be separately detected from the other terms by experimental measurements. Second, $\boldsymbol{P}^{\mathrm{B}}_{G}$ allows for distinguishing $\boldsymbol{\chi}^{\mathrm{IQB}}$ and $\boldsymbol{\chi}^{\mathrm{B}}$. We have argued in Chap. 7 and Figure 7.5 that $\boldsymbol{\chi}^{\mathrm{IQB}}$ and $\boldsymbol{\chi}^{\mathrm{B}}$ have different physical meanings and different dependence on the optical geometry. The role of $\boldsymbol{\chi}^{\mathrm{IQB}}$ has been increasingly recognized in the SFG spectroscopy [12, 13]. The former definition is convenient to examine the effects of $\boldsymbol{\chi}^{\mathrm{IQB}}$ and $\boldsymbol{\chi}^{\mathrm{B}}$ separately.

Bibliography

1. Adler E (1964) Nonlinear optical frequency polarization in a dielectric. Phys Rev 134:A728–A733
2. Bloembergen N, Chang RK, Jha SS, Lee CH (1968) Optical second-harmonic generation in reflection from media with inversion symmetry. Phys Rev 174:813–822
3. Bloembergen N, Pershan PS (1962) Light waves at the boundary of nonlinear media. Phys Rev 128:606–622
4. Boyd RW (2003) Nonlinear optics. Academic, Amsterdam
5. Byrnes SJ, Geissler PL, Shen YR (2011) Ambiguities in surface nonlinear spectroscopy calculations. Chem Phys Lett 516:115–124
6. Guyot-Sionnest P, Chen W, Shen YR (1986) General considerations on optical second-harmonic generation from surfaces and interfaces. Phys Rev B 33:8254–8263
7. Guyot-Sionnest P, Shen YR (1987) Local and nonlocal surface nonlinearities for surface optical second-harmonic generation. Phys Rev B 35:4420–4426
8. Guyot-Sionnest P, Shen YR (1988) Bulk contribution in surface second-harmonic generation. Phys Rev B 38:7985–7989
9. Hardy GH (1949) Divergent series. Clarendon Press, Oxford
10. Heinz TF (1991) Second-order nonlinear optical effects at surfaces and interfaces. In: Ponath H-E, Stegeman GI (eds) Nonlinear surface electromagnetic phenomena. Elsevier, Amsterdam, pp 353–416
11. Held H, Lvovsky AI, Wei X, Shen YR (2002) Bulk contribution from isotropic media in surface sum-frequency generation. Phys Rev B 66:205110
12. Kawaguchi T, Shiratori K, Henmi Y, Ishiyama T, Morita A (2012) Mechanisms of sum frequency generation from liquid benzene: symmetry breaking at interface and bulk contribution. J Phys Chem C 116:13169–13182
13. Kundu A, Tanaka S, Ishiyama T, Ahmed M, Inoue K, Nihonyanagi S, Sawai H, Yamaguchi S, Morita A, Tahara T (2016) Bend vibration of surface water investigated by heterodyne-detected sum frequency generation and theoretical study: dominant role of quadrupole. J Phys Chem Lett 7:2597–2601
14. Morita A (2004) Toward computation of bulk quadrupolar signals in vibrational sum frequency generation spectroscopy. Chem Phys Lett 398:361–366
15. Neipert C, Space B, Roney AB (2007) Generalized computational time correlation function approach: quantifying quadrupole contributions to vibrationally resonant second-order interface-specific optical spectroscopies. J Phys Chem C 111:8749–8756
16. Pershan PS (1963) Nonlinear optical properties of solids: energy considerations. Phys Rev 130:919–929
17. Shiratori K, Morita A (2012) Theory of quadrupole contributions from interface and bulk in second-order optical processes. Bull Chem Soc Jpn 85:1061–1076
18. Sipe JE, Mizrahi V, Stegeman GI (1987) Fundamental difficulty in the use of second-harmonic generation as a strictly surface probe. Phys Rev B 35:9091–9094
19. Sipe JE, Moss DJ, van Driel HM (1987) Phenomenological theory of optical second- and third-harmonic generation from cubic centrosymmetric crystals. Phys Rev B 35:1129–1141
20. Terhune RW, Maker PD, Savage CM (1962) Optical harmonic generation in calcite. Phys Rev Lett 8:404–406
21. Wei X, Hong SC, Goto T, Shen YR (2000) Nonlinear optical studies of liquid crystal alignment on a rubbed polyvinyl alcohol surface. Phys Rev E 62:5160–5172
22. Wilson MA, Pohorille A, Pratt LR (1989) Comment on "study on the liquid-vapor interface of water. I. Simulation results of thermodynamic properties and orientational structure". J Chem Phys 90:5211–5213

Chapter 8
Other Topics

Abstract This chapter introduces some recent topics in the SFG spectroscopy, particularly the $\chi^{(3)}$ effect for the charged interfaces and the chiral applications of SFG spectroscopy. The significance of $\chi^{(3)}$ effect is now recognized in the SFG and SHG spectra when interfaces are charged, such as the electrolyte interfaces and charged solid-liquid interfaces. Understanding of this effect is indispensable to analyze the spectra of these systems. Chiral application of the SFG spectroscopy has been also advanced recently, though chiral systems are known to be SFG active for long time. It could offer a new probe technique of chiral species at interfaces. The present chapter discuss basic features of chiral applications, which are somewhat different from the features of ordinary applications to interfaces.

Keywords $\chi^{(3)}$ effect · Surface charge · Chiral SFG

Recent development of SFG spectroscopy has brought new areas of applications as interface probe. Here we treat some recent topics in the SFG spectroscopy. Section 8.1 deals with the "$\chi^{(3)}$ effect", which is pertinent to liquids in contact to charged substrates, such as charged solid (membrane)-liquid interfaces and electrode-solution interfaces. The $\chi^{(3)}$ effect could have substantial contributions to the SFG/SHG spectra from these charged interfaces. It becomes an additional source of SFG/SHG, and could compete with the intrinsic signal from the interface. Precise understanding of this effect is critical to interpret the SFG/SHG spectra of charged interfaces. Section 8.2 treats chiral applications of SFG spectroscopy. The SFG is allowed in chiral systems in principle, as the inversion symmetry is broken. Actually the SFG as well as SHG from non-centrosymmetric materials is widely utilized in optical conversion devices. However, applications of SFG as a spectroscopic tool of probing chiral species have been relatively less explored. Recently the SFG/SHG could be utilized as a chiral probe at interfaces. The present section summarizes fundamental features of chiral SFG/SHG spectroscopy.

A. Morita, *Theory of Sum Frequency Generation Spectroscopy*,
Lecture Notes in Chemistry 97, https://doi.org/10.1007/978-981-13-1607-4_8

8.1 $\chi^{(3)}$ Effect at Charged Interfaces

In the SFG or SHG spectra from liquids at charged interfaces, another mechanism of SFG/SHG emission arises besides the usual one, called "$\chi^{(3)}$ effect". The charge at interface generates static electric field E that penetrates into the interface and bulk of the liquid, and thereby induces additional nonlinear polarization in the liquid. In such cases, the induced polarization of sum frequency $\Omega = \omega_1 + \omega_2$ by two incident light fields of frequencies ω_1 and ω_2 is represented as follows,

$$P_p^{(2)}(\Omega) = \sum_{q,r}^{x\sim z} \chi_{pqr}^{(2)}(\Omega, \omega_1, \omega_2) E_q(\omega_1) E_r(\omega_2) + \sum_{q,r,s}^{x\sim z} \chi_{pqrs}^{(3)}(\Omega, \omega_1, \omega_2, 0) E_q(\omega_1) E_r(\omega_2) E_s(0), \tag{8.1}$$

where $E_s(0)$ denotes the static electric field. This formula (8.1) is considered to be an extension of Eq. (1.4) in Chap. 1, and the additional second term of Eq. (8.1) refers to the $\chi^{(3)}$ effect.[1] In what follows, we formulate the SFG spectroscopy without losing generality, as SHG is a special case of SFG.

The $\chi^{(3)}$ effect in SHG spectroscopy was pointed out by Eisenthal and co-workers in silica-water interface with varying pH [23]. They discovered that the intensity of SHG signal from water strongly depends on the pH. Figure 8.1 demonstrates that the SHG signal is remarkably enhanced with increasing pH of the solution. The silica surface contains silanol groups (−SiOH) which undergo acid-base equilibrium, $-\text{SiOH} \rightleftharpoons -\text{SiO}^-$, and thus negative charge density at the silica surface increases with increasing pH. They argued that the SHG signal clearly correlates with the charge density, which is indicative of the $\chi^{(3)}$ effect as we will discuss below. Since then the $\chi^{(3)}$ effect has been intensively investigated in SHG and SFG for a variety of aqueous solutions in contact with charged interfaces [7–9, 12, 15, 18, 25]. In this section we clarify the fundamental properties and roles of the $\chi^{(3)}$ effect in SFG/SHG spectroscopy.

8.1.1 Properties of $\chi^{(3)}$ Tensor

First, we notice that the $\chi_{pqrs}^{(3)}$ in Eq. (8.1) is a fourth-rank tensor. Consequently, $\chi^{(3)}$ is not necessarily zero (allowed) in an isotropic matter for symmetry reason, whereas the third-rank tensor of $\chi^{(2)}$ is inevitably zero (forbidden). This important property of $\chi^{(3)}$ is intuitively understood by inspecting the role of $\chi^{(3)}$ in Eq. (8.1).

[1] Note that $P^{(2)}$ in Eq. (8.1) includes the third-order polarization of $\chi^{(3)}$. We use the notation $P^{(2)}$ for the SFG source polarization to show the correspondence to Eq. (1.4).

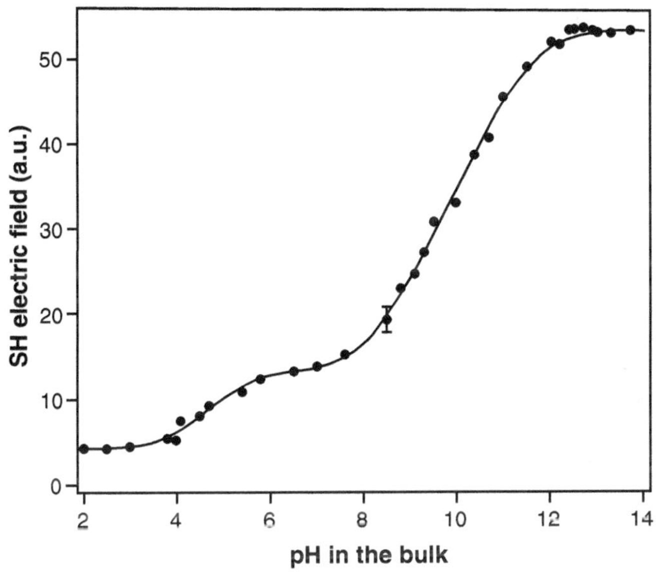

Fig. 8.1 Observed SHG signal from silica-water interface as a function of pH [23]. (Reprinted from Ref. [23], Copyright 1992, with permission from Elsevier)

The sum-frequency polarization in Eq. (8.1) is represented in the following form,

$$P_p^{(2)}(\Omega) = \sum_{q,r} \left[\chi_{pqr}^{(2)} + \sum_s \chi_{pqrs}^{(3)} E_s(0) \right] E_q(\omega_1) E_r(\omega_2), \tag{8.2}$$

where the quantity in the bracket corresponds to the SFG source term of $\chi^{(2)}$ including the static field $E(0)$. Accordingly, $\chi^{(3)}$ is represented with the derivative of $\chi^{(2)}$ with respect to the field $E(0)$,

$$\chi_{pqrs}^{(3)}(\Omega, \omega_1, \omega_2, 0) = \left(\frac{\partial \chi_{pqr}^{(2)}(\Omega, \omega_1, \omega_2)}{\partial E_s(0)} \right)_{E(0)=0}. \tag{8.3}$$

Equation (8.3) indicates that $\chi^{(3)}$ accounts for the induced SFG/SHG by the static electric field. The field-induced SFG/SHG is allowed in an isotropic media, since the imposed field $E(0)$ would break the isotropy of the system and thereby induce the SFG/SHG response. In charged solid-liquid interfaces, the SFG/SHG signal from $\chi^{(3)}$ stems from the liquid region as long as the electrostatic field penetrates into the place. The length of the region is comparable to the Debye screening length, which could be larger than the molecular scale as we discuss later.

Second issue is the microscopic origin of $\chi^{(3)}$. As we mentioned above, $\chi^{(3)}$ in Eq. (8.3) is understood as the perturbation on $\chi^{(2)}$ by the field. Microscopic origin of the perturbation can be twofold, orientational and electronic mechanisms. The former, orientational mechanism of $\chi^{(3)}$ means that the static electric field perturbs the orientational structure of molecules and thereby changes the $\chi^{(2)}$ of that material. The latter, electronic mechanism means that the electronic state of the material has the intrinsic third-order response to the electric fields. Relative significance of the two mechanisms in $\chi^{(3)}$ may depend on the system in question. In liquid water or dilute aqueous solutions, the orientational mechanism is confirmed to be dominant in $\chi^{(3)}$ by direct computation of the two mechanisms [16].

Third issue is the distinction of the $\chi^{(3)}$ effect from the quadrupole terms discussed in Chap. 7. Both the $\chi^{(3)}$ in the present chapter and the quadrupole terms are fourth-rank tensors and thus allowed in isotropic bulk media, though their mechanisms are fully distinct. The $\chi^{(3)}$ effect arises only when the electrostatic field is imposed on the system, while the quadrupole contributions are present in the SFG/SHG even in a system under no electrostatic field. If we take account of the quadrupole contributions in Eqs. (8.2) and (8.3), the $\chi^{(2)}$ in these equations should be replaced with the extended susceptibility $\chi^{(2)}_{qG}$ in Eq. (7.50) that includes both the dipole and quadrupole contributions.

The $\chi^{(3)}$ formula of Eq. (8.3) allows for straightforward calculation of the $\chi^{(3)}$ tensor by adopting the computational methods of $\chi^{(2)}$ in Chap. 4. One can calculate $\chi^{(2)}$ of a system with imposing a finite electrostatic field $E(0)$, and take the derivative of the calculated $\chi^{(2)}$ with respect to the applied electrostatic field. Such MD calculation of $\chi^{(3)}$ is conducted using isotropic bulk systems which are free from interfaces [16]. The calculated derivative of $\chi^{(2)}$ derives the pure $\chi^{(3)}$ term as a bulk property, since the isotropic systems are essentially free from the intrinsic $\chi^{(2)}$. Calculated result of $\chi^{(3)}$ for liquid water is shown in Fig. 8.2b, and compared with the intrinsic $\chi^{(2)}$ for water surface in panel (a).

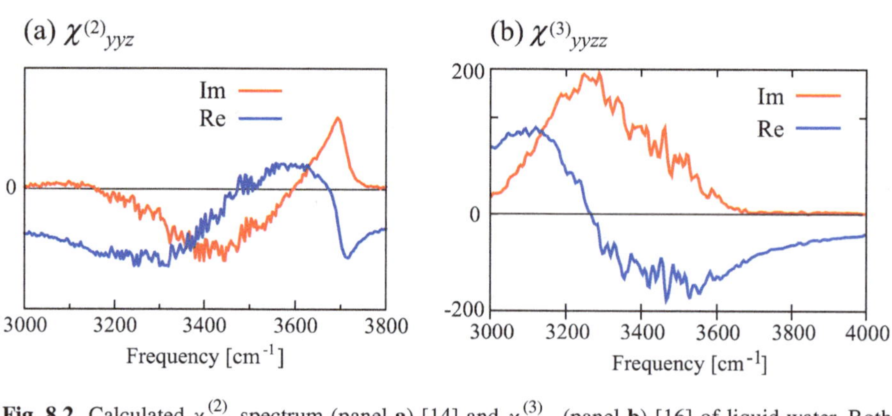

Fig. 8.2 Calculated $\chi^{(2)}_{yyz}$ spectrum (panel **a**) [14] and $\chi^{(3)}_{yyzz}$ (panel **b**) [16] of liquid water. Both panels are obtained using the same CRK model of water. Their real and imaginary parts are shown in blue and red, respectively. (Reprinted with permission from Ref. [14]; Copyright 2009, American Institute of Physics. Reproduced from Ref. [16] with permission from the PCCP Owner Societies)

8.1.2 *Role of* $\chi^{(3)}$ *in Electrolyte Solution*

Then we discuss the $\chi^{(3)}$ effect in electrolyte solutions near solid-liquid interfaces. In the conventional picture of electrochemistry, the structure of electrolyte solutions near a solid (or electrode) surface consists of the contact (Stern) layer and the diffuse (Guoy-Chapman) layer [2, 3]. The former indicates the region of a few monolayers at the solid-liquid interface, where specific interfacial structure, such as specific adsorption, is constructed. The latter, diffuse layer extends to the solution over the Debye screening length, where the counter ions of charged surface are distributed and the net electric field is penetrated. The $\chi^{(3)}$ signal arises from the diffuse layer of electrolyte solution, while the contact layer is considered to be the source of interface-specific SFG/SHG signal by $\chi^{(2)}$. Here we formulate the $\chi^{(3)}$ contribution from the diffuse layer based on the Gouy-Chapman theory [1–3].

Guoy-Chapman theory Suppose the normal coordinate z to the solid-liquid interface. The interface is located at $z = 0$ and has charge density σ per unit area, while the electrolyte solution with dielectric constant ε is extended in $z < 0$. (See the scheme of Fig. 1.2 or Fig. 8.3, where the medium α is regarded as solid and the medium β as the solution.[2]) The electrolyte in the solution consists of N_i kinds of ions of charge Z_i and number density n_i ($i = 1, \cdots, N_i$). The static electric field $E_z(0; z)$ is generated in the diffuse layer of solution along the z axis, and is related to the electrostatic potential $\Phi(z)$ at z by

$$E_z(0; z) = -\frac{d\Phi(z)}{dz} \quad \text{or} \quad \Phi(z) = -\int_{-\infty}^{z} E_z(0; z)dz, \tag{8.4}$$

where the origin of the potential is set in the bulk liquid, $\Phi(z \to -\infty) = 0$.

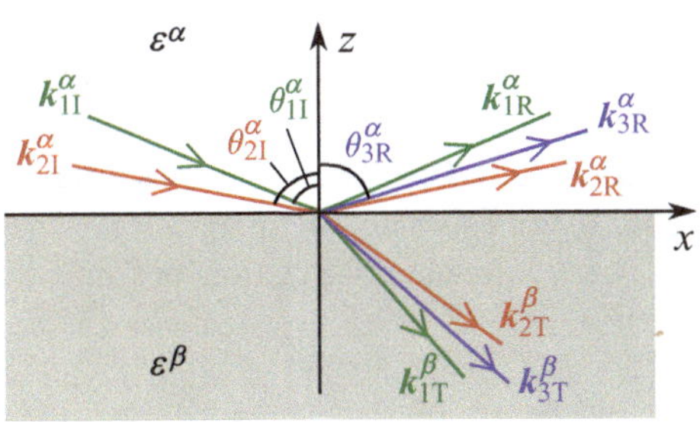

Fig. 8.3 Definition of parameters used for the surface SFG. The subscripts $f = 1, 2, 3$ denote the two incident lights and SFG, respectively. e.g. $\boldsymbol{k}^{\alpha}_{1I} = \boldsymbol{k}^{\alpha}_{I}(\omega_1)$, $\boldsymbol{k}^{\alpha}_{2R} = \boldsymbol{k}^{\alpha}_{R}(\omega_2)$, $\theta^{\alpha}_{3R} = \theta^{\alpha}_{R}(\Omega)$. (Same as Fig. 7.1 in Chap. 7)

---

[2]This convention is taken so as to be consistent to Chaps. 2 and 7, where the medium β ($z < 0$) is the source of polarization.

[Problem 8.1] Using the above notations and the Poisson-Boltzmann equation, derive the following formula of the electric field $E_z(0; z)$ in $z \leq 0$,

$$E_z(0; z)^2 = \left(\frac{d\Phi(z)}{dz}\right)^2 = \frac{8\pi k_B T}{\varepsilon} \sum_{i=1}^{N_i} n_i \left\{\exp\left(-\frac{Z_i e\Phi(z)}{k_B T}\right) - 1\right\}. \tag{8.5}$$

In a case of $Z : Z$ electrolyte solution ($N_i = 2$, $Z_1 = -Z_2 = Z$, $n_1 = n_2 = n$), show that Eq. (8.5) is simplified to be

$$E_z(0; z) = -\frac{d\Phi(z)}{dz} = -\sqrt{\frac{32\pi k_B T n}{\varepsilon}} \sinh\left(\frac{Ze\Phi(z)}{2k_B T}\right). \tag{8.6}$$

(*Hint*) Note that $\Phi(z)$ is related to the charge density $\rho(z)$ by the one-dimensional Poisson equation,

$$\frac{d^2\Phi(z)}{dz^2} = -\frac{4\pi}{\varepsilon}\rho(z), \tag{8.7}$$

while $\rho(z)$ is assumed to be given by the Boltzmann distribution of ions,

$$\rho(z) = \sum_{i=1}^{N_i} Z_i e n_i \exp\left(-\frac{Z_i e\Phi(z)}{k_B T}\right). \tag{8.8}$$

Coupled solution of Eqs. (8.7) and (8.8) leads to Eq. (8.5).

Surface charge and potential Using the static electric field $E_z(0; z)$ in the diffuse layer, the $\chi^{(3)}$ contribution to the SFG/SHG signal is estimated by integrating it over the z coordinate. The following formulas are analogous with those for the bulk quadrupole in Chap. 7, in the sense that the overall contributions from deep region ($z \ll 0$) are estimated by integration.

The nonlinear polarization per unit area $\boldsymbol{P}^S$ is given by integrating Eq. (8.2) along z,

$$\begin{aligned} P_p^S &= \int_{-\infty}^{0} dz \sum_{q,r} \left[\chi_{pqr}^{(2)} + \chi_{pqrz}^{(3)} E_z(0)\right] E_q(\omega_1) E_r(\omega_2) \\ &\approx \sum_{q,r} \int_{-\infty}^{0} dz \left[\chi_{pqr}^{(2)} \delta(z - 0_-) + \chi_{pqrs}^{(3)} E_z(0; z)\right] E_q(\omega_1) E_r(\omega_2) \\ &\approx \sum_{q,r} \left[\chi_{pqr}^{(2)} - \chi_{pqrs}^{(3)} \Phi(0)\right] E_q(\omega_1) E_r(\omega_2). \end{aligned} \tag{8.9}$$

In the final expression of Eq. (8.9), $\chi^{(2)}$ is defined as the second-order susceptibility *per unit area*, and $\chi^{(3)}$ as the third-order susceptibility *per unit volume*. Since $\chi^{(3)}$ is a bulk property, the definition per unit volume is feasible for a uniform material. Equation (8.9) includes the surface potential $\Phi(0) = -\int_{-\infty}^{0} E_z(0; z)dz$, which affects the relative amplitude of $\chi^{(3)}\Phi(0)$ term to $\chi^{(2)}$.[3]

The surface potential $\Phi(0)$ is related to the surface charge density σ. By applying the Gauss divergence theorem (see Sect. 2.1.2) to the solid-liquid interface, where the surface charge σ is located and the electric field E (or electric displacement D) is along the z direction, one leads to

$$4\pi\sigma = -\varepsilon E_z(0; z = 0_-).$$

In the case of Z:Z electrolyte solution, the result of Guoy-Chapman theory in Eq. (8.6) is further applied to the above equation to obtain

$$\sigma = \sqrt{\frac{2k_B T n\varepsilon}{\pi}} \sinh\left(\frac{Ze\Phi(0)}{2k_B T}\right) \tag{8.10}$$

or

$$\Phi(0) = \frac{2k_B T}{Ze} \sinh^{-1}\left(\sqrt{\frac{\pi}{2k_B T n\varepsilon}}\sigma\right). \tag{8.11}$$

Therefore, the surface charge density σ gives rise to the surface potential $\Phi(0)$, and thereby the $\chi^{(3)}$ signal of SFG/SHG in Eq. (8.9).

Phase factor In the derivation of Eq. (8.9), we have simply integrated Eq. (8.2) with omitting the phase of light electric fields. The validity of this treatment depends on the thickness of the range that the static electric field $E_z(0; z)$ penetrates. If the range of $E_z(0; z)$ is sufficiently shorter than the wavelengths of incident or emitted light fields, their phase factors are regarded as constant over the integral range of z and therefore can be neglected. Otherwise we need to explicitly take account of the phase factors in the integral of Eq. (8.9) [11, 19, 25]. We discuss this issue here, and provide the modified formula if necessary.

We first examine the range of static electric field $E_z(0; z)$ based on the Guoy-Chapman theory. Equation (8.6) for the Z : Z electrolyte solution can be analytically solved for $\Phi(z)$. This Eq. (8.6) is modified to

$$\frac{d}{dz}\left\{\tanh\left(\frac{Ze\Phi(z)}{4k_B T}\right)\right\} = \frac{1}{r_D} \tanh\left(\frac{Ze\Phi(z)}{4k_B T}\right),$$

where

$$r_D = \sqrt{\frac{\varepsilon k_B T}{8\pi Z^2 e^2 n}} \tag{8.12}$$

[3]The original literature of Eisenthal et al. [23] adopted a positive sign to $\Phi(0)$ in Eq. (8.9). This difference stems from definition of the direction of z axis.

Table 8.1 Estimated Debye length r_D in Eq. (8.12) for 1:1 electrolyte solutions with varying concentration. $T = 298.15$ K and $\varepsilon = 78.5$ are employed. (Note that this estimate assumes unit activity coefficient)

Concentration	Debye length
1 M	3.0 Å
0.1 M	9.6 Å
0.01 M (10 mM)	3.0 nm
0.001 M (1 mM)	9.6 nm
0.0001 M (0.1 mM)	30 nm

is called the Debye length for the $Z : Z$ electrolyte solution. The solution of $\Phi(z)$ is

$$\Phi(z) = \frac{4k_BT}{Ze}\tanh^{-1}\left\{\tanh\left(\frac{Ze\Phi(0)}{4k_BT}\right)\exp\left(\frac{z}{r_D}\right)\right\} \quad (z \le 0). \tag{8.13}$$

In a sufficiently dilute solution where $Ze\Phi(0)/4k_BT \ll 1$, this equation is linearized using $\tanh x \approx x$ ($x \ll 1$) to be

$$\Phi(z) \simeq \Phi(0)\exp\left(\frac{z}{r_D}\right) \quad (z \le 0). \tag{8.14}$$

Equations (8.13) or (8.14) indicates that the range of electrostatic field is determined by the Debye length r_D. We roughly estimate the spatial scale of r_D for 1:1 electrolyte solutions with varying concentration in Table 8.1. The table shows that the Debye length increases with lower concentration.

Then we derive the formula of $\chi^{(3)}$ contribution with explicitly considering the phase factors. Here we consider the SFG geometry in Fig. 8.3, where the incident lights emit from the medium α ($z > 0$) and polarization is generated in β ($z < 0$). And we use the same notations for other geometric properties as those defined in Chap. 7. With considering the phase factors, the nonlinear polarization per unit area $\boldsymbol{P}^S$ in Eq. (8.9) is modified to

$$P_p^S = \int_{-\infty}^{0} dz \exp\left(-ik_{G,z}^{\beta}(\Omega)z\right)\sum_{q,r}\left[\chi_{pqr}^{(2)}\delta(z-0_-) + \chi_{pqrz}^{(3)}E_z(0;z)\right]$$
$$\cdot L_{I,q}(\omega_1)L_{I,r}(\omega_2)E_{I,q}^{\alpha}(\omega_1)E_{I,r}^{\alpha}(\omega_2)\exp\left(ik_{T,z}^{\beta}(\omega_1)z\right)\exp\left(ik_{T,z}^{\beta}(\omega_2)z\right). \tag{8.15}$$

The derivation of this equation is in parallel with that of Eqs. (7.35) and (7.36) in Chap. 7. In this equation, $k_{G,z}^{\beta}(\Omega)$ is the z component of the SFG wavevector in medium β ($z < 0$) in the propagating direction G ($G = T$ for transmission and R for reflection). $\boldsymbol{E}_I^{\alpha}(\omega_f)$ is the incident electric field in medium α at frequency ω_f ($f = 1, 2$), and $\boldsymbol{L}_I(\omega_f)$ is the optical factor in Eqs. (5.29) and (5.30) to convert the incident light to transmission one. (Note that the local field factors in Eq. (7.36) are incorporated in $\chi^{(2)}$ and $\chi^{(3)}$.)

Equation (8.15) is integrated over z,

$$P_p^S=\sum_{q,r}\left[\chi^{(2)}_{pqr}+\chi^{(3)}_{pqrz}\int_{-\infty}^{0} dz E_z(0;z)\exp\left\{i(k^{\beta}_{T,z}(\omega_1)+k^{\beta}_{T,z}(\omega_2)-k^{\beta}_{G,z}(\Omega))z\right\}\right]$$
$$\cdot L_{I,q}(\omega_1)L_{I,r}(\omega_2)E^{\alpha}_{I,q}(\omega_1)E^{\alpha}_{I,r}(\omega_2). \tag{8.16}$$

The $\chi^{(3)}$ term including the z integral in Eq. (8.16) is further simplified as follows.

$$\begin{aligned}(\chi^{(3)}\text{ term}) &= \chi^{(3)}_{pqrz}\int_{-\infty}^{0} dz\, E_z(0;z)\exp\left\{i(k^{\beta}_{T,z}(\omega_1)+k^{\beta}_{T,z}(\omega_2)-k^{\beta}_{G,z}(\Omega))z\right\}\\ &= \chi^{(3)}_{pqrz}\int_{-\infty}^{0} dz\left(-\frac{d\Phi(z)}{dz}\right)\exp\left(\frac{iz}{l_G}\right)\\ &= \chi^{(3)}_{pqrz}\left\{-\Phi(0)+\frac{i}{l_G}\int_{-\infty}^{0} dz\,\Phi(z)\exp\left(\frac{iz}{l_G}\right)\right\},\end{aligned} \tag{8.17}$$

where $l_G = 1/(k^{\beta}_{T,z}(\omega_1)+k^{\beta}_{T,z}(\omega_2)-k^{\beta}_{G,z}(\Omega))$ is the coherence length in the direction G in Eq. (7.37). In the dilute solution where $\Phi(z)$ is expressed with Eq. (8.14),

$$\begin{aligned}(\chi^{(3)}\text{ term}) &\simeq \chi^{(3)}_{pqrz}\left\{-\Phi(0)+\frac{i}{l_G}\int_{-\infty}^{0} dz\,\Phi(0)\exp\left(\frac{z}{r_D}\right)\exp\left(\frac{iz}{l_G}\right)\right\}\\ &= -\chi^{(3)}_{pqrz}\Phi(0)\frac{1}{1+i\dfrac{r_D}{l_G}}.\end{aligned} \tag{8.18}$$

The result of Eq. (8.18) deviates from $-\chi^{(3)}_{pqrz}\Phi(0)$ when the condition of $r_D \ll l_G$ breaks down. This indicates that the phase factor in $\chi^{(3)}$ contribution becomes significant in dilute solutions where the Debye length r_D is comparable to the coherent length l_G of SFG.

8.1.3 *Calibrating $\chi^{(3)}$ Effect in SFG Spectra*

Attempts to estimate the $\chi^{(3)}$ contribution in observed SFG/SHG spectra of charged interfaces have been done experimentally and theoretically. Experimental estimate of $\chi^{(3)}$ requires decomposition of observed signal into $\chi^{(2)}$ and $\chi^{(3)}$ contributions. The decomposition was performed by changing ionic strength (concentration) of the electrolyte solution, with an assumption that the changing ionic strength affects the surface potential $\Phi(0)$ but does not affect $\chi^{(2)}$ from the interface layer [25]. Though the validity of such assumption should be carefully examined, the results of $\chi^{(3)}$ are generally consistent to the direct calculation of $\chi^{(3)}$ shown in Fig. 8.2b, which supports the validity of the estimate.

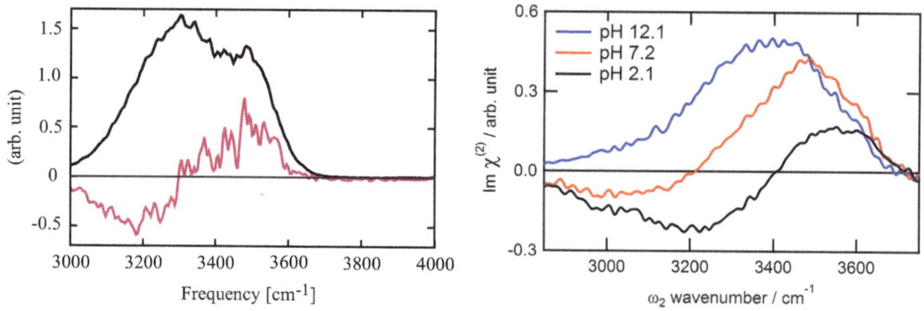

Fig. 8.4 (Left panel) calculated Im[$\chi^{(2)}$] spectrum (black) and the intrinsic spectrum after calibrating $\chi^{(3)}$ (purple) for charged α-quartz/H_2O interface [16]. (Right panel) experimentally observed Im[$\chi^{(2)}$] spectra for silica glass/HOD solution with varying pH [18]. (Reprinted with the permission from Ref. [18]; Copyright 2016 American Chemical Society)

One of the important concerns is the relative amplitude of $\chi^{(2)}$ and $\chi^{(3)}\Phi(0)$ in Eq. (8.9), which depends on the surface potential $\Phi(0)$. This estimate can be carried out with the calculated $\chi^{(2)}$ and $\chi^{(3)}$ spectra on the same footing, using the same time correlation function formula and the common molecular model, in Fig. 8.2. The MD results show that the amplitudes of $|\chi^{(2)}|$ and $|\chi^{(3)}\Phi(0)|$ in the O–H stretching region of liquid water become comparable at $\Phi(0) \sim 0.02$ V [16]. Quantitative accuracy of this estimate may have some room for improvement, but this estimate provides the rough idea on the relative amplitudes between $|\chi^{(2)}|$ and $|\chi^{(3)}\Phi(0)|$. In ordinary silica surfaces in contact with basic solutions, $\Phi(0)$ can be the order of 0.1 V, where the $\chi^{(3)}\Phi(0)$ contribution becomes comparable to or outweighs the $\chi^{(2)}$ contribution in observed SFG spectra. In electrode-solution surfaces with controlling surface potential, $\chi^{(3)}\Phi(0)$ can readily be a dominant contribution.

Another concern is to extract the intrinsic $\chi^{(2)}$ spectra related to the interfacial structure from the observed SFG spectra by calibrating the $\chi^{(3)}$ contribution. Since the $\chi^{(3)}$ is a bulk property and its spectra are known, its contribution can be subtracted from an observed (calculated) SFG spectrum from charged surface. Such attempt is feasible by MD simulation. Figure 8.4 shows that SFG spectrum of charged silica-water interface with and without the calibration of $\chi^{(3)}$ signal. The intrinsic $\chi^{(2)}$ spectrum after calibrating $\chi^{(3)}$ contribution clearly exhibits the bands assigned to hydrogen-bonding O–H to surface silanol [16]. The intrinsic Im[$\chi^{(2)}$] spectrum after the calibration is similar to experimentally observed Im[$\chi^{(2)}$] spectrum in acid solution where surface silanol is not deprotonated [18].

8.2 Chiral Elements of $\chi^{(2)}$

The present section discusses the applications of SFG spectroscopy to chiral systems. Chirality is another source of SFG in addition to the interface, since both are pertinent to the lack of inversion symmetry, the necessary condition for the SFG to be dipole allowed. In the preceding chapters we have focused on the SFG response from interfaces. Chiral materials were known to be SFG active in 1960s [10], though the chiral vibrational SFG spectroscopy has not been fully explored until recently, mainly because the chiral response is weak in ordinary conditions. However, there is a promising avenue to exploit both the interface and chiral sensitivities in future SFG spectroscopy. In what follows, we summarize fundamental features of SFG from chiral systems, which are somewhat distinct from the features of SFG from the interfaces [5, 13].

8.2.1 Symmetry

For a chiral system, the inversion or reflection operation changes its chirality. Accordingly, the $\boldsymbol{\chi}^{(2)}$ elements associated to chiral properties should not be invariant by the inversion or reflection. The above reasoning from the symmetry requirement leads to the fact that the following six elements of $\chi^{(2)}_{pqr}$ could be pertinent to the chirality:

$$\chi^{(2)}_{xyz}, \quad \chi^{(2)}_{xzy}, \quad \chi^{(2)}_{yxz}, \quad \chi^{(2)}_{yzx}, \quad \chi^{(2)}_{zxy}, \quad \chi^{(2)}_{zyx}. \tag{8.19}$$

This is a general conclusion for third-rank tensor properties. Any other tensor elements $\chi^{(2)}_{pqr}$ than in Eq. (8.19) miss either of x, y or z in the suffixes pqr. For example, $\chi^{(2)}_{zxx}$ does not have y in the suffixes zxx. We can readiy understand that those other tensor elements $\chi^{(2)}_{pqr}$ except for those in Eq. (8.19) do not satisfy the symmetry requirement of inversion or reflection. For example, let us consider the element $\chi^{(2)}_{zxx}$. It is responsible to the second-order polarization $P^{(2)}_z(\Omega)$ induced by $E_x(\omega_1)$ and $E_x(\omega_2)$,

$$P^{(2)}_z(\Omega) = \chi^{(2)}_{zxx}(\Omega, \omega_1, \omega_2)E_x(\omega_1)E_x(\omega_2).$$

This equation can never be associated to the chirality. If we operate the reflection at the plane perpendicular to the y axis, the vector elements of $P^{(2)}_z$, $E_x(\omega_1)$, $E_x(\omega_2)$ are unaffected. Consequently, the $\chi^{(2)}_{zxx}$ should not change to satisfy the relation, which means that $\chi^{(2)}_{zxx}$ is not a chiral property. The same argument holds for the other elements of $\chi^{(2)}_{pqr}$ except for the above six elements.

Some of the six $\boldsymbol{\chi}^{(2)}$ elements in Eq. (8.19) may be related when the material has a certain spatial symmetry. We treat here two cases, (i) isotropic bulk and (ii) interface of C_∞ symmetry.

(i) Istropic bulk In the case of an isotropic material, the following relations hold for the six elements,

$$\chi^{(2)}_{xyz} = -\chi^{(2)}_{xzy} = -\chi^{(2)}_{yxz} = \chi^{(2)}_{yzx} = \chi^{(2)}_{zxy} = -\chi^{(2)}_{zyx} \left(\equiv \chi^{(2)}_{\text{chiral}}\right). \tag{8.20}$$

These relations are readily confirmed from the fact that $\boldsymbol{\chi}^{(2)}$ for an isotropic material should be invariant by arbitrary rotation around an arbitrary axis. For example, if we rotate the system by 90° around the z axis, the coordinates should change as follows.

$$x \to y, \quad y \to -x, \quad z \to z \qquad \text{(rotation around } z\text{)}.$$

Accordingly, $\chi^{(2)}_{xyz} = -\chi^{(2)}_{yxz}$ is required in order that the rotation does not affect the material properties. The similar argument holds to derive the other relations in Eq. (8.20).

Equation (8.20) has one independent variable, $\chi^{(2)}_{\text{chiral}}$. It is given by the average,

$$\chi^{(2)}_{\text{chiral}} = \frac{1}{6} \sum_{p,q,r}^{x\sim z} \varepsilon_{pqr} \chi^{(2)}_{pqr}, \tag{8.21}$$

where ε_{pqr} is the Levi-Civita antisymmetric tensor (see Appendix A.2),

$$\varepsilon_{pqr} = \begin{cases} 1 & (pqr = xyz,\ yzx,\ zxy), \\ -1 & (pqr = yxz,\ zyx,\ xzy), \\ 0 & \text{(otherwise)}. \end{cases} \tag{7.132}$$

Equation (8.21) clarifies the invariance of $\chi^{(2)}_{\text{chiral}}$ to arbitrary rotation, since the Levi-Civita tensor is invariant as shown in Eq. (7.137) in Appendix A.2. Therefore, the expression of Eq. (8.21) is valid for the molecule-fixed coordinates as well.

(ii) Interface Then, we consider the situation that the chiral material forms an interface of C_∞ symmetry.

[Problem 8.2] Suppose a chiral interface system of azimuthal C_∞ symmetry. What relations are required among the six $\boldsymbol{\chi}^{(2)}$ elements of chiral origin in Eq. (8.19)?

In a chiral system with interface, the nonvanishing $\boldsymbol{\chi}^{(2)}$ elements of this system are attributed to either interface or chiral origin. We find from Sect. 3.3.3 and Problem 8.2 that the $\chi^{(2)}_{pqr}$ elements of chiral origin are distinct from the $\chi^{(2)}_{pqr}$ elements of interface origin. The nonvanishing elements of $\chi^{(2)}_{pqr}$ are summarized below.

- $\boldsymbol{\chi}^{(2)}$ of interface origin (achiral)—

$$\chi^{(2)}_{xxz} = \chi^{(2)}_{yyz}, \qquad \chi^{(2)}_{xzx} = \chi^{(2)}_{yzy}, \qquad \chi^{(2)}_{zxx} = \chi^{(2)}_{zyy}, \qquad \chi^{(2)}_{zzz}. \tag{3.48}$$

- $\boldsymbol{\chi}^{(2)}$ of chiral origin—

$$\chi^{(2)}_{xyz} = -\chi^{(2)}_{yxz}, \qquad \chi^{(2)}_{xzy} = -\chi^{(2)}_{yzx}, \qquad \chi^{(2)}_{zxy} = -\chi^{(2)}_{zyx}. \tag{8.22}$$

These tensor elements are related to the following polarization combinations of sum frequency, visible and infrared fields, as discussed in Sect. 3.3.3.

- $\chi^{(2)}$ of interface origin (achiral)—SSP, SPS, PSS, PPP
- $\chi^{(2)}$ of chiral origin—SPP, PSP, PPS

(8.23)

The above polarization combinations are mutually exclusive, which allows for distinguishing the SFG signals of chiral origin and of interface origin by choosing proper polarization combinations. The selection rules in Eq. (8.23) are valid for whether the chiral $\chi^{(2)}$ elements actually originate from the isotropic bulk (8.20) or from the interface (8.22). The SFG spectroscopy is capable of detecting the chiral signal in background-free condition. This feature of SFG as a chiral probe is in contrast to other chiral probe techniques based on difference spectra by different circular polarizations, such as circular dichroism and Raman optical activity.

8.2.2 Intensity

The $\chi^{(2)}$ elements of chiral origin are also represented in the perturbation and time dependent formulas in Chap. 3 in the same manner with those of interface origin. Therefore, the computational methods of chiral SFG signals are common with those of interface SFG described in the preceding chapters. However, there are some differences to describe the $\chi^{(2)}$ elements for chiral SFG.

The $\chi^{(2)}$ of chiral origin also consists of the vibrationally resonant and nonresonant terms in Eq. (3.33), $\chi^{(2)} = \chi^{(2),\text{res}} + \chi^{(2),\text{nonres}}$. In the case of isotropic bulk, $\chi^{(2)}_{\text{chiral}}$ in Eq. (8.21) is accordingly given by

$$\chi^{(2)}_{\text{chiral}} = \chi^{(2),\text{res}}_{\text{chiral}} + \chi^{(2),\text{nonres}}_{\text{chiral}} = \frac{1}{6}\sum_{p,q,r}^{x\sim z} \varepsilon_{pqr}\left(\chi^{(2),\text{res}}_{pqr} + \chi^{(2),\text{nornes}}_{pqr}\right), \tag{8.21}$$

where $\chi^{(2),\text{res}}_{\text{chiral}}$ and $\chi^{(2),\text{nonres}}_{\text{chiral}}$ are represented from Eqs. (3.36) and (3.38), respectively, to be

$$\begin{aligned}
\chi^{(2),\text{res}}_{\text{chiral}}(\Omega,\omega_1,\omega_2) &= \frac{1}{6}\sum_{p,q,r}^{x\sim z} \varepsilon_{pqr}\chi^{(2),\text{res}}_{pqr} \\
&= -\frac{1}{6\hbar}\sum_{p,q,r}^{x\sim z}\varepsilon_{pqr}\sum_{g,m}\left(\rho^{(0)}_g - \rho^{(0)}_m\right)\langle g|\alpha_{pq}(\Omega)|m\rangle\frac{\langle m|\mu_r|g\rangle}{\omega_2-\omega_{mg}+i\Gamma_{mg}} \\
&= -\frac{1}{6\hbar}\sum_{g,m}\left(\rho^{(0)}_g - \rho^{(0)}_m\right)\frac{1}{\omega_2-\omega_{mg}+i\Gamma_{mg}}\Big[\langle g|\left\{\alpha_{yz}(\Omega)-\alpha_{zy}(\Omega)\right\}|m\rangle\langle m|\mu_x|g\rangle \\
&\quad +\langle g|\left\{\alpha_{zx}(\Omega)-\alpha_{xz}(\Omega)\right\}|m\rangle\langle m|\mu_y|g\rangle+\langle g|\left\{\alpha_{xy}(\Omega)-\alpha_{yx}(\Omega)\right\}|m\rangle\langle m|\mu_z|g\rangle\Big]
\end{aligned} \tag{8.24}$$

and

$$
\begin{aligned}
\chi_{\text{chiral}}^{(2),\text{nonres}} &= \frac{1}{6}\sum_{p,q,r}^{x\sim z} \varepsilon_{pqr}\chi_{pqr}^{(2),\text{nonres}} \\
&= -\frac{1}{6\hbar}\sum_{p,q,r}^{x\sim z} \varepsilon_{pqr}\sum_{g,m}\left(\rho_g^{(0)}-\rho_m^{(0)}\right)\langle g|\alpha_{pr}(\Omega)|m\rangle \frac{\langle m|\mu_q|g\rangle}{\omega_1-\omega_{mg}+i\Gamma_{mg}} \\
&= \frac{1}{6\hbar}\sum_{g,m}\left(\rho_g^{(0)}-\rho_m^{(0)}\right)\frac{1}{\omega_1-\omega_{mg}+i\Gamma_{mg}}\Big[\langle g|\left\{\alpha_{yz}(\Omega)-\alpha_{zy}(\Omega)\right\}|m\rangle\langle m|\mu_x|g\rangle \\
&\quad +\langle g|\left\{\alpha_{zx}(\Omega)-\alpha_{xz}(\Omega)\right\}|m\rangle\langle m|\mu_y|g\rangle+\langle g|\left\{\alpha_{xy}(\Omega)-\alpha_{yx}(\Omega)\right\}|m\rangle\langle m|\mu_z|g\rangle\Big].
\end{aligned}
\tag{8.25}
$$

The above formulas of Eqs. (8.24) and (8.25) include the antisymmetric components of the Raman tensor $\left\{\alpha_{pq}-\alpha_{qp}\right\}$. Therefore, the isotropic average $\chi_{\text{chiral}}^{(2)}$ in Eq. (8.21) should vanish if the Raman tensor is symmetric. The antisymmetric components of the Raman tensor is generally minor in electronically off-resonant conditions, as discussed in Sect. A.3. This is the main reason why the chiral SFG has not been fully exploited for a long time, though it is allowed for symmetry reason. We also notice in passing that Eqs. (8.24) and (8.25) lead to $\chi_{\text{chiral}}^{(2),\text{res}} = -\chi_{\text{chiral}}^{(2),\text{nonres}}$ for $\omega_1 = \omega_2$. This indicates that SHG yield no signal from isotropic bulk materials.[4]

However, the intensity of chiral SFG signals is remarkably augmented when Ω becomes close to the electronic resonant condition [4, 6]. This is because the antisymmetric components of the Raman tensor become signficantly strong in electronically resonant or near-resonant conditions. Figure 8.5 displays an example of chiral SFG spectra from R-1,1'-bi-2-naphtol with varying sum-frequency photon energy $\hbar\Omega$ [4]. These spectra demonstrate that the intensity of chiral SFG signal is augmented by 10^5 in the near resonance condition with an electronic excited state. Such remarkable enhancement makes the chiral SFG a practically useful tool to detect chiral vibrational spectra with monolayer sensitivity.

8.2.3 *Future Development*

The vibrational SFG spectroscopy of chiral probe has many applications, particularly to biological molecules. It could attain monolayer sensitivity to detect chiral species. Yan et al. argued that the chiral SFG could have interface selectivity in the electronically off-resonant conditions that the chiral signal from isotropic bulk vanishes [26].

---

[4]Note that the SHG from chiral interface is allowed [17, 24].

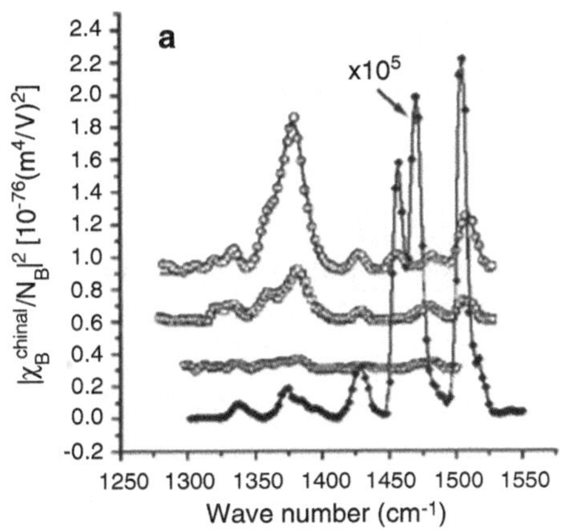

Fig. 8.5 Chiral (SPP) spectra from R-1,1'-bi-2-naphtol solution in acetone with varying sum-frequency photon energy ($\hbar\Omega$ = 3.70, 3.65, 3.59, 2.51 eV from top to the bottom). The baselines are vertically shifted for clarity, and the 2.51 eV spectrum is multiplied by 10^5. (Reprinted with permission from Ref. [4]. Copyright 2003 by American Physical Society.)

Another promising direction of chiral SFG is to combine the heterodyne measurement [20–22]. The heterodyne measurement can expand the applicability of chiral SFG. First, it allows for distinguishing enantiomers from the sign of the $\mathrm{Im}[\chi^{(2)}]$ amplitude. It is advantageous over conventional intensity measurement to detect small signals by amplifying weak signals with a local oscillator, while the conventional measurement detects the square of the small amplitude. Therefore, the heterodyne measurement allows for detecting the chiral signal in ordinary conditions of electronically off-resonance without resort to electronic and vibrational double resonance conditions.

In summary, the SFG spectroscopy is capable of detecting the signal of chiral origin selectively by choosing proper combination of polarizations. The chiral signal from isotropic bulk should be vanishingly small in electronically off-resonant condition, but its intensity is remarkably enhanced in the electronically (near) resonant condition. The chiral SFG signal from the surface is possibly allowed even in the electronically off-resonant condition, which offers an intriguing possibility of interface-specific chiral probe technique. The heterodyne measurement will greatly expand the range of detection and application of chiral SFG spectroscopy.

8.3 Solutions to Problems

8.3.1 Guoy-Chapman Theory

[Problem 8.1] Using the above notations and the Poisson-Boltzmann equation, derive the following formula of the electric field $E_z(0; z)$ in $z \le 0$,

$$E_z(0;z)^2 = \left(\frac{d\Phi(z)}{dz}\right)^2 = \frac{8\pi k_B T}{\varepsilon} \sum_{i=1}^{N_i} n_i \left\{ \exp\left(-\frac{Z_i e \Phi(z)}{k_B T}\right) - 1 \right\}. \tag{8.5}$$

In a case of $Z : Z$ electrolyte solution ($N_i = 2$, $Z_1 = -Z_2 = Z$, $n_1 = n_2 = n$), show that Eq. (8.5) is simplified to be

$$E_z(0; z) = -\frac{d\Phi(z)}{dz} = -\sqrt{\frac{32\pi k_B T n}{\varepsilon}} \sinh\left(\frac{Ze\Phi(z)}{2k_B T}\right). \tag{8.6}$$

(*Hint*) Note that $\Phi(z)$ is related to the charge density $\rho(z)$ by the one-dimensional Poisson equation,

$$\frac{d^2\Phi(z)}{dz^2} = -\frac{4\pi}{\varepsilon}\rho(z), \tag{8.7}$$

while $\rho(z)$ is assumed to be given by the Boltzmann distribution of ions,

$$\rho(z) = \sum_{i=1}^{N_i} Z_i e n_i \exp\left(-\frac{Z_i e\Phi(z)}{k_B T}\right). \tag{8.8}$$

Coupled solution of Eqs. (8.7) and (8.8) leads to Eq. (8.5).

Equations (8.7) and (8.8) are analytically solved as follows. $d^2\Phi/dz^2$ in Eq. (8.7) is represented using $E_z = -d\Phi/dz$ by

$$\frac{d^2\Phi}{dz^2} = -\frac{d}{dz}E_z(0; z) = -\frac{dE_z}{d\Phi}\frac{d\Phi}{dz} = \frac{dE_z}{d\Phi}E_z = \frac{1}{2}\frac{d}{d\Phi}E_z^2,$$

using a functional relation of Φ and z. Therefore, the equation to be solved is

$$\frac{1}{2}\frac{d}{d\Phi}E_z^2 = -\frac{4\pi}{\varepsilon}\sum_{i=1}^{N_i} Z_i e n_i \exp\left(-\frac{Z_i e\Phi}{k_B T}\right). \tag{8.26}$$

By integrating Eq. (8.26) from $\Phi = 0$ to Φ (or $z = -\infty$ to z),

$$\frac{1}{2}\left\{E_z(0; z)^2 - E_z(0; z = -\infty)\right\} = \frac{1}{2}E_z(0; z)^2$$

$$= \frac{4\pi}{\varepsilon}\left[\sum_{i=1}^{N_i} k_B T n_i \exp\left(-\frac{Z_i e\Phi}{k_B T}\right)\right]_{\Phi=0}^{\Phi} = \frac{4\pi}{\varepsilon}\sum_{i=1}^{N_i} k_B T n_i \left\{\exp\left(-\frac{Z_i e\Phi}{k_B T}\right) - 1\right\},$$

where the boundary condition $E_z(0; z \to -\infty) = 0$ is used. The above relation leads to Eq. (8.5).

In the case of $Z : Z$ electrolyte solution, Eq. (8.5) is further simplified. By assuming $N_i = 2$, $Z_1 = -Z_2 = Z$, $n_1 = n_2 = n$, Eq. (8.5) becomes

$$E_z(0;z)^2 = \frac{8\pi k_B T}{\varepsilon} n \left\{ \exp\left(-\frac{Ze\Phi}{k_B T}\right) + \exp\left(\frac{Ze\Phi}{k_B T}\right) - 2 \right\}$$
$$= \frac{8\pi k_B T}{\varepsilon} n \left\{ 2\sinh\left(\frac{Ze\Phi}{2k_B T}\right) \right\}^2$$

and thus by taking the square root to obtain Eq. (8.6). The minus sign of Eq. (8.6) is determined to satisfy the boundary condition, $\Phi(z \to -\infty) = 0$, since this condition requires $E_z = -d\Phi/dz < 0$ if $\Phi(z) > 0$, or $E_z = -d\Phi/dz > 0$ if $\Phi(z) < 0$.

8.3.2 *Chiral $\chi^{(2)}$ Components*

[Problem 8.2] Suppose a chiral interface system of azimuthal C_∞ symmetry. What relations are required among the six $\chi^{(2)}$ elements of chiral origin in Eq. (8.19)?

For a system of C_∞ symmetry, the $\chi^{(2)}$ property should be invariant by arbitrary rotation around the normal axis z. For example, if we rotate the coordinate system by 90° around the z axis, the coordinates should change as

$$x \to y, \qquad y \to -x, \qquad z \to z$$

Therefore, the following three conditions should hold among the chiral components,

$$\chi^{(2)}_{xyz} = -\chi^{(2)}_{yxz}, \qquad \chi^{(2)}_{xzy} = -\chi^{(2)}_{yzx}, \qquad \chi^{(2)}_{zxy} = -\chi^{(2)}_{zyx}. \tag{8.22}$$

Bibliography

1. Adamson AW, Gast AP (1997) Physical chemistry of surfaces, 6th edn. Wiley, New York
2. Bagotsky VS (2006) Fundamentals of electrochemistry, 2nd edn. Wiley, Hoboken
3. Bard AJ, Faulkner LR (2001) Electrochemical methods: fundamentals and applications, 2nd edn. Wiley, New York
4. Belkin MA, Shen YR (2003) Doubly resonant IR-UV sum-frequency vibrational spectroscopy on molecular chirality. Phys Rev Lett 91:213907
5. Belkin MA, Shen YR (2005) Non-linear optical spectroscopy as a novel probe for molecular chirality. Int Rev Phys Chem 24:257–299
6. Belkin MA, Shen YR, Harris RA (2004) Sum-frequency vibrational spectroscopy of chiral liquids off and close to electronic resonance and the antisymmetric raman tensor. J Chem Phys 120:10118–10126
7. Darlington AM, Jarisz TA, DeWalt-Kerian EL, Roy S, Kim S, Azam MS, Hore DK, Gibbs JM (2017) Separating the pH-dependent behavior of water in the Stern and diffuse layers with varying salt concentration. J Phys Chem C 121:20229–20241

8. Eftekhari-Bafrooei A, Borguet E (2011) Effect of electric fields on the ultrafast vibrational relaxation of water at a charged solid-liquid interface as probed by vibrational sum frequency generation. J Phys Chem Lett 2:1353–1358
9. Geiger FM (2009) Second harmonic generation, sum frequency generation, and $\chi^{(3)}$: dissecting environmental interfaces with a nonlinear optical Swiss army knife. Annu Rev Phys Chem 60:61–83
10. Giordmaine JA (1965) Nonlinear optical properties of liquids. Phys Rev 138:A1599–1606
11. Gonella G, Lütgebaucks C, de Beer AG, Roke S (2016) Second harmonic and sum-frequency generation from aqueous interfaces is modulated by interference. J Phys Chem C 120:9165–9173
12. Gragson DE, McCarty BM, Richmond GL (1997) Ordering of interfacial water molecules at the charged air/water interface observed by vibrational sum frequency generation. J Am Chem Soc 119:6144–6152
13. Haupert LM, Simpson GJ (2009) Chirality in nonlinear optics. Annu Rev Phys Chem 60:345–365
14. Ishiyama T, Morita A (2009) Analysis of anisotropic local field in sum frequency generation spectroscopy with the charge response kernel water model. J Chem Phys 131:244714
15. Jena KC, Covert PA, Hore DK (2011) The effect of salt on the water structure at a charged solid surface: differentiating second- and third-order nonlinear contributions. J Phys Chem Lett 2:1056–1061
16. Joutsuka T, Hirano T, Sprik M, Morita A (2018) Effect of third-order susceptibility in sum frequency generation spectroscopy: molecular dynamics study in liquid water. Phys Chem Chem Phys 20:3040–3053
17. Kriech MA, Conboy JC (2003) Label-free chiral detection of melittin binding to a membrane. J Am Chem Soc 125:1148–1149
18. Myalitsin A, Urashima S, Nihonyanagi S, Yamaguchi S, Tahara T (2016) Water structure at the buried silica/aqueous interface studied by heterodyne-detected vibrational sum-frequency generation. J Phys Chem C 120:9357–9363
19. Ohno PE, Saslow SA, Wang H-F, Geiger FM, Eisenthal KB (2016) Phase-referenced nonlinear spectroscopy of the α-quartz/water interface. Nat Commun 7:13587
20. Okuno M, Ishibashi T (2014) Chirality discriminated by heterodyne-detected vibrational sum frequency generation. J Phys Chem Lett 5:2874–2878
21. Okuno M, Ishibashi T (2015) Heterodyne-detected achiral and chiral vibrational sum frequency generation of proteins at air/water interface. J Phys Chem C 119:9947–9954
22. Okuno M, Ishibashi T (2015) Sensitive and quantitative probe of molecular chirality with heterodyne-detected doubly resonant sum frequency generation spectroscopy. Anal Chem 87:10103–10108
23. Ong S, Zhao X, Eisenthal KB (1992) Polarization of water molecules at a charged interface: second harmonic studies of the silica/water interface. Chem Phys Lett 191:327–335
24. Petralli-Mallow T, Wong TM, Byers JD, Yee HI, Hicks JM (1993) Circular dichroism spectroscopy at interfaces: a surface second harmonic generation study. J Phys Chem 97:1383–1388
25. Wen Y-C, Zha S, Liu X, Yang S, Guo P, Shi G, Fang H, Shen YR, Tian C-S (2016) Unveiling microscopic structures of charged water interfaces bysurface-specific vibrational spectroscopy. Phys Rev Lett 116:016101
26. Yan ECY, Fu L, Wang Z, Liu W (2014) Biological macromolecules at interfaces probed by chiral vibrational sum frequency generation spectroscopy. Chem Rev 114:8471–8498

Chapter 9
Applications: Aqueous Interfaces

Abstract This chapter presents recent applications of the computational SFG analysis for aqueous interfaces. Aqueous interfaces are relevant to a variety of fields in chemistry and engineering, and have been extensively investigated by SFG spectroscopy. Nevertheless, their hydrogen-bonding network and complicated vibrational coupling hinder simple intuitive interpretation of the observed spectra. The aid of MD simulation to analyze the complicated SFG spectra is actually quite powerful. This chapter introduces the results of analysis for various aqueous interfaces led by the author's group and others, including water, ice, electrolyte aqueous solutions, water-oil and water-membrane interfaces.

Keywords O–H stretching · Vibrational coupling · Interface thickness · Ion segregation · Electric double layer

Water and aqueous interfaces have been extensively studied by the SFG spectroscopy, and computational analysis of their SFG spectra has been particularly developed [32, 36, 61, 76, 96]. Water has its characteristic three-dimensional network structure of hydrogen bonds, and the hydrogen-bond structure gives rise to a number of peculiar properties of liquid water [3, 39]. Therefore, it is an intriguing issue to understand the hydrogen-bonding structure of water surface, and to reveal the structural difference from that of bulk water and its implications [44]. The interface structure is of fundamental importance to interfacial properties and heterogeneous reactions at water surface.

One important advantage of applying the SFG spectroscopy to aqueous interfaces is the availability of the O–H stretching vibrational band. The O–H stretching band in 3000–3800 cm^{-1} region is fairly easy to measure technically by SFG spectroscopy, and its frequency and intensity are sensitive indicators of the hydrogen-bond strength. Selective detection of the O–H stretching band of interfacial water provides quite useful information on the interfacial structure of water.

On the other hand, SFG spectra of aqueous interfaces pose a challenge to interpretation. The O–H stretching vibration of water involves various kinds of vibrational couplings, both intramolecular and intermolecular, which complicate the

A. Morita, *Theory of Sum Frequency Generation Spectroscopy*,
Lecture Notes in Chemistry 97, https://doi.org/10.1007/978-981-13-1607-4_9

spectral assignment. Just as an example, two equivalent O–H vibrations of a single isolated water molecule are coupled and split into symmetric and antisymmetric vibrational modes, which have orthogonal transition dipole moments. Therefore, the single water molecule exhibits two different directions of transition moments. We could not determine the orientation of the molecule from the observed direction of transition moment unless we know the character of the vibrational mode. In condensed phase, the water O–H moieties are coupled by the hydrogen bonds, and consequently the vibrational modes are delocalized among intermolecular O–H moieties. Such delocalized vibrations are often beyond our intuitive understanding. To properly assign the observed O–H vibrations in relation to the interface structure, theoretical analysis is particularly required.

Here we summarize the current status of the theoretical analysis of SFG spectra of water and aqueous interfaces. We discuss the surfaces of liquid water, ice, and electrolyte solutions. Besides the air-aqueous interfaces, liquid/liquid (water/oil) and water/monolayer interfaces are also treated. The knowledge described below should be regarded as fundamentals for future progress. In this chapter, we discuss the SFG spectra of O–H stretching region in the SSP polarization, the most commonly employed geometry, unless otherwise noted.

9.1 Water Surface

The vibrational spectra of water surface have been computationally analyzed by a number of researchers [4, 10, 13, 29, 52, 56, 57, 62, 73, 74, 80, 81, 103, 110]. Here we briefly show some basic information of water surface derived from the computational analysis of the SFG spectra.

Historical perspective The SFG spectrum of water surface in the O–H stretching region was first reported by Shen and co-workers [14], and subsequently studied by other groups including Shultz and Richmond [88, 94] in the pioneer stage of SFG spectroscopy. Since then the water surface is one of the most intensively studied surfaces by SFG [9, 93]. The spectral shape has been shown in Fig. 1.1 in Chap. 1, which apparently consists of two bands, a sharp band at $3700\,\mathrm{cm}^{-1}$ and a broad one at $3000{\sim}3600\,\mathrm{cm}^{-1}$. We note that the spectral lineshapes reported in early period [14, 88, 94] were noticeably different from each other. This problem of disagreement has been resolved by the progress of spectroscopy, and Fig. 1.1 is currently considered to be the established intensity spectrum of liquid water in the O–H stretching region.

First MD calculation of the water SFG spectrum was carried out by Morita and Hynes on the basis of the energy representation in Sect. 4.1 [57]. The calculated SFG spectrum of water surface reproduced the two-band structure in the O–H stretching region. The computation predicted the $\mathrm{Im}[\chi^{(2)}]$ spectrum that the sharp band at $3700\,\mathrm{cm}^{-1}$ has a positive amplitude of $\mathrm{Im}[\chi^{(2)}]$ while the broad band at $3000{\sim}3600\,\mathrm{cm}^{-1}$ has a negative sign (see Figs. 8.2a and 9.1). As we have argued in

Sect. 4.2, the sign of the Im[$\chi^{(2)}$] spectrum in the SSP polarization is indicative of the orientation of the transition dipole for the O–H stretching vibration. The positive sign at 3700 cm^{-1} indicates upward O–H orientation with H pointing to the vapor, while the negative sign at 3000∼3600 cm^{-1} implies downward O–H orientation. Subsequently the phase-sensitive or heterodyne-detected SFG measurement confirmed the band shape of the Im[$\chi^{(2)}$] spectrum [40, 66, 105].[1] The prediction of the Im[$\chi^{(2)}$] spectral features is the first successful demonstration of the computational analysis in the SFG spectroscopy.

Thickness of surface sensitivity One important and general concern in the SFG spectroscopy is to find the depth of detection in the interface region. As we have argued in Chap. 1, the SFG is sensitive to the interface where the inversion symmetry is broken. The SFG-active region of interface may depend on the system in question. In molecular liquids such as water, the broken symmetry arises from anisotropic molecular orientation near the interface, whereas the bulk material has random and isotropic orientation. The MD simulation is able to investigate molecular orientation with varying depth, and thereby answer the question about the spatial origin of surface sensitivity.

The detected depth by SFG spectroscopy can be readily examined with MD simulation. As illustrated in Fig. 9.1, we introduce a depth coordinate $\hat{z}$ which is normal to the interface and that the interface is located at $\hat{z} \approx 0$. We tentatively set a region at the surface $z > z^{\text{thres}}$, and calculate a $\chi^{(2)}$ spectrum of the

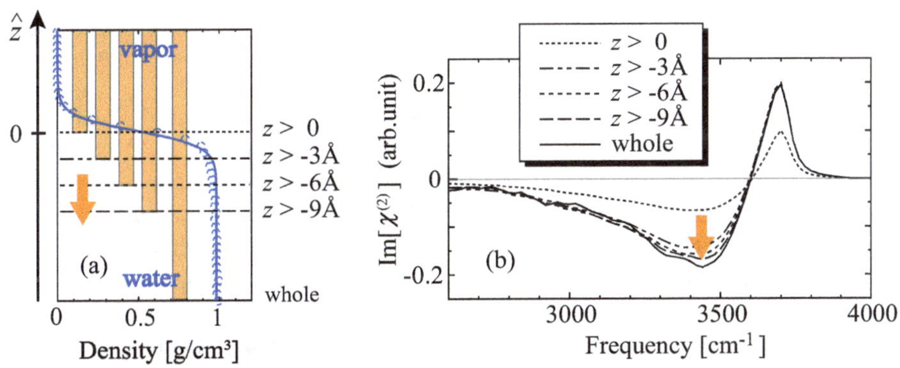

Fig. 9.1 MD analysis of surface thickness in SFG spectroscopy. (Left panel **a**) density profile of water surface as a function of $\hat{z}$ in blue. The varying surface regions $\hat{z} > z^{\text{thres}}$ are shown with orange bars. (Right panel **b**) convergence behavior of calculated Im[$\chi^{(2)}$] spectrum with expanding the interface region $\hat{z} > z^{\text{thres}}$ [57]. (Reprinted from Ref. [57], Copyright 2012, with permission from Elsevier)

[1]Early phase-sensitive SFG experiments reported a positive tail of Im[$\chi^{(2)}$] at around 3000 cm^{-1} region [40, 66, 105]. That feature is considered to be an artifact of phase calibration at present [67, 114].

system in that restricted region. This calculation is feasible by MD simulation. By gradually lowering the threshold z^{thres} and expanding that region, we can observe the convergence behavior of the calculated $\chi^{(2)}$. Figure 9.1 shows the results of such analysis. The figure tells us that the positive band at 3700 cm^{-1} originates from the region of $\hat{z} > -3$ Å, since the band amplitude reaches the convergence up to the region. This result evidences that this positive band at 3700 cm^{-1} comes from the top monolayer of the water surface, again supporting the free O–H exposed to the air. On the other hand, the negative band at 3000∼3600 cm^{-1} converges at about $\hat{z} > -9$ Å, indicating that the band of hydrogen-bonded O–H is attributed to a few top monolayers of the water surface. This analysis confirms the remarkably acute selectivity of SFG spectroscopy in the monolayer scale.

Analysis of hydrogen-bonded O–H band: The broad band at 3000∼3600 cm^{-1} is attributed to hydrogen-bonded O–H, due to the substantial red shift of frequency. However, further detailed assignment of this band has invoked a number of studies and often confusions. As seen in the experimental SFG spectrum of water (left panel of Fig. 1.1), this broad band appears to have two sub-bands, one at about 3200 cm^{-1} and the other at about 3400 cm^{-1}. The two sub-bands are often called "ice-like" and "liquid-like" bands, respectively, in analogy with these spectra of the corresponding bulk materials. The O–H band at 3200 cm^{-1} is obviously seen in the infrared and Raman spectra of ice [79], while the band at 3400 cm^{-1} is seen in liquid water [16]. However, the physical origin of these two sub-bands in the SFG spectrum is still controversial. An intuitive understanding of the two sub-bands comes from the picture that the water surface is a mixture of ice-like water and liquid-like water. This idea of two-state mixture model has a long history in bulk water [16], though there is no consensus to support this idea with microscopic investigation by MD simulation or other means.

Bonn and co-workers proposed an alternative assignment of the two-band structure that the two sub-bands are due to Fermi splitting of O–H stretching vibrational states by the H–O–H bending overtone [98, 99]. Their argument is supported by the experiment of H-D isotope dilution. By replacing H_2O with HOD, they observed that the two sub-bands merge into one. This implies that the two-band structure should originates from vibrational coupling rather than the structure, since the isotope dilution little affects on the structure of nuclei.

As we argued in the beginning of this chapter, a main challenge in analyzing O–H vibration stems from extensive intra- and inter-molecular vibrational couplings, which delocalize the O–H vibrations and complicates the relation to the molecular orientation. To disentangle the O–H vibrations, the H-D isotope dilution offers a useful means. The isotope dilution preserves the structure of nuclei, while it effectively eliminates the intra- and inter-molecular couplings among nearby O–H bonds. Therefore, the observed spectrum approaches that of assembly of isolated O–H bonds in dilute conditions, and the ideal picture about the relation of O–H orientation and Im[$\chi^{(2)}$] in Sect. 4.2.1 becomes increasingly reliable in the spectral analysis.

The two sub-bands at 3200 and 3400 cm^{-1} are widely seen in the O–H bands of a variety of aqueous systems, though they exhibits varying spectral lineshapes and relative intensities. These O–H band shapes of various aqueous systems are likely influenced by inhomogeneous species as well as vibrational couplings. Previous time-resolved SFG measurements confirmed that the O–H band includes comparable homogeneous and inhomogeneous widths[8, 25, 51]. Comprehensive understanding of those band shapes is one of the principal aims of the further SFG analysis of water surface.

Bend vibration Most SFG studies on water surface have dealt with the O–H stretching vibrations. The O–H stretching vibration is easy to measure and provides rich information on hydrogen-bonding environment, whereas the spectral analysis is complicated. There are some attempts to analyze other modes of the SFG spectrum, such as bend and libration [46, 60, 64, 75]. Here we discuss the bend mode in the SFG spectrum of water.

The SFG measurement of the bend band was first carried out by Benderskii and co-workers [108], and subsequently the heterodyne measurement of the bend band was reported by Kundu et al. [47]. The measured $\mathrm{Im}[\chi^{(2)}]$ spectrum revealed a positive band over the bend frequency region, though a previous simulation by Nagata et al. had predicted a bipolar shape of the $\mathrm{Im}[\chi^{(2)}]$ band [60]. The qualitative disagreement between the simulation and experiment was elucidated by further theoretical analysis [47], concluding that the bend band of SFG is actually dominated by the χ^{IQB} term of bulk quadrupole origin discussed in Chap. 7. The sign of $\mathrm{Im}[\chi^{(2)}]$ in the bend region is therefore insensitive to the molecular orientation at surface.

We note in passing that the role of quadrupole contribution on SFG spectra was also found in benzene [45, 50], as discussed in Sect. 10.2. Systematic investigation of the quadrupole effect on various systems is desired.

9.2 Ice Surface

The structure of ice surface has been drawing wide attention for more than a century. Faraday suggested premelting of ice surface [15] below the freezing temperature, and subsequently a number of experiments using optical, magnetic or electrical means confirmed that the premelting layer is developed in several tens of nanometers at about $T \gtrsim -10\,°\mathrm{C}$ [79]. The SFG spectroscopy is powerful to selectively probe the microscopic hydrogen bonding environment at the ice surface. The experimental SFG spectrum of ice Ih basal surface is shown in Fig. 9.2. We notice the remarkably intense band at about 3200 cm^{-1} in the ice spectrum, much stronger than that of the water spectrum. The intensity further augments with lowering temperature [19, 113]. The origin of this remarkable band is understood with the help of MD analysis.

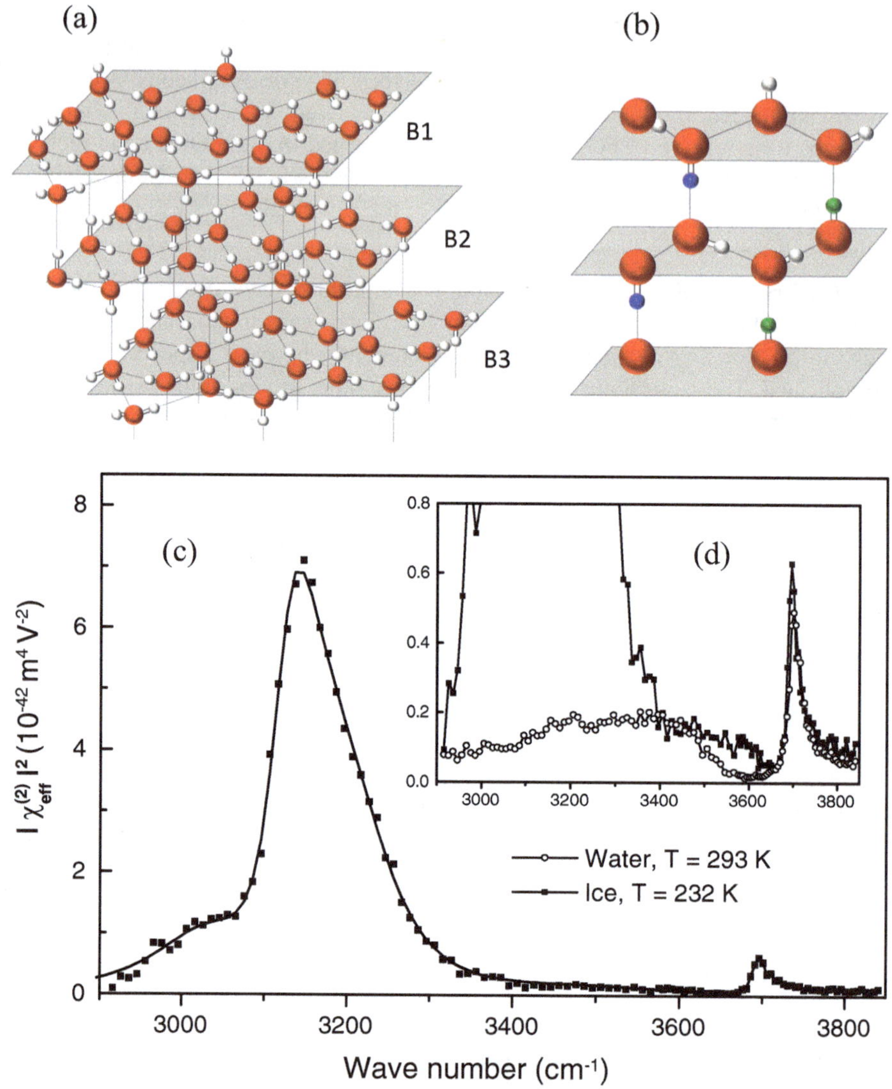

Fig. 9.2 (**a**, **b**) Schematic pictures of basal face of ice surface [71], where red and white symbols denote oxygen and hydrogen, and the bilayers are labeled with B1, B2, ..., in order from the topmost layer to the bulk. In the magnified picture (**b**), upward and downward hydrogens of bilayer-stitching hydrogen bonds are colored in green and blue, respectively. (**c**, **d**) Experimental ice spectrum of SSP polarization at 232 K [113]. The inset (**d**) compares the ice and water spectra. (Reprinted with permission from Ref. [113]. Copyright 2001 by American Physical Society)

The intense band and its temperature dependence of the ice surface were reproduced by MD simulation [31, 35] using QM/MM calculation instead of the classical polarizable MD simulation. These authors found that the band intensity at

3200 cm^{-1} is substantially augmented by the charge transfer, and accordingly the classical model that omits the charge transfer effect could not sufficiently reproduce the remarkable intensity. The band intensity is sensitive to the local disorder of the tetrahedral ice structure as well as the charge transfer. The ordered ice structure facilitates delocalized O–H vibrations, which couple with the charge transfer and augment the band intensity. The large temperature dependence of the band intensity is elucidated with the sensitivity to the local disorder.

The intense band is mainly attributed to the bilayer-stitching O–H vibrations between the first (B1) and second (B2) bilayers of the ice surface in Fig. 9.2b. By looking at the panel (b), one may think that the signals from the upward O–H (green) and downward O–H (blue) vibrations should cancel each other in the ideal ice lattice. However, the MD analysis showed that the upward and downward O–H bonds between B1 and B2 layers are actually inequivalent because of more structural disorder in the B1 bilayer than that in B2, and the broken symmetry near the surface causes the strong SFG signal [71]. The larger disorder in the B1 layer is indicative of the surface premelting in the atomic level, and the premelting develops toward deeper bilayers with increasing temperature [92].

The above interpretation of the intense SFG band should be examined in comparison with experimental measurement of the Im[$\chi^{(2)}$] spectrum. However, experimentally reported lineshapes of the Im[$\chi^{(2)}$] spectrum are under serious controversy at present [68, 71, 97], though those of the SFG intensity spectrum agree. Otsuki et al. [71] and Smit et al. [97] reported a negative main band of the Im[$\chi^{(2)}$] spectrum, which is in accord with previous theoretical studies [11, 31, 35, 112], whereas Nojima et al. [68] experimentally reported a positive band. This problem of Im[$\chi^{(2)}$] spectrum should be resolved to establish the spectrum. There remain some important issues to be elucidated in relation to the SFG spectrum of ice, including the proton ordering near the ice surface [68, 102, 112] and the contribution of bulk signal in SFG [95, 112].

9.3 Electrolyte Solution Surfaces

Understanding of electrolyte aqueous solution surfaces has been remarkably advanced in this century [43, 89, 115]. In early days of the twentieth century, people believed that water surface is generally void of ions, on the basis of surface tension measurements [21, 84] and the theory of dielectrics [70]. This picture appears to be consistent to intuitive idea that ions prefer to be strongly hydrated in the interior of bulk rather than to expose themselves to the air. This intuitive picture was challenged in 2001 with MD simulation by Jungwirth and Tobias [42]. They predicted that some anions, such as I^- or Br^-, rather prefer to be exposed to the air as shown in Fig. 9.3. This prediction stimulated experimental studies of electrolyte solution surfaces by various means, including SFG, SHG and photoelectron spectroscopies [43, 72, 77]. The SFG spectroscopy played one of

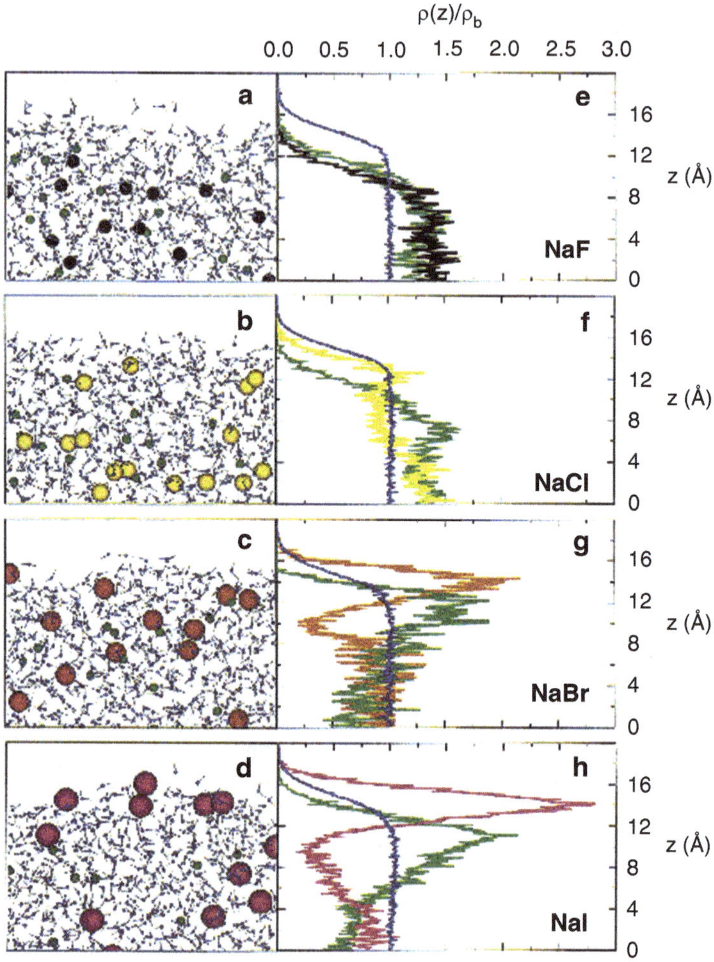

Fig. 9.3 MD snapshots (left) and density profiles (right) of NaX (X = F, Cl, Br, I) aqueous solution surfaces [42]. (Reprinted with the permission from Ref. [42]. Copyright 2001 American Chemical Society)

the leading roles to unveil the surface structures in combination with MD analysis [18, 32, 36, 41]. Such studies became a prototype of close collaboration of SFG experiment and MD simulation. Here we focus on the findings brought by the MD analysis of SFG spectroscopy.

9.3.1 *Halide Ions: Surface Segregation*

The MD simulation predicts the order of surface preference to be $I^- > Br^- > Cl^- > F^-$, indicating that larger anions (e.g. I^-) exhibit surface preference while smaller ones (F^-) are buried into the bulk liquid (see Fig. 9.3). In response to the MD prediction, SFG measurements of NaX (X = F^-, Cl^-, Br^-, I^-) aqueous solutions were carried out in 2004 by Liu et al. [49] and Raymond et al. [85]. While the two groups reported analogous spectra of NaX solutions, their interpretations were contradictory about the essential issue whether the observed SFG spectra support the surface preference of I^- and Br^-. This ambiguity clearly indicates the need of MD analysis to draw definite interpretation from the SFG spectra.

The MD simulation of NaI solution predicts the surface preference of I^-, and also reproduces the perturbed SFG spectrum of NaI solution from that of pure water. The role of NaI on the spectral perturbation is manifested in the Im[$\chi^{(2)}$] spectrum more clearly than the intensity spectrum ($\sim |\chi^{(2)}|^2$). Figure 9.4a shows the Im[$\chi^{(2)}$] spectrum of NaI solution [40], where the amplitude of Im[$\chi^{(2)}$] is shifted to the positive direction compared to that of the pure water in 3100∼3500 cm^{-1}, and the perturbed Im[$\chi^{(2)}$] spectrum is well reproduced by the MD simulation in Fig. 9.4b [28].

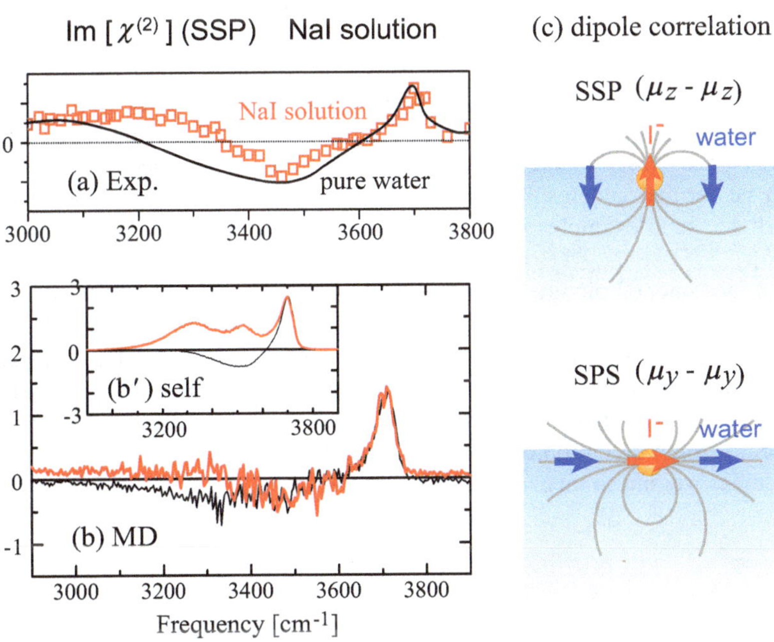

Fig. 9.4 (**a**) Experimental Im[$\chi^{(2)}$] (SSP) spectra of 2 M NaI solution (red) and pure water (black) [40], (**b**) Calculated Im[$\chi^{(2)}$] spectra [28] by Eq. (5.27), and (**b′**) the inset shows the self parts by Eq. (5.27′). (**c**) Dipole-dipole correlation schemes of I^- and water at surface. $\mu_z - \mu_z$ (SSP) cancels the Im[$\chi^{(2)}$] amplitude, while $\mu_y - \mu_y$ (SPS) augments. (Reprinted with permission from Ref. [40]; Copyright 2008 by American Physical Society. Reprinted with the permission from Ref. [28]; Copyright 2007 American Chemical Society)

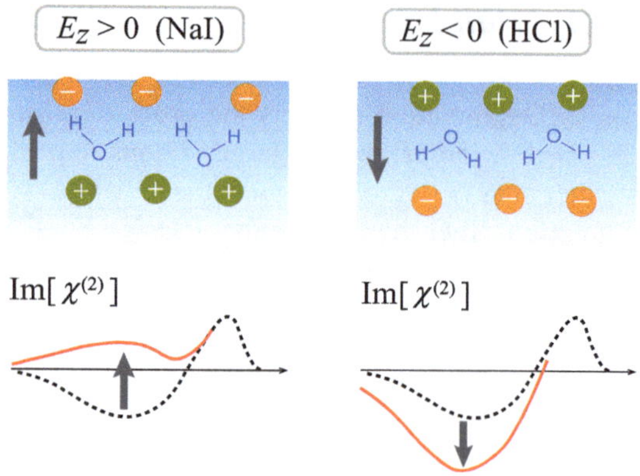

Fig. 9.5 Two types of electric double layer formation at electrolyte interfaces and their perturbation on Im[$\chi^{(2)}$] spectra

Electric double layer The perturbation of electrolyte on the surface structure and SFG spectra is generally understood by the electric double layer picture, as illustrated in Fig. 9.5. This idea was applied to the SFG analysis by Shultz and co-workers [94], and has been widely utilized as a basic picture for qualitative understanding of spectral perturbation. In the case of NaI solution, the I^- anion comes closer to the surface than the Na^+ cation, and thus these ions form an electric double layer near the surface with an upward electric field (left panel of Fig. 9.5). Consequently the water molecules in the double layer orient their dipole upward, which gives rise to the positive perturbation on the Im[$\chi^{(2)}$] spectra. In fact, the positive perturbation of Im[$\chi^{(2)}$] was observed by experiment and MD simulation, indicative of the surface preference of I^- more than Na^+. On the other hand, if cations comes closer to the surface than anions, a reverse electric double layer should be formed and result in negative perturbation on Im[$\chi^{(2)}$] (right panel). We will see such examples in other electrolyte solutions, including HCl solution [26, 36].

This fundamental picture of spectral perturbation is widely valid for electrolyte solution surfaces, since the electric double layer is generally formed at electrified interfaces. Previous MD studies have shown that even a small charge separation of ions causes sensitive perturbation on the SFG spectra [23, 38]. We occasionally encounter SFG spectra of electrolyte solutions which are not readily interpreted with this picture. Such exceptional cases offer further specific insight into the spectra and structure, as we find some examples below.

Validity of $\boldsymbol{\chi}^{(2)} \approx N \cdot \overline{\boldsymbol{\alpha}^{(2)}}$: In the above paragraph we argued that the positive perturbation of $\text{Im}[\chi^{(2)}]$ is indicative of the surface preference of I^- more than Na^+. When we evaluate the perturbation on the spectral amplitude quantitatively, further insight is obtained into the mechanism of SFG spectra for electrolyte solution surfaces. The amplitude of $\chi^{(2)}$ is conventionally presented with Eq. (3.41),

$$\chi^{(2)}_{pqr} \approx \sum_{l=1}^{N} \alpha^{(2)}_{l,pqr} = N \cdot \overline{\alpha^{(2)}_{pqr}}, \tag{3.41}$$

on the basis of an assumption that the nonlinear susceptibility $\boldsymbol{\chi}^{(2)}$ is the sum of hyperpolarizabilities $\boldsymbol{\alpha}^{(2)}$ of constituent molecules. Equation (3.41) implies that $\boldsymbol{\chi}^{(2)}$ is governed by two factors, number density N and orientational order $\overline{\boldsymbol{\alpha}^{(2)}}$. The electric double layer orients the water molecules, as illustrated in Fig. 9.5, and thus enhances the orientational average $\overline{\alpha^{(2)}_{pqr}}$. This interpretation is qualitatively in accord with the above picture of electric double layer.

However, the actual enhancement of the $\chi^{(2)}$ amplitude is significantly smaller than that predicted by Eq. (3.41) in the SSP polarization. $\chi^{(2)}$ is expressed by Eq. (5.27),

$$\chi^{(2),\text{res}}_{pqr}(\Omega, \omega_1, \omega_2) = \frac{i\omega_2}{k_B T} \int_0^\infty dt \left\langle A_{\text{eff},\,pq}(t) M_r(0) \right\rangle \exp(i\omega_2 t), \tag{5.27}$$

where

$$A_{\text{eff},pq}(t) = \sum_{l=1}^{N} \alpha_{\text{eff},pq}(l,t) \quad M_r(0) = \sum_{m=1}^{N} \mu_{\text{eff},r}(m,0). \tag{5.28}$$

On the other hand, Eq. (3.41) represents the $\boldsymbol{\chi}^{(2)}$ as the sum of molecular hyperpolarizabilities $\boldsymbol{\alpha}^{(2)}_l$, and accordingly corresponds to

$$\chi^{(2),\text{res}}_{pqr}(\Omega, \omega_1, \omega_2) \approx \sum_{l=1}^{N} \alpha^{(2)}_{l,pqr} = \sum_{l=1}^{N} \frac{i\omega_2}{k_B T} \int_0^\infty dt \left\langle \alpha_{\text{eff},pq}(l,t) \mu_{\text{eff},r}(l,0) \right\rangle \exp(i\omega_2 t) \tag{5.27$'$}$$

in the form of time correlation function. We notice that Eq. (5.27′) includes only the self correlation ($l = m$) of Eq. (5.27). Equation (5.27′) should be regarded as an approximation of Eq. (5.27) by omitting the intermolecular correlation ($l \neq m$).

The calculated $\text{Im}[\chi^{(2)}]$ spectra of NaI solution by Eqs. (5.27) and (5.27′) are compared in panels (b) and (b′) of Fig. 9.4, respectively. The perturbation of NaI (red lines) from pure water (black) exhibits noticeable differences in the two panels. The self correlation of $\text{Im}[\chi^{(2)}]$ spectrum in panel (b′) exhibits a remarkable positive amplitude (red line), because of the orientational order of water molecules induced by the electric double layer. However, the Im[$\chi^{(2)}$] amplitude of NaI (red line) in panel (b) shows a rather modest perturbation to the positive direction. The difference between (b) and (b′) indicates that Eq. (5.27′) or Eq. (3.41) overestimates the positive perturbation on $\chi^{(2)}$. The deviation is obviously attributed to the cross correlation among neighbor molecules ($l \neq m$) in Eq. (5.27). The cross correlation is illustrated in Fig. 9.4c in the case of SSP polarization, where the z-component dipole is relevant through the $\chi^{(2)}_{yyz}$ element (see Eq. (3.49)). In the NaI solution surface, the induced z-component dipoles of different molecules tend to correlate anti-parallel, which thereby suppresses the total amplitude of dipole. This correlation effect becomes obvious in SFG spectra including surface-active and very polarizable species, such as I^-.

SPS polarization The SPS spectra of aqueous systems could provide complementary information to the SSP spectra. However, the SPS spectra have been less explored than SSP, because the signal intensity is generally weak and the analysis is less intuitive. The SPS spectra are associated to the $\chi^{(2)}_{yzy}$ element, which involves the y-component dipole. Since the relevant dipole is parallel to the interface, the mechanism of SPS spectra is not interpreted in terms of up/down dipole orientation of surface species. To understand the SPS spectra even in qualitative sense, the MD analysis is often required.

In the SPS polarization, the correlation effect discussed above has an opposite influence on the $\chi^{(2)}$ amplitude, as illustrated in Fig. 9.4c. In contrast to the z-component dipole in SSP, the induced y-component dipoles in SPS tend to correlate parallel in the electric double layer and thus enhance the total $\chi^{(2)}$ amplitude for the SPS polarization. The constructive effect of dipole correlation in the SPS case has been demonstrated by MD calculation and consistently elucidated the experimental spectrum [28]. The SPS spectrum of NaI solution provides another decisive evidence for the electric double layer formation by Na^+ and I^-.

9.3.2 *Buried Ions: F^-, SO_4^{2-}*

In contrast to the ions in the preceding subsection, some other ions are repelled from the water surface and buried, in accord with the traditional picture of interfacial ions. MD simulation predicts that F^- and SO_4^{2-} are typical examples of such buried ions. One may expect that such buried electrolytes little perturb the surface structure, since these ions do not penetrate into the topmost layer. Yet the SFG spectroscopy can report perturbed SFG spectra for some of these electrolyte solutions, which implies the water surface is still perturbed by the ions. The mechanism of spectral perturbation and surface structure are elucidated with the help of MD analysis.

Figure 9.6a shows the computational [23] and experimental [17, 49] SFG spectra of NaF and Na_2SO_4 solutions. We find that the NaF (blue) and Na_2SO_4 (red)

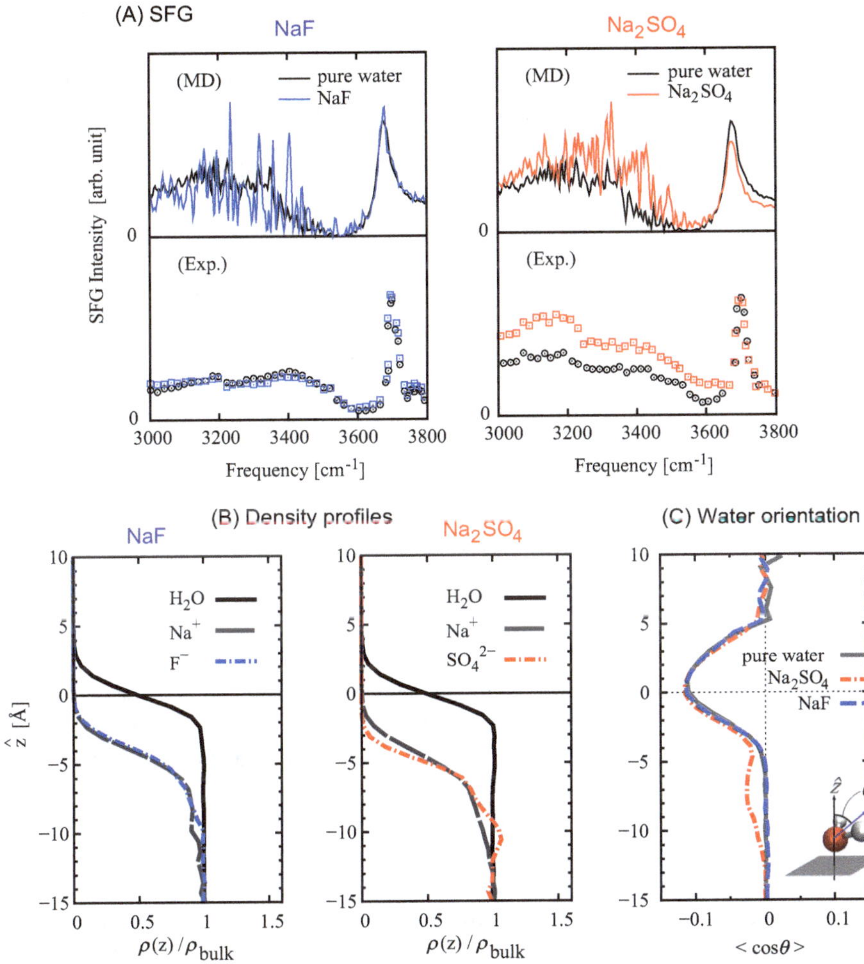

Fig. 9.6 (**a**) SFG spectra of 0.016 mole fraction (≈0.9 M) NaF (blue) and 1M Na_2SO_4 (red) solutions in comparison with that of pure water (black). Upper panels are calculated results [23], while lower panels are experimental [17, 49]. (**b**) Density profiles of ions and water along the depth coordinate $\hat{z}$. The density of each species is normalized with that in the bulk. (**c**) $\langle \cos\theta \rangle$ profiles of water orientation for pure water (gray), Na_2SO_4 (red) and NaF solutions (blue). The definition of the tilt angle θ is illustrated in the inset. (Reprinted with the permission from Refs. [17, 23, 49]. Copyright 2012, 2004, 2005 American Chemical Society)

solution spectra are noticeably different. NaF little perturbs the SFG spectrum of neat water (black), whereas Na_2SO_4 obviously enhances the SFG intensity. These differences illustrate that specific effects on the surface are present among such buried ions. The reason of the spectral difference is understood as follows.

Panels (b) and (c) show the calculated density profiles and water orientation, respectively, of the NaF and Na_2SO_4 solution surfaces. The density profiles in (b) confirm that all the present ions of Na^+, F^- and SO_4^{2-} are repelled from the topmost layer of the surface. By comparing the density profiles of NaF and Na_2SO_4 solutions in panel (b), we find that the NaF solution (left) exhibits almost overlapping profiles of Na^+ and F^- ions, whereas the Na_2SO_4 solution (right) shows slight but significant difference in the Na^+ and SO_4^{2-} profiles. In the Na_2SO_4 solution, SO_4^{2-} is more repelled from the surface than Na^+, arguably because SO_4^{2-} is a divalent ion. Consequently, charge separation between Na^+ and SO_4^{2-} generates an electric double layer in a deep region by a few monolayers from the surface ($\hat{z} \sim -5$ Å).

The effect of electric double layer in the Na_2SO_4 solution is manifested in the orientational profile of water in panel (c). This panel displays the $\langle \cos\theta \rangle$ profile as a function of the depth coordinate $\hat{z}$, where θ is the tilt angle of the water dipole from the surface normal. Near the Gibbs dividing surface of water ($\hat{z} \approx 0$ Å), all the $\langle \cos\theta \rangle$ profiles of pure water (black), Na_2SO_4 solution (red), and NaF solution (blue) show negative and nearly identical shapes. This feature means that the water orientation of the top layer is little perturbed by the buried ions. However, we see a negative $\langle \cos\theta \rangle$ region in a deeper region $\hat{z} \approx -5 \sim -10$ Å for the Na_2SO_4 solution (red dashed). This feature is a consequence of the electric double layer of Na^+ and SO_4^{2-} formed in that region. Further MD analysis confirmed that the enhanced SFG intensity in the Na_2SO_4 solution originates from the perturbed water orientation of the negative $\langle \cos\theta \rangle$ in that deep region, and the perturbed $\mathrm{Im}[\chi^{(2)}]$ band of Na_2SO_4 solution was confirmed by the heterodyne detected SFG measurement [107].

In summary, the SFG signals of water originate from the surface region where the isotropic orientation is broken, and the electric double layer formation of electrolytes is a typical cause to perturb the water orientation. The above cases of NaF and Na_2SO_4 solutions exemplify that the SFG spectroscopy is sensitive to the perturbed water orientation induced by slight charge separation, even when the charge separation arises from a somewhat deep region from the topmost layer. The perturbed SFG spectra of electrolyte solutions may reflect the structural change in a deeper region than the topmost layer of the water surface.

9.3.3 Acid

In the surface of acid solutions, the excess hydronium (H_3O^+) cations preferentially reside on the topmost surface of water. This microscopic behavior was predicted by MD simulation [58, 78]. It has been long known experimentally that the surface tension of acid solutions becomes smaller than that of neat water [84], implying a positive surface excess from the thermodynamic view. The relation to the microscopic surface structure of acid solutions and their SFG spectra is analyzed in the following.

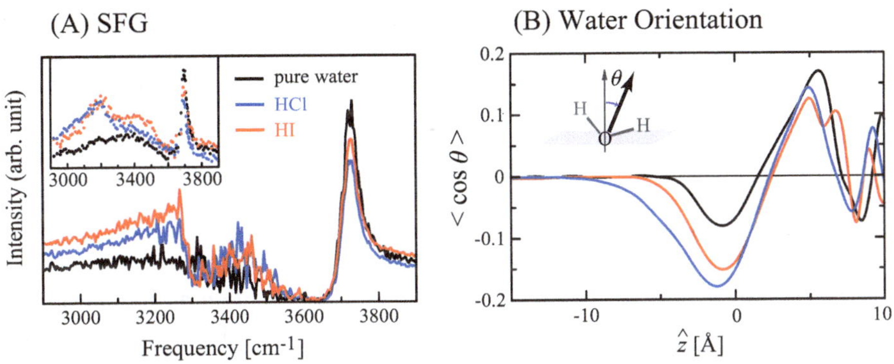

Fig. 9.7 SFG spectra and orientation for acid solution surfaces. (**a**) calculated SFG spectra of pure water (black), 1.1 M HCl (blue) and 1.1M HI (red) [26]. The inset shows the experimental spectra by Mucha et al. [58]. (**b**) ⟨cos θ⟩ profile of water orientation. (Reprinted with the permission from Refs. [26, 58]. Copyright 2007, 2005 American Chemical Society)

HCl, HI (strong acid) solution surfaces Figure 9.7 displays the calculated and experimental SFG spectra of strong acid solutions, HCl and HI [26, 58].[2] The SFG spectra of acid solutions are generally characterized with two features: (i) reduced intensity of the free O–H band at about 3700 cm^{-1} and (ii) enhanced intensity of hydrogen-bonding O–H, particularly in 3200 cm^{-1} region. These characters are consistently elucidated by the MD analysis of SFG spectra.

(i) The amplitude of the free O–H band is reduced in the acid solutions, essentially because the H_3O^+ covers the water surface and decreases the density of free O–H at the topmost layer. (ii) The increased intensity of hydrogen-bonding O–H band is understood from the electric double layer picture. As illustrated in the right panel of Fig. 9.5, the H_3O^+ layer at the topmost surface and the counter anions located below form an electric double layer. The double layer orients the water molecules toward the bulk liquid, resulting in negative perturbation on $\langle \cos\theta \rangle$. This pertubation augments the negative $\langle \cos\theta \rangle$ at water surface, since the pure water surface has intrinsic negative $\langle \cos\theta \rangle$ orientation at the top layer $\hat{z} = -3 \sim 0$ Å (see right panel of Fig. 9.7). Therefore, the enhanced water orientation of negative $\langle \cos\theta \rangle$ by the acid perturbation augments the negative $\mathrm{Im}[\chi^{(2)}]$ amplitude and the SFG intensity in the hydrogen-bonded O–H frequency region.

H_2SO_4 solution surface Sulfuric acid solution surface is relevant to heterogeneous atmospheric chemistry, as it is the main chemical component of sulfate aerosols, ubiquitously present in troposphere and stratosphere. The SFG measurement of sulfuric acid solution was performed in early stage of SFG spectroscopy [2, 83]. The observed spectra showed that the SFG intensity decreases in concentrated

[2]Note that the present MD simulation employed the point polarizable model [27] instead of CRK. Therefore, the spectra and structure may not coincide with those of the CRK model in other parts.

solutions (mole fraction $x \gtrsim 0.2$) and the whole O–H band eventually vanishes with increasing x. This behavior posed confusing interpretations for the surface structure.

This system poses an additional challenge in interpretation besides the strong acid solutions discussed above. Sulfuric acid undergoes two-step ionization:

$$H_2SO_4 \overset{pKa_1}{\longleftrightarrow} HSO_4^- \overset{pKa_2}{\longleftrightarrow} SO_4^{2-}. \tag{9.1}$$

The first ionization is considered to be strong enough ($pKa_1 \ll 0$), whereas the second ionization ($pKa_2 \simeq 1.9$ [20]) is not as strong as the first one. In such cases that the acid dissociation is not complete, the local composition of the ion species at the interface could be different from that in the bulk, since the interface is a less polar environment than the bulk. The ion dissociation at interface may be sensitive to the depth or other conditions, and coupled to the interface structure and SFG spectra. These uncertainties make such surfaces challenging to understand. The collaboration of SFG spectra and MD analysis can provide valuable information on the surface structure and ion composition.

Since we could not determine the local ion composition a priori, we assumed typical cases (a–c) about acid dissociation (9.1), where

(a) Same $pKa_{1,2}$ in Eq. (9.1) holds for surface as well, (HSO_4^- dominant, SO_4^{2-} present)
(b) Second ionization is suppressed at surface ($pKa_2 \gg 0$), (no SO_4^{2-})
(c) First ionization is also suppressed ($pKa_1 \gg 0$), (only neutral H_2SO_4)

and predicted the SFG spectra in each case by MD simulation. Figure 9.8 displays the MD results in comparison with the experiments [2, 33] to see whether the experimental features in O–H and S–O bands are consistently reproduced by MD predictions. We first find that case (a) would lead to excessively strong hydrogen bonding O–H signal, arguably because the presence of divalent anions (SO_4^{2-}) too much augments the electric double layer. Therefore, case (a) is not likely to be the proper case. On the other hand, case (c) shows the main S–O band at $1150\,cm^{-1}$ in the SSP spectrum, which is assigned to the neutral H_2SO_4, while the experimental main S–O band as well as cases (a) and (b) arises at $1050\,cm^{-1}$ from HSO_4^-. Therefore, case (c) is not the proper case either. The combination of SFG measurement and MD simulation thereby conclude that the first dissociation is facile at the surface, whereas the second dissociation to form SO_4^{2-} is substantially more suppressed at the surface than in the bulk [30, 33, 53].

9.3.4 Base

The surface preference of hydroxide (OH^-) anion remains a controversial and active issue both theoretically and experimentally [1]. Previous MD simulation studies reported different conclusions about its surface preference, which arguably depend on the methods and conditions. Direct experimental measurement of surface OH^-

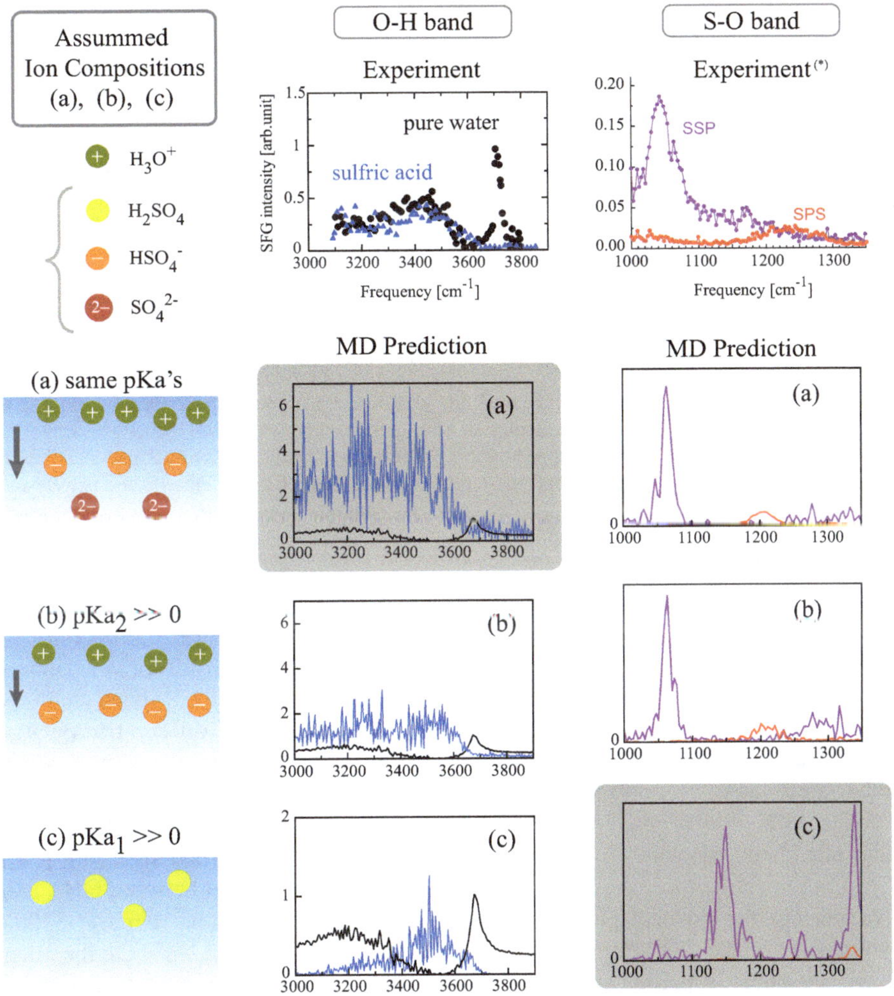

Fig. 9.8 Predicted SFG spectra of O–H and S–O bands of 0.2x sulfuric acid solution in comparison with experimental ones (top panels [2, 33]). In O–H spectra (center column), blue symbols refer to sulfuric acid solution, and black ones for pure water. In S–O spectra (right coloumn), SSP (purple) and SPS (red) combinations are displayed. MD calculations were performed in each case of three assumptions (**a**)–(**c**) about local acid dissociation, discussed in the text. Shadowed panels show obviously inconsistent spectra with experiments. (Reproduced from Ref. [33] with permission from the PCCP Owner Societies. Reprinted with the permission from Ref. [2]; Copyright 1997 American Chemical Society.) (*)Top right panel shows experimental S–O spectra for 0.29x solution

species is also challenging. What information can be extracted from the SFG spectra of basic solutions [104, 106]? Here we summarize the implications drawn from the MD analysis of the Im[$\chi^{(2)}$] spectrum.

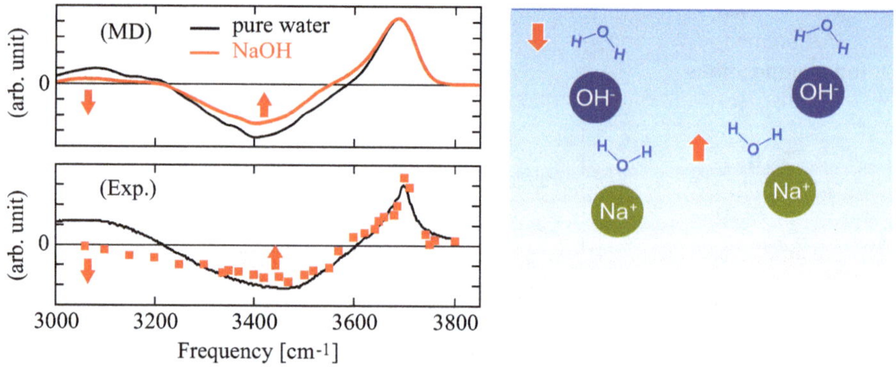

Fig. 9.9 (Left) calculated and experimental Im[$\chi^{(2)}$] spectra of pure water (black) and 1.2 M NaOH solution (red). Both MD calculation [24] and experiment [106] show the opposite perturbations on the Im[$\chi^{(2)}$] amplitude at about 3400 and 3100 cm^{-1} regions. (Right) illustration of the first solvation shell of OH^- and the electric double layer. (Reprinted with the permission from Refs. [24, 106]. Copyright 2014, 2008 American Chemical Society)

Figure 9.9 displays the calculated and experimental Im[$\chi^{(2)}$] spectrum of NaOH solution in comparison with that of neat water. The Im[$\chi^{(2)}$] spectrum of NaOH solution shows remarkable features of perturbation. The Im[$\chi^{(2)}$] band at 3300–3600 cm^{-1} shifts to the positive direction, whereas the Im[$\chi^{(2)}$] band at 3000–3200 cm^{-1} to negative [106]. These opposite perturbations in different frequency regions are not interpreted with the electric double layer picture, and imply some other mechanism beyond the double layer picture.

The MD simulation of NaOH solution reproduces the opposite perturbations, and elucidates the whole mechanisms by analyzing the perturbed spectrum [24]. To summarize the mechanisms, the electric double layer formed between OH^- and Na^+ brings the positive perturbation to the main O–H stretching band at 3300–3600 cm^{-1}, since OH^- comes slightly closer to the surface than Na^+. On the other hand, the negative perturbation at 3000–3200 cm^{-1} originates from the water in the first solvation shell (FSS) of OH^-, as we discuss below.

This contribution of FSS is general in the electrolyte solutions, and we briefly explain the mechanism in Fig. 9.10. In a case of an anion, the FSS includes water molecules that orient their dipoles toward the anion, as illustrated in panel (a), and consequently the upward and downward orientations co-exist in the whole FSS. When the ions with their FSS are distributed in the surface region (panel (b)), the net contribution of the topmost, downward contribution remains while the other contributions cancel each other. This mechanism is common with the χ^{IQB} mechanism of the quadrupole contribution in Chap. 7 (see detailed discussion in

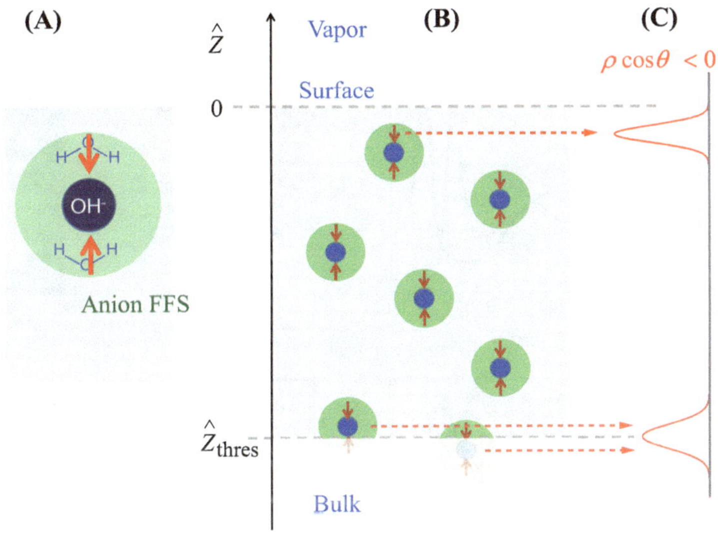

Fig. 9.10 (**a**) Illustration of the first solvation shell (FSS, green) around an anion OH^-. (**b**) Distributed ions with FSS in the surface region. (**c**) Net dipole orientation generated from the FSS water [24]. One may consider that the topmost, downward orientation remains while the other contributions cancel. Alternatively, if one considers a FSS as a quadrupole of opposite dipoles, the net downward contribution emerges at an arbitrary threshold $\hat{z}_{thres}$

Appendix A.1).[3] We note that the FSS is regarded to form a quadrupole with a pair of opposite dipoles.

The mechanism of FSS is clearly manifested in OH^-, since the FSS component appears in a particularly low-frequency region at 3000–3200 cm^{-1} and is separated from the main O–H band. From the above discussion, we can readily understand that the FSS of anions have generally negative Im[$\chi^{(2)}$] contributions while the FSS of cations have positive Im[$\chi^{(2)}$] contribution. The feature of the FSS evidences that the OH^- anions retain the first solvation shell and do not preferentially expose themselves at the surface.

9.4 Oil/Water Interfaces

Oil/water interfaces are relevant to various phenomena, such as micelle formation, extraction, membrane transport, sensors, and phase transfer catalysis. Microscopic understanding of oil/water interfaces has been pursued with various experimental

[3]One may wonder that the net dipole in the FSS cancel and thus no signal is generated. Even though one considers the FSS of an ion as a quadrupole consisting of opposite dipoles, as illustrated in Fig. 9.10b, the net negative contribution still remains at an arbitrary threshold $\hat{z}_{thres}$. This mechanism is same with the χ^{IQB} term of the quadrupole contribution in Appendix A.1.

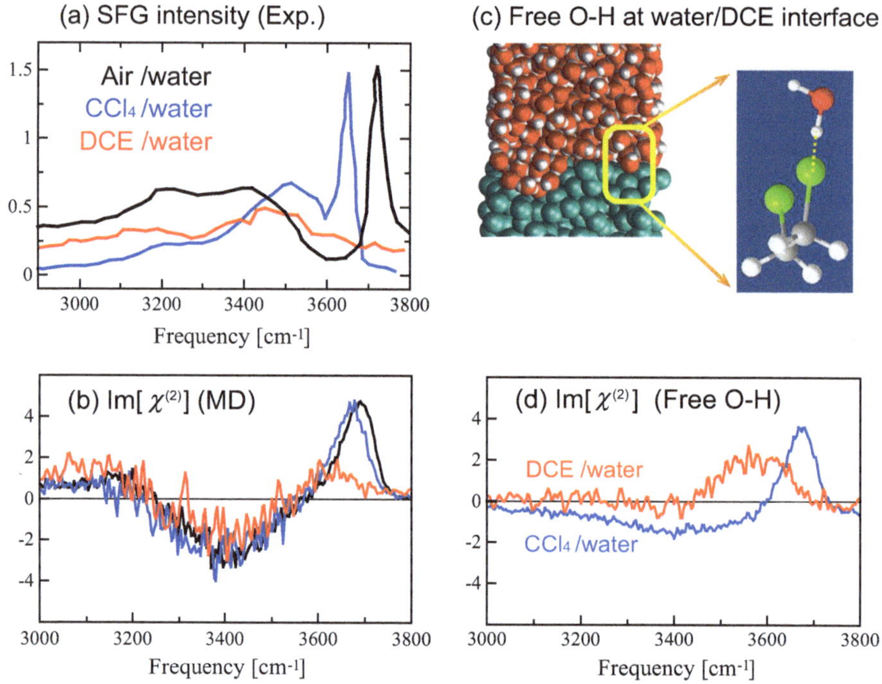

Fig. 9.11 (**a**) Experimental SFG spectra of neat water/vapor (black), water/CCl_4 (blue) and water/1,2-dichloroethane (DCE, red) interfaces [55, 109, 111]. (**b**) Calculated Im[$\chi^{(2)}$] spectra of the three interfaces [34]. (**c**) Illustration of free O–H at water/DCE interface. (**d**) Calculated Im[$\chi^{(2)}$] spectra by restricting the water molecules having free O–H bonds at the interfaces. The free O–H band at CCl_4/water is little perturbed, while that at DCE/water interface is more red shifted and broadened. (Reprinted with the permission from Ref. [34]. Copyright 2012 American Chemical Society)

techniques [48, 55] as well as MD simulation [5, 6, 12, 82]. A main challenge of investigating oil/water interfaces lies in difficulties of selective detection of interfacial molecules with sufficient spatial resolution. The interface-sensitive nonlinear spectroscopy is able to probe buried oil/water interfaces with excellent selectivity as long as it is accessible by light. The SHG was applied to reveal detailed polarity environment in the vicinity of the liquid-liquid interfaces [100, 101]. The collaboration of vibrational SFG spectroscopy and MD analysis is powerful to reveal the details of the oil/water interfaces.

Typical and fundamental examples of oil/water interfaces were studied by Richmond and co-workers [55, 109, 111]. Figure 9.11a shows the SFG spectra of carbon tetrachloride (CCl_4)/water and 1,2-dichloroethane (DCE)/water interfaces in comparison to that of air/water interface. We find that the CCl_4/water spectrum retains the two-band structure that resembles air/water, while the spectrum of DCE/water interface becomes structureless and weaker. The apparent spectral difference may allow various interpretations. The MD analysis can clarify the interpretation of the spectra in relation to the structure of oil/water interfaces.

The MD simulation of the SFG spectra was performed [34], which well reproduced the distinct SFG spectra of CCl_4/water and DCE/water interfaces. The MD simulation allows for direction observation of interfacial water structure. However, the MD showed rather similar structure of interfacial water at CCl_4 and DCE in terms of the density profile or orientation of water molecules, irrespective of the apparently distinct SFG spectra reproduced. The mechanism of the noticeable spectral differences becomes clearer in the calculated Im[$\chi^{(2)}$] spectra in Fig. 9.11b. Comparing the Im[$\chi^{(2)}$] spectra of the two oil/water interfaces, we find that the spectra in the H-bonding region below 3600 cm^{-1} not much different, which is consistent to the direct MD observation that the density and orientation of surface water are analogous at CCl_4/water and DCE/water interfaces. However, we find that the free O–H band in 3600–3700 cm^{-1} is particularly suppressed in the DCE/water interface.

The perturbation on the free O–H band is understood in the following manner. The free O–H of water at oil/water interface actually interacts with adjacent oil molecules, as illustrated in Fig. 9.11c. The "free" O–H at DCE/water interface is more perturbed than that at CCl_4/water because of larger polarity of DCE than CCl_4. As a consequence, the positive Im[$\chi^{(2)}$] band of "free" O–H is substantially red shifted and broadened for the DCE/water interface, as evidenced in Fig. 9.11d. The red-shifted free O–H band of DCE/water interface overlaps with the negative Im[$\chi^{(2)}$] band of the H-bonding O–H, and cancel the intensity. The apparent spectral difference between CCl_4/water and DCE/water interfaces is attributed to the local interaction of water and oil molecules at the interfaces, rather than qualitatively distinct structure of molecular orientation.

9.5 Water at Monolayers

Amphiphilic molecules tend to form various self assembled structures in/on water, such as Langmuir monolayer, micelle and lipid bilayers. Such structures generally include interfaces of water and amphiphilic molecules, and their interfaces govern the stability of these structures. The interfaces of phospholipid membranes have been drawing particular attention by SFG spectroscopy [41], as the lipid membranes define the boundary of cells, control mass transport, and thereby play vital roles of living functions [7, 63]. A number of MD studies in relation to the SFG spectroscopy have been performed to aim at selective detection and understanding of water structure in contact with those amphiphilic monolayers [37, 38, 59, 69, 86, 87, 90, 91].

One of the basic concepts of the water structure is the flip-flop model of water orientation in Fig. 9.12. The orientational structure of water molecules is determined by the net charges of the monolayer molecules. When the monolayer molecules are negatively charged, such as sodium dodecyl sulfate (SDS, $C_{12}H_{25}SO_4^- \cdot Na^+$), the water molecules take upward orientation and leads to positive Im[$\chi^{(2)}$] band. On the other hand, if the monolayer is positively charged, such as cetyltrimethylammonium bromide (CTAB, $C_{16}H_{33}N^+(CH_3)_3 \cdot Br^-$), the water takes downward orientation

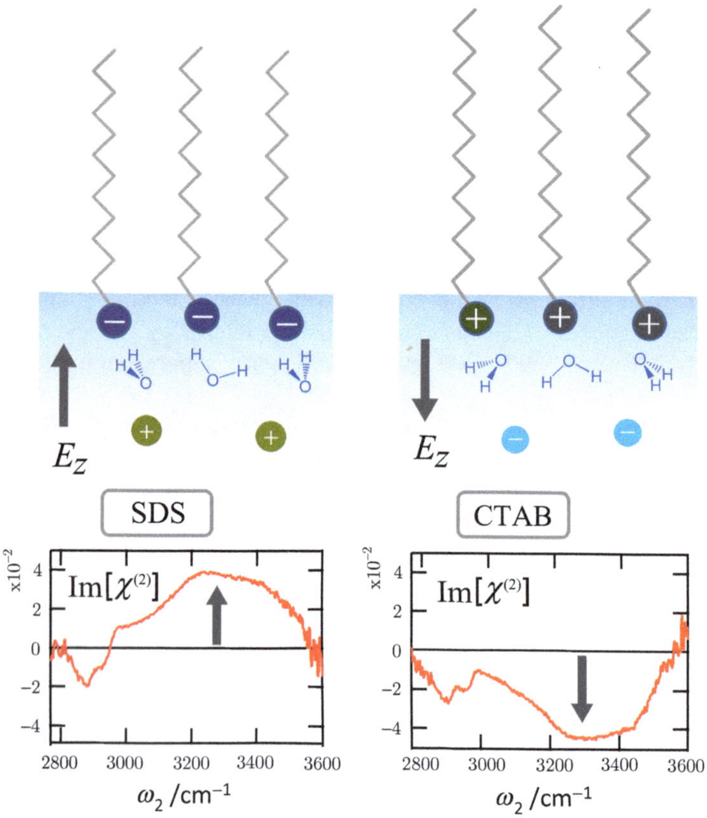

Fig. 9.12 Water structure near negatively and positively charged monolayer. The O–H band of $\mathrm{Im}[\chi^{(2)}]$ spectrum changes its sign [65]. (Reprinted with permission from Ref. [65]. Copyright 2009, American Institute of Physics)

and shows negative $\mathrm{Im}[\chi^{(2)}]$ band. The reversal of the $\mathrm{Im}[\chi^{(2)}]$ band of water O–H stretching is experimentally verified in Fig. 9.12 [65]. We notice that this mechanism is essentially common with the electric double layer picture in Fig. 9.5.

In a case that the amphiphilic molecules are neutral, the SFG spectra of water interface are not amenable to the simple flip-flop model. Biological membranes include various neutral but zwitterionic phospholipid molecules, such as phosphatidylcholine (POPC) and dipalmitoylphosphatidylcholine (DPPC). Figure 9.13 displays an illustration of the water/POPC interface (left panel) and the experimental [54] and calculated [37] $\mathrm{Im}[\chi^{(2)}]$ spectra (right). The polar head group of POPC includes a positively charged choline ($-N^{+}(CH_3)_3$) and a negatively charged phosphate ($-PO_4^-$), which form a zwitterionic molecule. The experimental $\mathrm{Im}[\chi^{(2)}]$ spectrum of the water/POPC interface shows a positive main band at 3300 cm^{-1} and a positive minor band at 3580 cm^{-1}. Another, related experimental $\mathrm{Im}[\chi^{(2)}]$ spectrum of water/DPPC also shows a quite similar lineshape [22]. The MD analysis

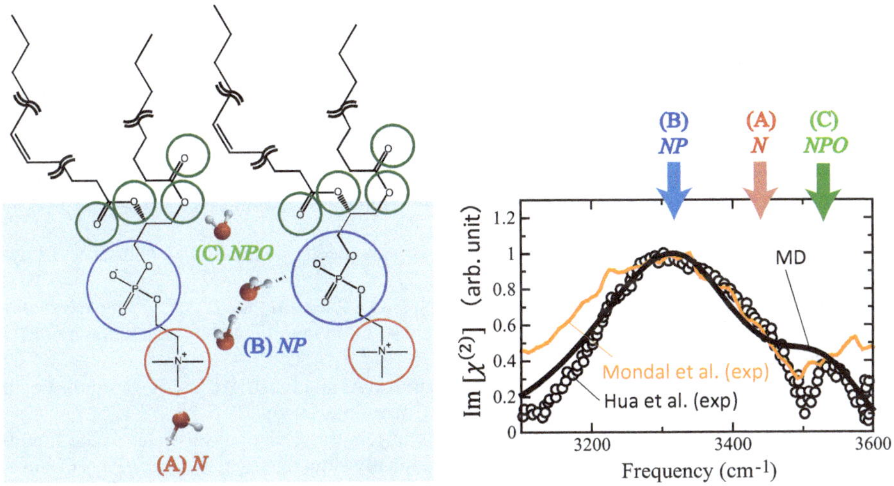

Fig. 9.13 (Left) Illustration of water/phosphatidylcholine (POPC) interface, where the polar head group of POPC is classified to choline (N, red), phosphate (P, blue) and ester oxygen (O, green). (Right) Experimental [54] (orange) and calculated [37] (black) Im[$\chi^{(2)}$] spectra of water/POPC interface. Experimental spectrum of water/DPPC (open circles) [22] is also shown. (Reprinted with the permission from Refs. [37, 54]; Copyright 2012, 2016 American Chemical Society. Reproduced from Ref. [22] by permission of John Wiley & Sons Ltd)

revealed that the main positive band at 3300 cm^{-1} is assigned to the water molecules between the choline and phosphate groups, labeled "*NP*". The positive sign of this "*NP*" band is understood from the electric double layer of the charged groups. The minor band at 3580 cm^{-1} is attributed to the water molecules penetrating to the ester group, labeled "*NPO*". This O–H band is located in a higher frequency region due to weaker hydrogen-bonding environment. The water molecules attached to the choline group, labeled "*N*", do not show up clearly in the Im[$\chi^{(2)}$] spectrum, but they contribute to the dip between the two positive bands.

Bibliography

1. Agmon N, Bakker HJ, Campen RK, Henchman RH, Pohl P, Roke S, Thämer M, Hassanali A (2016) Protons and hydroxide ions in aqueous systems. Chem Rev 116:7642–7672
2. Baldelli S, Schnitzer C, Shultz MJ, Campbell DJ (1997) Sum frequency generation investigation of water at the surface of H_2O/H_2SO_4 binary systems. J Phys Chem B 101:10435–10441
3. Ball P (2008) Water – an enduring mystery. Nature 452:291–292
4. Benjamin I (1994) Vibrational spectrum of water at the liquid/vapor interface. Phys Rev Lett 73:2083–2086
5. Benjamin I (1995) Theory and computer simulations of solvation and chemical reactions at liquid interfaces. Acc Chem Res 28:233–239

6. Benjamin I (2015) Reaction dynamics at liquid interfaces. Annu Rev Phys Chem 66:165–188
7. Berg JM, Tymoczko JL, Gregory J, Gatto J, Stryer L (2015) Biochemistry, 8th edn. W. H. Freeman, New York
8. Bonn M, Bakker HJ, Ghosh A, Yamamoto S, Sovago M, Campen RK (2010) Structural inhomogeneity of interfacial water at lipid monolayers revealed by surface-specific vibrational pump-probe spectroscopy. J Am Chem Soc 132:14971–14978
9. Bonn M, Nagata Y, Backus EHG (2015) Molecular structure and dynamics of water at the water-air interface studied with surface-specific vibrational spectroscopy. Angew Chem Int Ed 54:5560–5576
10. Buch V (2005) Molecular structure and OH-stretch spectra of liquid water surface. J Phys Chem B 109:17771–17774
11. Buch V, Tarbuck T, Richmond GL, Groenzin H, Li I, Shultz MJ (2007) Sum frequency generation surface spectra of ice, water, and acid solution investigated by an exciton model. J Chem Phys 127:204710
12. Chang T, Dang LX (1996) Molecular dynamics simulations of CCl_4-H_2O liquid-liquid interface with polarizable potential models. J Chem Phys 104:6772–6783
13. Das B, Sharma B, Chandra A (2018) Effects of tert-Butyl alcohol on water at the liquid-vapor interface: structurally bulk-like but dynamically slow interfacial water. J Phys Chem C 122:9374–9388
14. Du Q, Superfine R, Freysz E, Shen YR (1993) Vibrational spectroscopy of water at the vapor/water interface. Phys Rev Lett 70:2313–2316
15. Faraday M (1850) On certain conditions of freezing water. The Athenaeum 1181:640
16. Franks F (ed) (1972) Water: a comprehensive treatise, vol 1. Plenum Press, New York
17. Gopalakrishnan S, Jungwirth P, Tobias DJ, Allen HC (2005) Air-liquid interfaces of aqueous solutions containing ammonium and sulfate: spectroscopic and molecular dynamics studies. J Phys Chem B 109:8861–8872
18. Gopalakrishnan S, Liu D, Allen HC, Kuo M, Shultz MJ (2006) Vibrational spectroscopic studies of aqueous interfaces: salts, acids, bases, and nanodrops. Chem Rev 106:1155–1175
19. Groenzin H, Li I, Buch V, Shultz MJ (2007) The single-crystal, basal face of ice Ih investigated with sum frequency generation. J Chem Phys 127:214502
20. Haynes WM (ed) (2012) CRC handbook of chemistry and physics, 93rd edn. CRC Press, Boca Raton
21. Heydweiller A (1910) Über physikalische Eigenschaften von Lösungen in ihrem Zusammenhang. II. Oberflächenspannung und elektrisches Leitvermögen wässeriger Salzlösungen. Ann Physik 4 33:145–185
22. Hua W, Verreault D, Allen HC (2015) Solvation of calcium-phosphate headgroup complex at the DPPC/aqueous interface. Chem Phys Chem 16:3910–3915
23. Imamura T, Mizukoshi Y, Ishiyama T, Morita A (2012) Surface structures of NaF and Na_2SO_4 aqueous solutions: specific effects of hard ions on surface vibrational spectra. J Phys Chem C 116:11082–11090
24. Imamura T, Ishiyama T, Morita A (2014) Molecular dynamics analysis of naoh aqueous solution surface and the sum frequency generation spectra: is surface OH^- detected by SFG spectroscopy? J Phys Chem C 118:29017–29027
25. Inoue K, Ishiyama T, Nihonyanagi S, Yamaguchi S, Morita A, Tahara T (2016) Efficient spectral diffusion at the air/water interface revealed by femtosecond time-resolved heterodyne-detected vibrational sum frequency generation spectroscopy. J Phys Chem Lett 7:1811–1815
26. Ishiyama T, Morita A (2007) Molecular dynamics analysis of interfacial structures and sum frequency generation spectra of aqueous hydrogen halide solutions. J Phys Chem A 111:9277–9285
27. Ishiyama T, Morita A (2007) Molecular dynamics study of gas-liquid aqueous sodium halide interfaces. I. Flexible and polarizable molecular modeling and interfacial properties. J Phys Chem C 111:721–737

28. Ishiyama T, Morita A (2007) Molecular dynamics study of gas-liquid aqueous sodium halide interfaces II. Analysis of vibrational sum frequency generation spectra. J Phys Chem C 111:738–748
29. Ishiyama T, Morita A (2009) Analysis of anisotropic local field in sum frequency generation spectroscopy with the charge response kernel water model. J Chem Phys 131:244714
30. Ishiyama T, Morita A (2011) Molecular dynamics simulation of sum frequency generation spectra of aqueous sulfuric acid solution. J Phys Chem C 115:13704–13716
31. Ishiyama T, Morita A (2014) A direct evidence of vibrationally delocalized response at ice surface. J Chem Phys 141:18C503
32. Ishiyama T, Morita A (2017) Computational analysis of vibrational sum frequency generation spectroscopy. Annu Rev Phys Chem 68:355–377
33. Ishiyama T, Morita A, Miyamae T (2011) Surface structure of sulfuric acid solution relevant to sulfate aerosol: molecular dynamics simulation combined with sum frequency generation measurement. Phys Chem Chem Phys 13:20965–20973
34. Ishiyama T, Sato Y, Morita A (2012) Interfacial structures and vibrational spectra at liquid/liquid boundaries: molecular dynamics study of water/carbon tetrachloride and water/1,2-dichloroethane interfaces. J Phys Chem C 116:21439–21446
35. Ishiyama T, Takahashi H, Morita A (2012) Origin of vibrational spectroscopic response at ice surface. J Phys Chem Lett 3:3001–3006
36. Ishiyama T, Imamura T, Morita A (2014) Theoretical studies of structures and vibrational sum frequency generation spectra at aqueous interfaces. Chem Rev 114:8447–8470
37. Ishiyama T, Terada D, Morita A (2016) Hydrogen-bonding structure at zwitterionic lipid/water interface. J Phys Chem Lett 7:216–220
38. Ishiyama T, Shirai S, Okumura T, Morita A (2018) Molecular dynamics study of structure and vibrational spectra at zwitterionoic lipid/aqueous KCl, NaCl, and $CaCl_2$ solution interfaces. J Chem Phys 148:222801
39. Jeffrey GA (1997) An introduction to hydrogen bonding. Oxford University Press, Oxford
40. Ji N, Ostroverkhov V, Tian CS, Shen YR (2008) Characterization of vibrational resonances of water-vapor interfaces by phase-sensitive sum-frequency spectroscopy. Phys Rev Lett 100:096102
41. Johnson CM, Baldelli S (2014) Vibrational sum frequency spectroscopy studies of the influence of solutes and phospholipids at vapor/water interfaces relevant to biological and environmental systems. Chem Rev 114:8416–8446
42. Jungwirth P, Tobias DJ (2001) Molecular structure of salt solutions: a new view of the interface with implications for heterogeneous atmospheric chemistry. J Phys Chem B 105:10468–10472
43. Jungwirth P, Finlayson-Pitts BJ, Tobias DJ (eds) (2006) Structure and chemistry at aqueous interfaces. Chem Rev 106:1137–1539
44. Jungwirth P et al (2008) Water – from interfaces to the bulk. Faraday Discuss 141:1–488
45. Kawaguchi T, Shiratori K, Henmi Y, Ishiyama T, Morita A (2012) Mechanisms of sum frequency generation from liquid benzene: symmetry breaking at interface and bulk contribution. J Phys Chem C 116:13169–13182
46. Khatib R, Hasegawa T, Sulpizi M, Backus EHG, Bonn M, Nagata Y (2016) Molecular dynamics simulations of SFG librational modes spectra of water at the water-air interface. J Phys Chem C 120:18665–18673
47. Kundu A, Tanaka S, Ishiyama T, Ahmed M, Inoue K, Nihonyanagi S, Sawai H, Yamaguchi S, Morita A, Tahara T (2016) Bend vibration of surface water investigated by heterodyne-detected sum frequency generation and theoretical study: dominant role of quadrupole. J Phys Chem Lett 7:2597–2601
48. Leich MA, Richmond GL (2005) Recent experimental advances in studies of liquid/liquid interfaces. Faraday Discuss 129:1–21, and the papers in this issue
49. Liu D, Ma G, Levering LM, Allen HC (2004) Vibrational spectroscopy of aqueous sodium halide solutions and air-liquid interfaces: observation of increased interfacial depth. J Phys Chem B 108:2252–2260

50. Matsuzaki K, Nihonyanagi S, Yamaguchi S, Nagata T, Tahara T (2013) Vibrational sum frequency generation by the quadrupolar mechanism at the nonpolar benzene/air interface. J Phys Chem Lett 4:1654–1658
51. McGuire JA, Shen YR (2006) Ultrafast vibrational dynamics at water interfaces. Science 313:1945–1948
52. Medders GR, Paesani F (2016) Dissecting the molecular structure of the air/water interface from quantum simulations of the sum-frequency generation spectrum. J Am Chem Soc 138:3912–3919
53. Miyamae T, Morita A, Ouchi Y (2008) First acid dissociation at an aqueous H_2SO_4 interface with sum frequency generation spectroscopy. Phys Chem Chem Phys 10:2010–2013
54. Mondal JA, Nihonyanagi S, Yamaguchi S, Tahara T (2012) Three distinct water structures at a zwitterionic lipid/water interface revealed by heterodyne-detected vibrational sum frequency generation. J Am Chem Soc 134:7842–7850
55. Moore FG, Richmond GL (2008) Integration or segregation: how do molecules behave at oil/water interfaces? Acc Chem Res 41:739–748
56. Morita A (2006) Improved computation of sum frequency generation spectrum of water surface. J Phys Chem B 110:3158–3163
57. Morita A, Hynes JT (2000) A theoretical analysis of the sum frequency generation spectrum of the water surface. Chem Phys 258:371–390
58. Mucha M, Frigato T, Levering LM, Allen HC, Tobias DJ, Dang LX, Jungwirth P (2005) Unified molecular picture of the surfaces of aqueous acid, base, and salt solutions. J Phys Chem B 109:7617–7623
59. Nagata Y, Mukamel S (2010) Vibrational sum-frequency generation spectroscopy at the water/lipid interface: molecular dynamics simulation study. J Am Chem Soc 132:6434–6442
60. Nagata Y, Hsieh C-S, Hasegawa T, Voll J, Backus EHG, Bonn M (2013) Water bending mode at the water-vapor interface probed by sum-frequency generation spectroscopy: a combined molecular dynamics simulation and experimental study. J Phys Chem Lett 4:1872–1877
61. Nagata Y, Ohto T, Backus EHG, Bonn M (2016) Molecular modeling of water interfaces: from molecular spectroscopy to thermodynamics. J Phys Chem B 120:3785–3796
62. Nagata Y, Pool R, Backus EHG, Bonn M (2012) Nuclear quantum effects affect bond orientation of water at the water-vapor interface. Phys Rev Lett 109:226101
63. Nelson DL, Cox MM (2013) Lehninger: principles of biochemistry, 6th edn. W. H. Freeman, New York
64. Ni Y, Skinner JL (2015) IR and SFG vibrational spectroscopy of the water bend in the bulk liquid and at the liquid-vapor interface, respectively. J Chem Phys 143:014502
65. Nihonyanagi S, Yamaguchi S, Tahara T (2009) Direct evidence for orientational flip-flop of water molecules at charged interfaces: a heterodyne-detected vibrational sum frequency generation study. J Chem Phys 130:204704
66. Nihonyanagi S, Ishiyama T, Lee T-K, Yamaguchi S, Bonn M, Morita A, Tahara T (2011) Unified molecular view of air/water interface based on experimental and theoretical $\chi^{(2)}$ spectra of isotopically diluted water surface. J Am Chem Soc 133:16875–16880
67. Nihonyanagi S, Kusaka R, Inoue K, Adhikari A, Yamaguchi S, Tahara T (2015) Accurate determination of complex $\chi^{(2)}$ spectrum of the air/water interface. J Chem Phys 143:124707
68. Nojima Y, Suzuki Y, Takahashi M, Yamaguchi S (2017) Proton order toward the surface of ice Ih revealed by heterodyne- detected sum frequency generation spectroscopy. J Phys Chem Lett 8:5031–5034
69. Ohto T, Backus EHG, Hsieh CS, Sulpizi M, Bonn M, Nagata Y (2015) Lipid carbonyl groups terminate the hydrogen-bond network of membrane-bound water. J Phys Chem Lett 6:4499–4503
70. Onsager L, Samaras NNT (1934) The surface tension of Debye-Hückel electrolytes. J Chem Phys 2:528–536
71. Otsuki Y, Sugimoto T, Ishiyama T, Morita A, Watanabe K, Matsumoto Y (2017) Unveiling subsurface hydrogen-bond structure of hexagonal water ice. Phys Rev B 96:115405

72. Ottosson N, Faubel M, Bradforth SE, Jungwirth P, Winter B (2010) Photoelectron spectroscopy of liquid water and aqueous solution: electron effective attenuation lengths and emission-angle anisotropy. J Elec Spec Rel Phen 177:60–70
73. Perry A, Ahlborn H, Space B, Moore PB (2003) A combined time correlation function and instantaneous normal mode study of the sum frequency generation spectroscopy of the water/vapor interface. J Chem Phys 118:8411–8419
74. Perry A, Neipert C, Kasprzyk CR, Green T, Space B, Moore PB (2005) A theoretical description of the polarization dependence of the sum frequency generation spectroscopy of the water/vapor interface. J Chem Phys 123:144705
75. Perry A, Neipert C, Ridley C, Space B, Moore PB (2005) Identification of a wagging vibrational mode of water molecules at the water/vapor interface. Phys Rev E 71:050601
76. Perry A, Neipert C, Space B, Moore PB (2006) Theoretical modeling of interface specific vibrational spectroscopy: methods and applications to aqueous interfaces. Chem Rev 106:1234–1258
77. Petersen PB, Saykally RJ (2006) On the nature of ions at the liquid water surface. Annu Rev Phys Chem 57:333–364
78. Petersen MK, Iyengar SS, Day TJF, Voth GA (2004) The hydrated proton at the water liquid/vapor interface. J Phys Chem B 108:14804–14806 (2004)
79. Petrenko VF, Whitworth RW (1999) Physics of ice. Oxford University Press, Oxford
80. Pieniazek PA, Tainter CJ, Skinner JL (2011) Interpretation of the water surface vibrational sum-frequency spectrum. J Chem Phys 135:044701
81. Pieniazek PA, Tainter CJ, Skinner JL (2011) Surface of liquid water: three-body ineractions and vibrational sum-frequency spectroscopy. J Am Chem Soc 133:10360–10363
82. Pohorille A, Wilson MA (1996) Excess chemical potential of small solutes across water-membrane and water-hexane interfaces. J Chem Phys 104:3760–3773
83. Radüge C, Pflumio V, Shen YR (1997) Surface vibrational spectroscopy of sulfuric acid-water mixtures at the liquid-vapor interface. Chem Phys Lett 274:140–144
84. Randles JEB (1977) Structure at the free surface of water and aqueous electrolyte solutions. Phys Chem Liq 7:107–179
85. Raymond EA, Richmond GL (2004) Probing the molecular structure and bonding of the surface of aqueous salt solutions. J Phys Chem B 108:5051–5059
86. Re S, Nishima W, Tahara T, Sugita Y (2014) Mosaic of water orientation structures at a neutral zwitterionic lipid/water interface revealed by molecular dynamics simulations. J Phys Chem Lett 5:4343–4348
87. Reddy SK, Thiraux R, Rudd BAW, Lin L, Adel T, Joutsuka T, Geiger FM, Allen HC, Morita A, Paesani F (2018) Bulk contributions modulate the sum-frequency generation spectra of interfacial water on model sea-spray aerosols. Chem. https://doi.org/10.1016/j.chempr.2018.04.007
88. Richmond GL (2002) Molecular bonding and interactions at aqueous surfaces as probed by vibrational sum frequency spectroscopy. Chem Rev 102:2693–2724
89. Richmond GL (ed) (2014) Special section: aqueous interfaces. Chem Rev 114:8361–8498
90. Roy S, Hore D (2012) Simulated structure and nonlinear vibrational spectra of water next to hydrophobic and hydrophilic solid surfaces. J Phys Chem C 116:22867–22877
91. Roy S, Gruenbaum SM, Skinner JL (2014) Theoretical vibrational sum-frequency generation spectroscopy of water near lipid and surfactant monolayer interfaces. J Chem Phys 141:18C502
92. Sánchez MA, Kling T, Ishiyama T, van Zadel M-J, Bisson PJ, Mezger M, Jochum MN, Cyran JD, Smit WJ, Bakker HJ, Shultz MJ, Morita A, Donadio D, Nagata Y, Bonn M, Backus EHG (2017) Experimental and theoretical evidence for bilayer-by-bilayer surface melting of crystalline ice. Proc Natl Acad Sci 114:227–232
93. Shen YR, Ostroverkhov V (2006) Sum-frequency vibrational spectroscopy on water interfaces: polar orientation of water molecules at interfaces. Chem Rev 106:1140–1154

94. Shultz MJ, Schnitzer C, Simonelli D, Baldelli S (2000) Sum frequency generation spectroscopy of the aqueous interfaces: ionic and soluble molecular solutions. Int Rev Phys Chem 19:123–153
95. Shultz MJ, Bisson P, Groenzin H, Li I (2010) Multiplexed polarization spectroscopy: measuring surface hyperpolarizability orientation. J Chem Phys 133:054702
96. Skinner JL, Pieniazek PA, Gruenbaum SM (2012) Vibrational spectroscopy of water at interfaces. Acc Chem Res 45:93–100
97. Smit WJ, Tang F, Nagata Y, Sánchez MA, Hasegawa T, Backus EHG, Bonn M, Bakker HJ (2017) Observation and identification of a new oh stretch vibrational band at the surface of ice. J Phys Chem Lett 8:3656–3660
98. Sovago M, Campen RK, Wurpel GWH, Müller M, Bakker HJ, Bonn M (2008) Vibrational response of hydrogen-bonded interfacial watger is dominated by intramolecular coupling. Phys Rev Lett 100:173901
99. Sovago M, Campen RK, Bakker HJ, Bonn M (2009) Hydrogen bonding strength of interfacial water determined with surface sum-frequency generation. Chem Phys Lett 470:7–12
100. Steel WH, Walker RA (2003) Measuring dipolar width across liquid-liquid interfaces with 'molecular rulers'. Nature 424:296–299
101. Steel WH, Walker RA (2003) Solvent polarity at an aqueous/alkane interface: the effect of solute identity. J Am Chem Soc 125:1132–1133
102. Su X, Lianos L, Shen YR, Somorjai GA (1998) Surface-induced ferroelectric ice on Pt(111). Phys Rev Lett 80:1533–1536
103. Sulpizi M, Salanne M, Sprik M, Gaigeot M (2013) Vibrational sum frequency generation spectroscopy of the water liquid.vapor interface from density functional theory-based molecular dynamics simulations. J Phys Chem Lett 4:83–87
104. Tarbuck TL, Ota ST, Richmond GL (2006) Spectroscopic studies of solvated hydrogen and hydroxidde ions at aqueous surfaces. J Am Chem Soc 128:14519–14527
105. Tian C-S, Shen YR (2009) Isotopic dilution study of the water/vapor interface by phase-sensitive sum-frequency vibrational spectroscopy. J Am Chem Soc 131:2790–2791
106. Tian CS, Ji N, Waychunas GA, Shen YR (2008) Interfacial structures of acidic and basic aqueous solutions. J Am Chem Soc 130:13033–13039
107. Verreault D, Hua W, Allen HC (2012) From conventional to phase-sensitive vibrational sum frequency generation spectroscopy: probing water organization at aqueous interfaces. J Phys Chem Lett 3:3012–3028
108. Vinaykin M, Benderskii AV (2012) Vibrational sum-frequency spectrum of the water bend at the air/water interface. J Phys Chem Lett 3:3348–3352
109. Walker DS, Richmond GL (2007) Depth profiling of water molecules at the liquid-liquid interface using a combined surface vibrational spectroscopy and molecular dynamics approach. J Am Chem Soc 129:9446–9451
110. Walker DS, Richmond GL (2007) Understanding the effects of hydrogen bonding at the vapor-water interface: vibrational sum frequency spectroscopy of H_2O/HOD/D_2O mixtures studied using molecular dynamics simultions. J Phys Chem C 111:8321–8330
111. Walker DS, Moore FG, Richmond GL (2007) Vibrational sum frequency spectroscopy and molecular dynamics simulation of the carbon tetrachloride-water and 1,2-dichloroethane-water interfaces. J Phys Chem C 111:6103–6112
112. Wan Q, Galli G (2015) First-principles framework to compute sum-frequency generation vibrational spectra of semiconductors and insulators. Phys Rev Lett 115:246404
113. Wei X, Miranda PB, Shen YR (2001) Surface vibrational spectroscopic study of surface melting of ice. Phys Rev Lett 86:1554–1557
114. Yamaguchi S (2015) Development of single-channel heterodyne-detected sum frequency generation spectroscopy and its application to the water/vapor interface. J Chem Phys 143:034202
115. Zhang Y, Cremer PS (2010) Chemistry of hofmeister anions and osmolytes. Annu Rev Phys Chem 61:63–83

Chapter 10
Applications: Organic Interfaces

Abstract Interfaces of organic species offer enormous opportunities to be studied by SFG spectroscopy. In this chapter we introduce recent advances of computational analysis of SFG in the organic interfaces. Here we mainly deal with the C–H vibrations, as they are ubiquitously seen in organic species and have been most widely investigated by the SFG spectroscopy to date. We discuss molecular modeling of organic species, and then unified assignment method of alkyl C–H stretching of SFG spectra in comparison with IR and Raman spectra. Related topics revealed by the SFG analysis are also discussed, including SFG from centrosymmetric molecules, quadrupole contribution, and polarization analysis of molecular orientation.

Keywords C–H stretching · Alkyl groups · Polarization analysis

Interfaces of organic molecules include surfaces of various organic solvents, films and polymers. These interfaces are relevant to many applications, such as membranes, batteries, adhesion, wetting, tribology, etc. The organic interfaces are often buried to form liquid-liquid or solid-liquid interfaces. The SFG spectroscopy has large potential to selectively investigate various chemical species and their structure at those interfaces. Among various vibrational bands of organic molecules, the C–H stretching vibrations have been most intensively measured by SFG spectroscopy to date, since the first vibrational SFG measurement in 1987 [14]. This is because the C–H stretching vibrations show generally strong intensity and are technically easy to measure, in addition to that they are ubiquitously included in organic species. Besides the C–H stretching, other vibrations have been investigated, such as C=O and $\mathrm{C} \equiv \mathrm{N}$ stretching bands, by SFG spectroscopy [3, 20, 27, 30, 33, 37, 40].

The SFG spectra of organic interfaces have their own challenge to theoretical analysis, which is different from that of O–H band of aqueous systems in Chap. 9. First, the organic interfaces include variety of chemical species and vibrations, and the theoretical analysis has to accurately describe these variety of species and interfaces. Reliability of MD calculations capable of reproducing the various SFG spectra is an integral requisite of the theoretical analysis. This requirement is

A. Morita, *Theory of Sum Frequency Generation Spectroscopy*,
Lecture Notes in Chemistry 97, https://doi.org/10.1007/978-981-13-1607-4_10

essentially an issue of molecular modeling for the SFG calculations, and the general theory of polarization in Chap. 6 is particularly valuable for this requirement. Second, even the C–H bands of organic species are actually quite complicated and challenging to the analysis of SFG spectra. General formulas of C–H vibrations in the SFG spectroscopy have been proposed by Hirose and co-workers in early SFG theory [11, 12], and sophisticated by Wang and co-workers [7, 32]. These formulas are useful to characterize the SFG signals of the C–H band in various polarizations. A remaining issue in the theoretical analysis is to advance the MD simulation toward reproducing and even predicting the observed SFG spectra of organic interfaces.

The MD analysis aims at elucidating observed SFG spectra of various species using realistic molecular properties of constituent species. Theoretical analysis of organic SFG spectra has been relatively less developed in comparison to those of aqueous systems at present. The reliable theoretical support is definitely called for in practical interpretation of SFG spectra of organic species, particularly when examples of related infrared or Raman vibrational spectra do not offer a useful clue to help understanding the spectral features of SFG. This chapter summarizes the basis of theoretical SFG analysis of organic interfaces with the help of MD simulation.

10.1 C–H Bands of Alkyl Groups

The C–H vibrational band of alkyl groups appears in 2800–3000 cm^{-1} range, and has been one of the most common targets in the SFG measurement to date, as well as the O-H band of aqueous systems. The C–H band is ubiquitous in organic molecules, and its band shapes offer detailed information on the interface. Precise theoretical analysis of the C–H band is particularly beneficial to the application of SFG spectroscopy.

10.1.1 C–H Modes

Conventional assignment of the C–H band is performed in comparison with the infrared and Raman spectra of the corresponding C–H vibrations. The components in the C–H band are usually assigned with the notations in Fig. 10.1. In a methyl (CH_3–) group, the symmetric and asymmetric C–H vibrations are labeled with r^+ and r^-, respectively. The asymmetric vibration consists of two modes, classified into in-plane (r^-_{IP}) and out-of-plane (r^+_{OP}) modes. In a methylene (–CH_2–) group, the symmetric and asymmetric C–H vibrations are labeled with d^+ and d^-, respectively. These C–H stretching modes are generally close to the overtone or combination of bending modes, and thus often split by the Fermi resonance. The satellite modes of the Fermi resonace are labeled with FR, such as r^+_{FR} and d^+_{FR}.

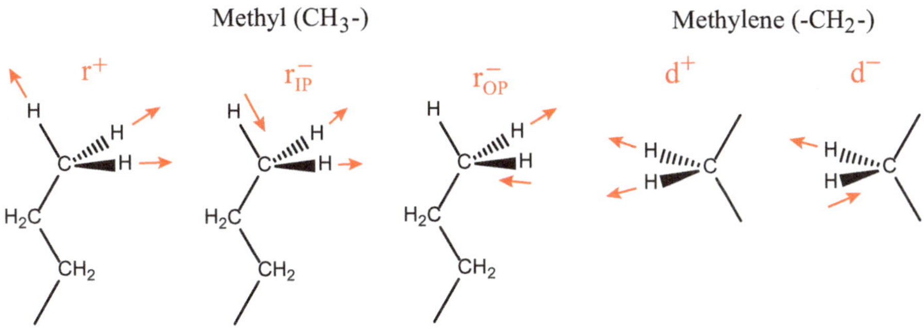

Fig. 10.1 C–H stretching modes of methyl (r^+, r^-_{IP}, r^-_{OP}) and methylene (d^+, d^-) groups

The C–H band of alkyl groups is complicated to interpret, essentially because that frequency range is congested with a number of overlapping components, i.e.

(a) different functional groups, such as methyl (CH_3–) and methylene (–CH_2–).
(b) different modes, such as symmetric and asymmetric stretching. These modes are often split by the Fermi resonance.
(c) different conformers, such as trans and gauche of alkyl chains.

One principal aim of the theoretical analysis is to disentangle the observed/calculated C–H band into these components and to establish the spectral assignment. The computational SFG analysis of C–H vibrations has been developed toward that aim. We briefly summarize the insight from the computational analysis below, in the case of C–H bands of methanol and ethanol, two smallest alcohols including methyl and methylene groups. Application to the species with longer alkyl chains is in progress.

10.1.2 Modeling of C–H

First we briefly discuss the importance of molecular modeling, as it is a critical factor of the SFG analysis. The computational SFG analysis requires two aspects of molecular modeling, (i) force field and (ii) polarization properties. The former requirement (i) is commonly relevant to all MD simulations, while the latter (ii) is specific to the SFG calculation. The charge response kernel (CRK) theory in Chap. 6 allows for general description of the (ii) polarization properties with retaining its ab initio quality. The universal modeling method is useful to the C–H vibrations as well.

A remaining challenge lies in the modeling of (i) force field of C–H stretching, particularly the intramolecular force field. The most challenging issue is to incorporate the Fermi resonance, since the Fermi resonance is essentially a quantum mechanical phenomenon and not straightforwardly compatible with classical MD simulation.[1] Full quantum molecular dynamics to allow for resonances is not feasible in most cases of condense phase simulation. One practical solution of this problem is to exploit the quantum-classical analogy of resonance phenomena [16]. Both classical and quantum mechanics can describe resonance phenomena of vibrations in a qualitatively analogous manner, that an anharmonic coupling Hamiltonian brings about significant mode mixing of nearly degenerate fundamental and overtone. However, the classical and quantum mechanics do not result in the same extent of resonance mode mixing [16]. Therefore, one can reasonably reproduce or mimic the Fermi resonance in the classical mechanics by using an effective coupling Hamiltonian including a quantum correction, so that the results of the resonance mixing by the classical mechanics are in accord with those by the quantum mechanics. This treatment of quantum correction is valid as long as we reproduce the features of Fermi resonance in the vibrational spectra by classical MD simulation. The following analysis of the Fermi resonance is based on the treatment.

10.1.3 Methanol C–H Vibrations

Methanol is the smallest alcohol containing only one methyl group, and the vibrational spectra of C–H band have been experimentally studied by SFG as well as infrared and Raman. The SFG spectrum in SSP polarization shows two components of C–H band in Fig. 10.2a, a major low-frequency one at $\sim$2830 cm^{-1} and a minor high-frequency one at $\sim$2950 cm^{-1}. It is noteworthy that similar two-band structure of the C–H band is also seen in the infrared and Raman spectra of liquid methanol as well, and the assignment of the structure has been argued to date [2, 4, 24, 29]. The widely accepted assignment of the two-band structure is distinct between the infrared and Raman spectra. The two bands in the IR spectrum are mostly assigned to the symmetric and asymmetric C–H stretching modes [2], whereas those in the Raman spectrum to the symmetric C–H stretching and the counterpart of its Fermi resonance [24, 29]. The different assignments of infrared and Raman spectra brought about confusion in the assignment of the SFG spectrum. This is a typical example that straightforward analogy to infrared or Raman spectra is not useful to interpret SFG spectra.

Band structure The MD analysis provides definite assignment of the SFG spectrum on a common basis to the infrared and Raman spectra, with also elucidating differences among three vibrational spectroscopies [16, 17]. The basic picture of vibrational levels of methanol is summarized in Fig. 10.2b. The C–H symmetric stretching of methyl ν_3 interact with bending overtone $2\nu_5$ and is split by the Fermi resonance, while the C–H asymmetric stretching ν_2, ν_9 overlap with the higher-frequency component of the Fermi splitting in the frequency domain. This scheme is common among the three vibrational spectroscopies in the case of methanol.

[1]The validity of classical mechanics to treat vibrational dynamics of molecules is based on the equivalence of quantum and classical descriptions of harmonic oscillators [1]. The Fermi resonance that stems from anharmonic coupling is essentially beyond the classical mechanics.

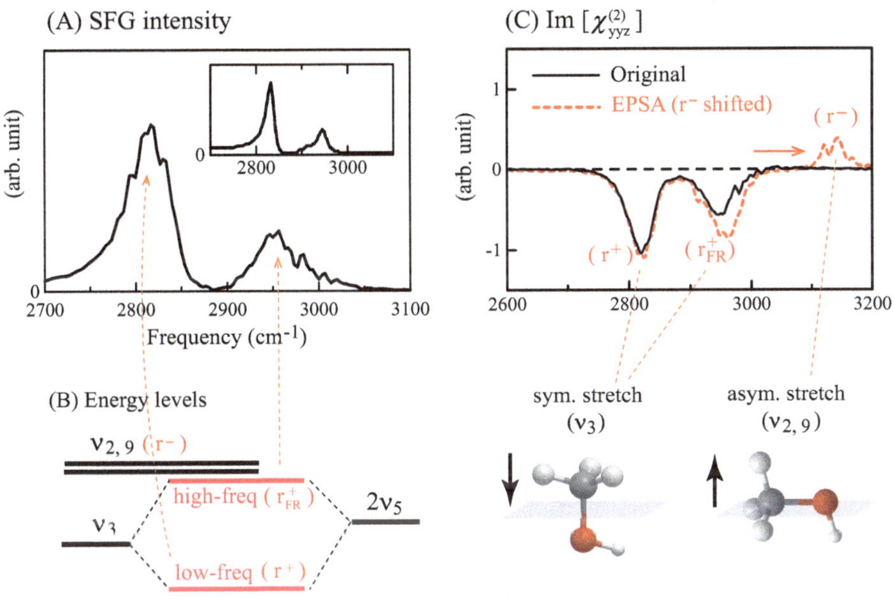

Fig. 10.2 (**a**) Calculated SFG intensity spectrum of liquid methanol surface [17], while the inset shows the experimental spectrum [24]. (**b**) Energy levels of C–H stretching modes of methanol. (**c**) Calculated Im[$\chi^{(2)}$] spectra. The black line is the original spectrum, while the red dashed line is the result of EPSA where the $\nu_{2,9}$ frequencies are tentatively blue shifted to disentangle the overlap. (Reprinted with permission from Ref. [17]; Copyright 2011, American Institute of Physics. Reprinted with the permission from Ref. [24]; Copyright 2003 American Chemical Society)

The different assignments stem from difference in relative intensities of the overlapping bands: the asymmetric stretching (r^-) and the high-frequency component of the Fermi resonance (r^+_{FR}). In the IR spectrum, the apparent high-frequency band is assigned to the asymmetric C–H stretching (r^-) since its intensity exceeds that of the Fermi resonance. On the other hand, the apparently same band of the Raman spectrum is assigned it to the Fermi resonance (r^+_{FR}) since the Raman intensity of asymmetric stretching is negligibly small. In the SFG spectrum, the asymmetric C–H stretching and the Fermi resonance overlap in the Im[$\chi^{(2)}$] amplitude with opposite signs, though the net sign of Im[$\chi^{(2)}$] is governed by the Fermi resonance. Therefore, the high-frequency component of SFG spectrum should be essentially assigned to the Fermi resonance, though the asymmetric stretching has substantial contribution to this component in a destructive manner.

Empirical parameter shift analysis (EPSA) This assignment is demonstrated by the empirical parameter shift analysis (EPSA). Figure 10.2c shows the calculated Im[$\chi^{(2)}$] spectrum, where both components of the two-band structure have negative sign. This is because the ν_3 mode (methyl symmetric C–H stretching) shows a negative amplitude when the methyl group points to the vapor, as we discussed in Sect. 4.2.2 and Fig. 3.2, and the Fermi splitting of the ν_3 mode distributes its negative amplitude to both components. As a result, the calculated Im[$\chi^{(2)}$] spectrum does not apparently show the $\nu_{2,9}$ modes. However, if we tentatively augmented the force constant of the $\nu_{2,9}$ modes with the other conditions intact, the MD simulation would yield the Im[$\chi^{(2)}$] spectrum with the red dashed line. This is a "thought experiment", feasible with MD simulation. Then a positive band of the asymmetric stretching modes emerges and moves to the blue, and consequently the positive band is separated from the original high-frequency negative component.[2] We also find that the original negative component at $\sim$2950 cm^{-1} increases its negative amplitude after the asymetric modes are separated out. This EPSA result of Panel (c) manifests the spectral overlap of the asymmetric stretching and the Fermi component depicted in Panel (b).

10.1.4 Ethanol C–H Vibrations

Ethanol contains both methyl and methylene groups, and also has trans and gauche conformers. Therefore, ethanol is a simplest model to incorporate all the complicating factors (a)–(c) in the alkyl C–H vibrations mentioned in Sect. 10.1.1; (a) different functional groups, (b) vibrational modes, (c) conformational isomers. An intensive experimental analysis of the ethanol C–H vibrations has been performed by Gan et al. [6, 8], who separated the methyl and methylene vibrations using normal and partially deuterated ethanol species, i.e. CH_3CH_2OH, CH_3CD_2OH, CD_3CH_2OH. The latter two species allow us to focus on the methyl and methylene C–H vibrations, respectively, in the ethanol molecule. The comprehensive set of experimental SFG spectra of ethanol, including the different isotopes and polarization combinations, were analyzed by the MD simulation with extending the above modeling of methanol.

MD calculation of liquid ethanol can reproduce and elucidate the C–H vibrational spectra of infrared, Raman and SFG in a unified manner [34, 35]. One of the important findings is that the trans and gauche conformers show rather different spectral components in the SFG spectra. In particular, methylene C–H vibrations near a gauche defect show a distinct feature in the vibrational SFG spectra. Accurate modeling of ethanol C–H vibrations is a significant step toward general simulation of alkyl molecules.

[2]The asymmetric C–H stretching mode has a positive amplitude when the molecule is tilted, as illustrated in Fig. 10.2c.

10.2 Benzene: SFG from Centrosymmetric Molecules

The SFG of benzene poses an intriguing issue. The liquid benzene surface generates intense C–H signal in the SFG spectrum, although each benzene molecule is centrosymmetric and thus it should have null hyperpolarizability [13, 25]. Elucidating the SFG mechanism of liquid benzene is pertinent to the fundamental mechanism of SFG and symmetry breaking.

First, let us recall the SFG from other ordinary non-centrosymmetric molecules, as we discussed in Sect. 3.3. These molecules have finite non-zero hyperpolarizabilities, and nevertheless the susceptibility of their bulk materials vanishes. This is because the molecular orientation is random and isotropic in the bulk, so that the contributions of hyperpolarizabilities cancel each other by taking the statistical average. On the other hand, the susceptibility of their interfaces is not necessarily zero due to anisotropic molecular orientation. However, such typical scenario for the SFG activity of interfaces does not elucidate the SFG signal from benzene, because the benzene itself has no hyperpolarizability regardless of its orientation. Consequently, there remain two possible scenarios in order to elucidate the SFG of benzene:

(i) **symmetry breaking:** the symmetry of benzene molecules is broken at the inhomogeneous environment at the interface so that the SFG becomes allowed there.
(ii) **quadrupole:** the quadrupole mechanism of SFG is invoked beyond the conventional dipole mechanism.

The MD simulation was fully utilized to examine these mechanisms [18], and found that both mechanisms are actually involved in the SFG from liquid benzene.

Regarding (i) symmetry breaking, the benzene molecule has six equivalent C–H groups, whose vibrations are degenerated in the zero-th order approximation. The C–H groups construct the normal modes of C–H stretching under the D_{6h} symmetry of benzene (i.e. ν_2 (A_{1g}), ν_7 (E_{2g}), ν_{13} (B_{1u}), and ν_{20} (E_{1u})), and these normal modes have nearly degenerate frequencies. Therefore, these C–H normal modes can be readily mixed each other by a small external perturbation. At the surface of liquid benzene, the nonuniform solvation environment acts as the perturbation to break the symmetry of the C–H modes. This mechanism implies that a centrosymmetric molecule can readily become SFG active at interface. Figure 10.3a shows the calculated result of $\mathrm{Im}[\chi^{(2)}]$ spectrum (red line) of benzene by the standard time correlation function formula of Eq. (5.27). The $\mathrm{Im}[\chi^{(2)}]$ spectrum shows a bipolar band at about 3100 cm^{-1} region. The positive and negative components are attributed to the upward and downward local C–H modes. These local modes are slightly split in frequency due to the distinct solvation environments.

(ii) The quadrupole mechanism turned out to be also significant in the SFG spectrum of benzene. The MD analysis was applied to examine all possible quadrupole contributions discussed in Chap. 7, and consequently revealed that the χ^{IQB} term has a particularly large contribution. Figure 10.3a shows the $\mathrm{Im}[\chi^{\mathrm{IQB}}]$

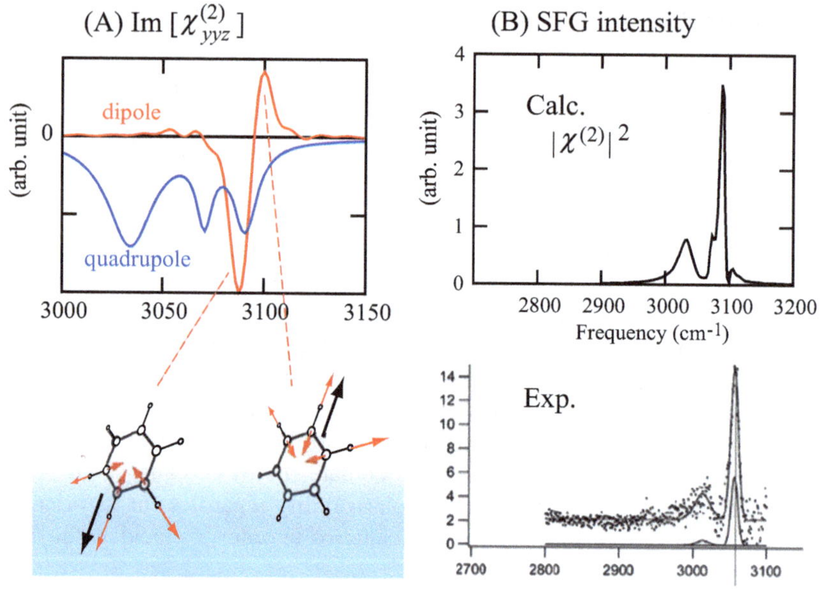

Fig. 10.3 (**a**) Calculated Im[$\chi^{(2)}$] spectrum of benzene surface. The red line denotes the χ^{ID} (dipole) term, while the blue line the χ^{IQB} term. The local modes to generate the bipolar band of χ^{ID} are also depicted. (**b**) Calculated SFG intensity spectrum (upper panel) [18] and the experimental one (lower panel) [13]. (Reprinted with the permission from Ref. [18]; Copyright 2012 American Chemical Society. Reproduced from Ref. [13] by permission of The Royal Society of Chemistry)

spectrum with a blue line in the same scale, which has an almost comparable amplitude to the dipole term (red line). The χ^{IQB} term is a bulk property and thus reflects the vibration in the bulk phase. Its spectral shape is analogous to the infrared spectrum of bulk liquid. By considering both mechanisms (i) and (ii), the experimental SFG intensity spectrum is successfully elucidated in Fig. 10.3b. We find that the satellite band of C–H stretching at $\sim$3030 cm^{-1} is of quadrupole character.

10.3 Molecular Orientation and Polarization Analysis

Molecular orientation at interfaces is one of the principal properties to characterize microscopic structure of interfaces. In principle, the interface structure is distinct from the bulk in terms of anisotropic molecular orientation. Therefore, molecular orientation at interfaces has been widely discussed by SFG spectroscopy. The SFG spectroscopy is able to investigate molecular orientation by the analysis of polarization, as we discussed in Sects. 3.3 and 4.2, and has been applied to the orientation of alkyl groups.

The procedure of polarization analysis is simply summarized into two steps as follows.

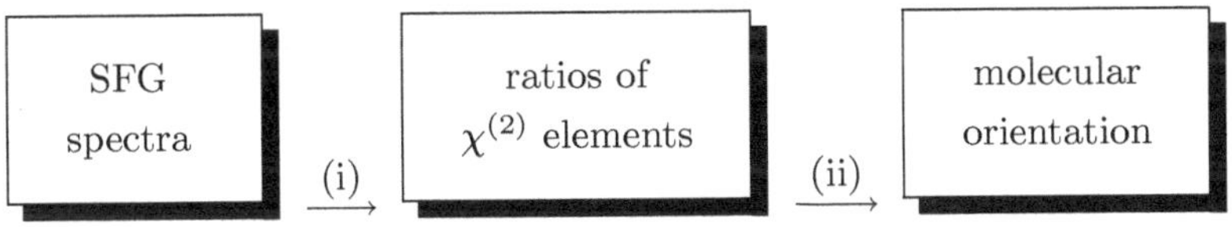

Essentially, the experimental measurements evaluate the ratios of different $\chi^{(2)}$ tensor elements, such as $B = \chi^{(2)}_{yyz}/\chi^{(2)}_{yzy}$ and $C = \chi^{(2)}_{zzz}/\chi^{(2)}_{yyz}$. There are some methods for the step (i) to evaluate the ratios of $\chi^{(2)}$ elements experimentally. A simple way is to compare the relative intensities of relevant polarization combinations. More sophisticated methods using interference of different polarizations have been also developed and utilized [5, 9, 10, 23, 30, 32, 38]. In any events, the obtained tensor elements in step (i) have to be analyzed in order to derive the microscopic molecular orientation. The theory of the analysis in the step (ii) is also an integral part of the polarization analysis of orientation, and we have depicted the theoretical procedure (ii) to derive the molecular orientation from the ratios of tensor elements in Sect. 4.2. The present section focuses on the analysis step (ii), and further examines the procedure.

When we apply the theoretical procedure (ii) to actual interfaces, it often requires some assumptions on microscopic properties of the systems. As we discuss in the following, the reliability of these assumptions may have significant influences on the results of molecular orientation. The MD simulation can help clarifying the relation between the tensor elements and the microscopic molecular orientation without resort to those assumptions, and help examining the reliability of the assumptions involved in the polarization analysis. In this section the polarization analysis of the methyl group is examined with two examples, methanol and acetonitrile, with the help of MD simulation.

10.3.1 Methanol

The SFG band of the methyl symmetric stretching has been discussed in Sect. 10.1. As illustrated in Fig. 10.2, the C–H band consists of the low-frequency component at $\sim$2830 cm^{-1} and high-frequency one at $\sim$2950 cm^{-1}, and the former is attributed to the methyl symmetric stretching, while the latter is a mixture of the Fermi splitting and the antisymmetric stretching. Therefore, the intensity of the former component becomes an index of methyl symmetric C–H stretching, which could be governed by the number density of methyl groups at interface and their orientation.

The experimental intensity of methyl symmetric stretching (ss) band shows an intriguing behavior when the methanol is diluted with water. In the methanol/water mixture solutions, the SFG intensity of this band shows a turn-over behavior as

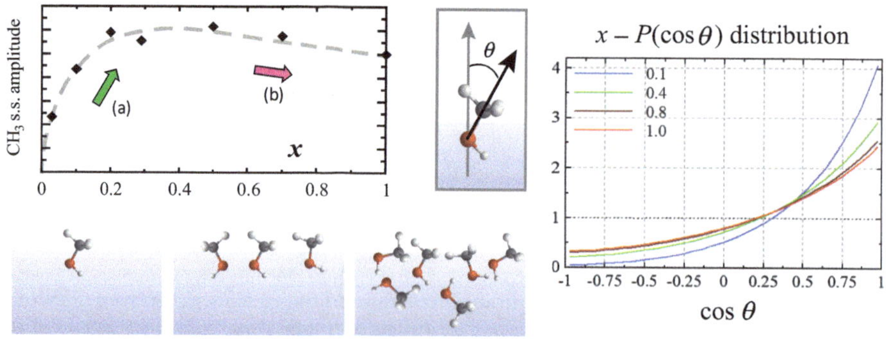

Fig. 10.4 (Left) Relative amplitude of methyl symmetric stretching band of the SSP-polarized SFG spectra for methanol/water mixture solutions as a function of the mole fraction of methanol x [31]. Schematic pictures of methanol distribution at the surface are also shown. (Right) Probability distribution $P(\cos\theta)$ of the methyl tilt angle θ with varying x [15]. The holizontal dashed line at $P = 1$ denotes the fully random orientation. (Reprinted with the permission from Ref. [15]. Copyright 2015 American Chemical Society)

a function of the mole fraction of methanol x, as shown in Fig. 10.4 [31]. This turn-over behavior is also reproduced by the MD simulation of SFG spectra for methanol/water solutions [15]. The MD simulation elucidated this behavior by two step mechanisms. In the low x region (a) the initial increase is understood with increasing number density of methanol at the surface, whereas in the high x region (b) the decreasing amplitude is attributed to the randomized orientation of surface methanol.

Sung et al. [31] also carried out the polarization analysis of the methanol/water mixture solutions, and found that the ratio of tensor elements $C = \chi^{(2)}_{zzz}/\chi^{(2)}_{yyz}$ is nearly invariant over the entire range of x. This experimental finding would have been interpreted by the ordinary theoretical analysis of polarization in Sect. 4.2 that the molecular orientation is also invariant over x. However, this conclusion appears contrary to the picture of (b) randomized orientation mentioned above. The MD simulation allows us to calculate the ratio C as well, and actually reproduced the nearly constant C over x, *irrespective* of the randomized orientation in the high x region [15].

The MD simulation of the present systems offers an instructive insight into the polarization analysis. The reason to solve this apparent inconsistency between the experimental tensor element ratio and the MD result of orientation lies in the wide distribution of the tilt angle θ of methyl groups at the surface, as shown in the right panel of Fig. 10.4. The MD simulation shows a quite wide probability distribution of $\cos\theta$ at the surface, which approaches to random orientation with increasing x. Such orientational distributions in Fig. 10.4 are not well approximated with a delta function or a Gaussian. The orientational distribution is found to be one of the crucial assumptions in the polarization analysis. The information on the distribution with the aid of MD simulation will help improving the reliability of the polarization analysis [15, 28].

10.3.2 Acetonitrile

Acetonitrile CH_3CN is an excellent model molecule to clarify the relation between the orientation and polarization, since it is a rod-like molecule with C_{3v} symmetry. Accordingly, the molecular orientation is readily described with the tilt angle θ, as discussed in Sect. 4.2.2. The experimental polarization analysis of acetonitrile in aqueous solutions was carried out by a number of groups so far [19, 28, 30, 39, 41], though their results of orientation appear inconsistent. Zhang et al. [39, 41] reported that the tilt angle θ of the methyl group of acetonitrile changes abruptly with increasing mole fraction of acetonitrile x at $x \simeq 0.07$, whereas Kim et al. [19] concluded that the orientation changes gradually with increasing x. Shultz et al. [30] and Saito et al. [28] reported nearly invariant orientation over x. These intriguing discrepancies should be resolved with the help of reliable theoretical analysis of polarization and orientation, and the MD simulation has been performed in collaboration with experimental measurement [28].

The MD simulation reproduced neary invariant orientation, and thus supported the recent conclusion by Shultz and Saito. The reason for discrepancies from the former previous studies is not clear at present, though the polarization analysis of acetonitrile may suffer from very weak SPS signal. (The recent experiments [28, 30] adopted the polarization angle null method to circumvent this problem.) The MD simulation of the polarization analysis revealed two significant problems in the quantitative accuracy. One issue is the orientational distribution, as we discussed in the case of methanol above. Caution should be required to apply the polarization analysis to an interface with an unknown distribution of orientation, particularly to liquid interfaces.

Another critical issue is the appropriate value of R,

$$R = \frac{(\partial\alpha_{\xi\xi}/\partial q_1)}{(\partial\alpha_{\zeta\zeta}/\partial q_1)} = \frac{(\partial\alpha_{\eta\eta}/\partial q_1)}{(\partial\alpha_{\zeta\zeta}/\partial q_1)}, \tag{4.10}$$

which is used in the theoretical analysis. The estimation of R appears to have considerable confusions so far. Shultz adopted $R = 0.58$ while Zhang used $R = 2.3$ in their previous studies. Table 10.1 results in $R = 0.2967/0.3731 = 0.80$ from the density functional theory (B3LYP/aug-cc-pVTZ) calculation. The value of R is often estimated experimentally from the Raman depolarization ratio of the bulk sample [21, 22, 26]. However, the Raman depolarization ratio allows dual solutions of R, and thus could not determine whether $R > 1$ or $R < 1$ by itself. From a simplified geometric argument (see illustration in Table 10.1), each methyl C–H bond of an acetonitrile molecule forms $\varphi \approx 70.5°$ ($= \cos^{-1}(1/3)$) from the ζ axis in the ideal tetrahedral geometry. Therefore, if each C–H bond possesses a Raman tensor component along the bond, the sum of three C–H bond polarizabilities should yield the ratio R to be

$$R \approx \frac{\sin^2\varphi + (-\frac{1}{2}\sin\varphi)^2 + (-\frac{1}{2}\sin\varphi)^2}{3\cos^2\varphi} = \frac{1}{2}\tan^2\varphi \approx 4. \tag{10.1}$$

Table 10.1 Calculated derivatives of dipole $\partial\mu_{r'}/\partial q_1$ and polarizability $\partial\alpha_{p'q'}/\partial q_1$ of acetonitrile by B3LYP/aug-cc-pVTZ [28]. Units: atomic units. q_1 denotes the normal mode coordinate of C–H symmetric stretching defined for the stretching direction with unit reduced mass. Right picture illustrate the configuration of acetonitrile in the molecule-fixed coordinates (ξ, η, ζ). (Same as Table 4.2 in Chap. 4)

$\partial\mu_{r'}/\partial q_1$	$r' = \xi$	η	ζ
	0	0	-0.005970
$\partial\alpha_{p'q'}/\partial q_1$	$q' = \xi$	η	ζ
$p' = \xi$	0.2967	0	0
η	0	0.2967	0
ζ	0	0	0.3731

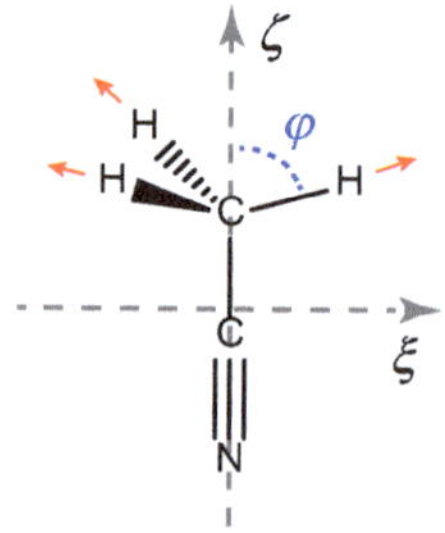

Table 10.2 Calculated ratios R in Eq. (4.10) for CH_3X molecules by B3LYP/aug-cc-pVTZ [28]

CH_3X	R[b]
CH_3H	2.1
CH_3CH_3[a]	1.2
CH_3F	1.8
CH_3Cl	0.94
CH_3Br	0.79
CH_3CN	0.80

[a]Staggered ethane
[b]The normal mode coordinate q_1 in Eq. (4.10) is assumed to be the symmetry adopted natural internal coordinate for the methyl symmetric stretching

All the ambiguities in R mentioned above cast large uncertainty in the estimated orientation from the polarization ratio B or C by Eqs. (4.14) or (4.15). The polarization analysis should be based on sufficiently accurate quantum chemical calculations of R unless a direct experimental value is available. The calculated value of $R = 0.80$ is smaller than unity, which is not in accord with the simple geometric estimate of Eq. (10.1) [36].

The quantum chemical calculations of related molecules showed that the R value of the methyl symmetric stretching is actually sensitive to the adjacent functional group X of the methyl group. Table 10.2 summarizes the calculated R values for various CH_3X molecules (X = H, CH_3, F, Cl, Br, CN). The electronically soft CN group reduces the R value due to the hyperconjugation of CH and CN π orbitals. Accurate evaluation of the R value that could depend on molecular species is of critical importance in the polarization analysis.

Bibliography

1. Bader JS, Berne BJ (1994) Quantum and classical relaxation rates from classical simulations. J Chem Phys 100:8359–8366
2. Bertie JE, Zhang SL, Eysel HH, Baluja S, Ahmed MK (1993) Infrared intensities of liquids XI: infrared refractive indices from 8000 to 2 cm^{-1}, absolute integrated intensities, and dipole moment derivatives of methanol at 25°C. Appl Spectrosc 47:1100–1144
3. Buck M, Himmelhaus M (2001) Vibrational spectroscopy of interfaces by infrared-visible sum frequency generation. J Vac Sci Technol A 19:2717–2736
4. Falk M, Whalley E (1961) Infreared spectra of methanol and deuterated methanols in gas, liquid, and solid phases. J Chem Phys 34:1554–1568
5. Gan W, Wu B-H, Chen H, Guo Y, Wang H-F (2005) Accuracy and sensitivity of determining molecular orientation at interfaces using sum frequency generation vibrational spectroscopy. Chem Phys Lett 406:467–473
6. Gan W, Zhang Z, Feng R-R, Wang H-F (2006) Identification of overlapping features in the sum frequency generation vibrational spectra of air/ethanol interface. Chem Phys Lett 423:261–265
7. Gan W, Wu B-H, Zhang Z, Guo Y, Wang H-F (2007) Vibrational spectra and molecular orientation with experimental configuration analysis in surface sum frequency generation (SFG). J Phys Chem C 111:8716–8725
8. Gan W, Zhang Z, Feng R-R, Wang H-F (2007) Spectral interference and molecular conformation at liquid interface with sum frequency generation vibrational spectroscopy (SFG-VS). J Phys Chem C 111:8726–8738
9. Groenzin H, Li I, Shultz MJ (2008) Sum frequency generation: polarization surface spectroscopy analysis of the vibrational surface modes on the basal face of ice Ih. J Chem Phys 128:214510
10. Heinz TF, Tom HWK, Shen YR (1983) Determination of molecular orientation of monolayer adsorbates by optical second-harmonic generation. Phys Rev A 28:1883–1885
11. Hirose C, Akamatsu A, Domen K (1992) Formulas for the analysis of surface sum-frequency generation spectrum by CH stretching modes of methyl and methylene groups. J Chem Phys 96:997–1004
12. Hirose C, Hiroyoshi Y, Akamatsu N, Domen K (1993) Orientation analysis by simulation of vibrational sum frequency generation spectrum: CH stretching bands of the methyl group. J Phys Chem 97:10064–10069
13. Hommel EL, Allen HC (2003) The air-liquid interface of benzene, toluene, *m*-xylene, and mesitylene: a sum frequency, Raman, and infrared spectroscopic study. Analyst 128:750–755
14. Hunt JH, Guyot-Sionnest P, Shen YR (1987) Observation of C–H stretch vibrations of monolayers of molecules optical sum-frequency generation. Chem Phys Lett 133:189–192
15. Ishihara T, Ishiyama T, Morita A (2015) Surface structure of methanol/water solutions via sum-frequency orientational analysis and molecular dynamics simulation. J Phys Chem C 119:9879–9889
16. Ishiyama T, Sokolov VV, Morita A (2011) Molecular dynamics simulation of liquid methanol. I. Molecular modeling including C–H vibration and Fermi resonance. J Chem Phys 134:024509
17. Ishiyama T, Sokolov VV, Morita A (2011) Molecular dynamics simulation of liquid methanol. II. Unified assignment of Infrared, Raman, and sum frequency generation vibrational spectra in methyl C–H stretching region. J Chem Phys 134:024510
18. Kawaguchi T, Shiratori K, Henmi Y, Ishiyama T, Morita A (2012) Mechanisms of sum frequency generation from liquid benzene: symmetry breaking at interface and bulk contribution. J Phys Chem C 116:13169–13182
19. Kim J, Chou KC, Somorjai GA (2003) Structure and dynamics of acetonitrile at the air/liquid interface of binary solutions studied by infrared-visible sum frequency generation. J Phys Chem B 107:1592–1596

20. Liu H, Tong Y, Kuwata N, Osawa M, Kawamura J, and Ye S (2009) Adsorption of propylene carbonate (PC) on the $LiCoO_2$ surface investigated by nonlinear vibrational spectroscopy. J Phys Chem C 113:20531–20534
21. Long DA (1977) Raman spectroscopy. McGraw-Hill, New York
22. Long DA (2002) The Raman effect. Wiley, New York
23. Lü R, Gan W, Wang H (2003) Novel method for accurate determination of the orientational angle of interfacial chemical groups. Chin Sci Bull 48:2183–2187
24. Ma G, Allen HC (2003) Surface studies of aqueous methanol solutions by vibrational broad bandwidth sum frequency generation spectroscopy. J Phys Chem B 107:6343–6349
25. Matsuzaki K, Nihonyanagi S, Yamaguchi S, Nagata T, Tahara T (2013) Vibrational sum frequency generation by the quadrupolar mechanism at the nonpolar benzene/air interface. J Phys Chem Lett 4:1654–1658
26. McQuarrie DA (2000) Statistical mechanics, 2nd edn. University Science Books, Sausalito
27. Nicolau BG, Garca-Rey N, Dryzhakov B, Dlott DD (2015) Interfacial processes of a model lithium ion battery anode observed, in situ, with vibrational sum-frequency generation spectroscopy. J Phys Chem C 119:10227–10233
28. Saito K, Peng Q, Qiao L, Wang L, Joutsuka T, Ishiyama T, Ye S, Morita A (2017) Theoretical and experimental examination on SFG polarization analysis at acetonitrile-water solution surfaces. Phys Chem Chem Phys 19:8941–8961
29. Schwartz M, Moradi-Araghi A, Koehler WH (1980) Fermi resonance in aqueous methanol. J Mol Struct 63:279–285
30. Shultz MJ, Bisson P, Groenzin H, Li I (2010) Multiplexed polarization spectroscopy: measuring surface hyperpolarizability orientation. J Chem Phys 133:054702
31. Sung J, Park K, Kim D (2005) Surfaces of alcohol-water mixtures studied by sum-frequency generation vibrational spectroscopy. J Phys Chem B 109:18507–18514
32. Wang H-F, Gan W, Lu R, Rao Y, Wu B-H (2005) Quantitative spectral and orientational analysis in surface sum frequency generation vibrational spectroscopy (SFG-VS). Int Rev Phys Chem 24:191–256
33. Wang J, Chen X, Clarke ML, Chen Z (2005) Detection of chiral sum frequency generation vibrational spectra of proteins and peptides at interfaces in situ. Proc Natl Acad Sci U S A 102:4978–4983
34. Wang L, Ishiyama T, Morita A (2017) Theoretical investigation of C–H vibrational spectroscopy. 1. Modeling of methyl and methylene groups of ethanol with different conformers. J Phys Chem A 121:6687–6700
35. Wang L, Ishiyama T, Morita A (2017) Theoretical investigation of C–H vibrational spectroscopy. 2. Unified assignment method of IR, Raman and SFG spectra of ethanol. J Phys Chem A 121:6701–6712
36. Wu H, Zhang W-K, Gan W, Cui Z-F, Wang H-F (2006) An empirical approach to the bond additivity model in quantitative interpretation of sum frequency generation vibrational spectra. J Chem Phys 125:133203
37. Yan ECY, Fu L, Wang Z, Liu W (2014) Biological macromolecules at interfaces probed by chiral vibrational sum frequency generation spectroscopy. Chem Rev 114:8471–8498
38. Zhang TG, Zhang CH, Wong GK (1990) Determination of molecular orientation in molecular monolayers by second-harmonic generation. J Opt Soc Am B 7:902–907
39. Zhang D, Gutow JH, Eisenthal KB, Heinz TF (1993) Sudden structural change at an air/binary liquid interface: sum frequency study of the air/acetonitrile-water interface. J Chem Phys 98:5099–5101
40. Zhang D, Gutow J, Eisenthal KB (1994) Vibrational spectra, orientations, and phase transitions in long-chain amphiphiles at the air/water interface: probing the head and tail groups by sum frequency generation. J Phys Chem 98:13729–13734
41. Zhang D, Gutow JH, Eisenthal KB (1996) Structural phase transition of small molecules at air/water interfaces. J Chem Soc Faraday Trans 92:539–543

Chapter 11
Summary

Abstract The present chapter summarizes the outline of the theory of SFG spectroscopy described in this book, including the fundamental and detailed theory of $\chi^{(2)}$ and application of computational SFG analysis for realistic interfaces. This chapter also shows future directions of the SFG spectroscopy, which will be greatly facilitated in collaboration of experimental advances and the computational analysis. Theoretical interpretation with the aid of reliable computation will be increasingly important in the future of the surface nonlinear spectroscopy.

Keywords Fundamental theory of $\chi^{(2)}$ · Modeling of polarization · Future directions

In this book, fundamental theory of the surface nonlinear spectroscopy and related principles are presented. The principal aim is toward comprehensive understanding of the vibrational SFG spectroscopy through close collaboration of experimental measurement and theoretical analysis. Here we briefly summarize the outline of the theory described in this book.

11.1 Outline of Theory

The theory of SFG consists of both macroscopic and microscopic aspects of the surface nonlinear spectroscopy, which were treated in Chaps. 2 and 3, respectively. The macroscopic aspect in Chap. 2 deals with the electromagnetic processes in which the nonlinear polarization is induced at the interface and thereby emits the radiation of sum frequency. On the other hand, the microscopic aspect in Chap. 3 defines material properties of the interface related to the nonlinear response.

Furthermore, Chap. 4 provides the microscopic formulations of the frequency-dependent nonlinear susceptibility of the interface $\chi^{(2)}(\Omega, \omega_1, \omega_2)$ by two routes, on the basis of the energy representation and the time-dependent representation. These two routes of formulations allow for calculating $\chi^{(2)}(\Omega, \omega_1, \omega_2)$ by molecular

A. Morita, *Theory of Sum Frequency Generation Spectroscopy*,
Lecture Notes in Chemistry 97, https://doi.org/10.1007/978-981-13-1607-4_11

simulation. In applying these formulas of $\chi^{(2)}$ to realistic interfaces, accurate treatment of electronic polarization and intermolecular couplings in the interface systems are of critical importance to provide reliable interpretation of the observed spectra. Therefore, the theory and modeling of these polarization properties were discussed in Chaps. 5 and 6. The knowledge summarized in Chaps. 4, 5 and 6 forms the necessary foundation of the computational analysis of SFG spectra.

Chapters 7 and 8 treat some advanced topics of SFG theory. We argued the quadrupole contributions to the SFG and SHG processes in Chap. 7 in details, which arise from the bulk region as well as from the interface. Chapter 8 discussed some other topics of SFG, including the $\chi^{(3)}$ effect and chiral applications of the SFG spectroscopy.

The theory and principles presented in this book not only provide the formal framework of the surface nonlinear spectroscopy, but allow us to understand experimentally observed SFG spectra from the molecular viewpoints of electronic properties and structure of the interfaces. Chapters 9 and 10 introduced some applications of the computational analysis pioneered by the author's group to aqueous and organic interfaces. These examples become prototypes of computational analysis of SFG spectroscopy for further applications. The theoretical analysis in combination of quantum chemical calculations and molecular dynamics simulations is growing to become a powerful means to interpret the experimental spectra.

11.2 Future Directions

The applications of computational analysis of SFG are currently expanding, with various modeling methods besides the CRK model [1, 4, 5, 12]. The future computational analysis will cover a variety of interfaces treated by experimental SFG measurement. These applications should include

- surfaces of water and aqueous solutions (water, ice, electrolyte solutions, etc.)
- liquid-liquid interfaces (water-oil, etc.)
- electrochemical interfaces (solution-electrode)
- molecules adsorbed on solid surfaces
- polymer surfaces
- Langmuir and Langmuir-Blodgett films, self-assemble monolayers
- biomolecules—cell membranes
- environmental interfaces—aerosol surface, water-mineral interface

⋮

The advance of theory is stimulated by the advance of experimental methods. Experimental techniques of surface nonlinear spectroscopy have been remarkably sophisticated since the first report of SFG spectrum. Just for example, the advancement of the heterodyne detected SFG provides experimental information of real (Re[$\chi^{(2)}$]) and imaginary (Im[$\chi^{(2)}$]) parts of the $\chi^{(2)}$ spectra, and allows unambiguous comparison with the calculated $\chi^{(2)}$ spectra. This progress has indeed a substantial impact on the collaboration of experiment and theory. We have discussed a number of such examples in the preceding chapters. Recent advancements of the experimental techniques worth noting should include

- heterodyne SFG
- time-resolved SFG, 2D-SFG
- high-resolution SFG
- sophisticated analysis of polarization
- chiral SFG
- double resonance SFG
- SFG microscope, imaging
- SFG scattering

⋮

These advancements have been achieved by the efforts of various experimental researchers. We could not detail these aspects in this book. Interested readers in these experimental aspects should refer to a recent excellent textbook by Shen [10] and the references therein.

In the future advances of SFG spectroscopy, one of the largest challenges lies in the unambiguous and comprehensive analysis of observed spectra. Although it is widely recognized that the SFG spectroscopy involves rich microscopic information on interfaces, it is not often straightforward to carry out precise assignment at the molecular level. Current experimental analysis of the observed spectra largely depends on empirical spectral fitting for assignment of band structure, though such empirical fitting suffers from large ambiguities and fitted bands are often hard to be interpreted. Reliable analysis methods with sound theoretical basis will help exploiting the potential of the surface nonlinear spectroscopy. In the next stage of analysis, detailed comparison of the spectra with molecular simulation will be widely utilized. Collaboration of experiment and theory will be of increasing importance in the surface nonlinear spectroscopy [2, 3, 6–9, 11, 13].

Bibliography

1. Auer BM, Skinner JL (2008) Vibrational sum-frequency spectroscopy of the liquid/vapor interface for dilute HOD in D_2O. J Chem Phys 129:214705
2. Ho J, Psciuk BT, Chase HM, Rudshteyn B, Upshur MA, Fu L, Thomson RJ, Wang H-F, Geiger FM, Batista VS (2016) Sum frequency generation spectroscopy and molecular dynamics simulations reveal a rotationally fluid adsorption state of alpha-pinene on silica. J Phys Chem C 120:12578–12589
3. Ishiyama T, Morita A, Miyamae T (2011) Surface structure of sulfuric acid solution relevant to sulfate aerosol: molecular dynamics simulation combined with sum frequency generation measurement. Phys Chem Chem Phys 13:20965–20973
4. Medders GR, Paesani F (2016) Dissecting the molecular structure of the air/water interface from quantum simulations of the sum-frequency generation spectrum. J Am Chem Soc 138:3912–3919

5. Nagata Y, Ohto T, Backus EHG, Bonn M (2016) Molecular modeling of water interfaces: from molecular spectroscopy to thermodynamics. J Phys Chem B 120:3785–3796
6. Nihonyanagi S, Ishiyama T, Lee T-K, Yamaguchi S, Bonn M, Morita A, Tahara T (2011) Unified molecular view of air/water interface based on experimental and theoretical $\chi^{(2)}$ spectra of isotopically diluted water surface. J Am Chem Soc 133:16875–16880
7. Otsuki Y, Sugimoto T, Ishiyama T, Morita A, Watanabe K, Matsumoto Y (2017) Unveiling subsurface hydrogen-bond structure of hexagonal water ice. Phys Rev B 96:115405
8. Reddy SK, Thiraux R, Rudd BAW, Lin L, Adel T, Joutsuka T, Geiger FM, Allen HC, Morita A, Paesani F (2018) Bulk contributions modulate the sum-frequency generation spectra of interfacial water on model sea-spray aerosols. Chem. https://doi.org/10.1016/j.chempr.2018.04.007
9. Sánchez MA, Kling T, Ishiyama T, van Zadel M-J, Bisson PJ, Mezger M, Jochum MN, Cyran JD, Smit WJ, Bakker HJ, Shultz MJ, Morita A, Donadio D, Nagata Y, Bonn M, Backus EHG (2017) Experimental and theoretical evidence for bilayer-by-bilayer surface melting of crystalline ice. Proc Natl Acad Sci 114:227–232
10. Shen YR (2016) Fundamentals of sum-frequency spectroscopy. Cambridge University Press, Cambridge
11. Stiopkin IV, Weeraman C, Pieniazek PA, Shalhout FY, Skinner JL, Benderskii AV (2011) Hydrogen bonding at the water surface revealed by isotopic dilution spectroscopy. Nature 474:192–195
12. Sulpizi M, Salanne M, Sprik M, Gaigeot M (2013) Vibrational sum frequency generation spectroscopy of the water liquid-vapor interface from density functional theory-based molecular dynamics simulations. J Phys Chem Lett 4:83–87
13. Walker DS, Richmond GL (2007) Understanding the effects of hydrogen bonding at the vapor-water interface: vibrational sum frequency spectroscopy of H_2O/HOD/D_2O mixtures studied using molecular dynamics simultions. J Phys Chem C 111:8321–8330

Correction to: Quadrupole Contributions from Interface and Bulk

Correction to:
Chapter 7 in: A. Morita, *Theory of Sum Frequency Generation Spectroscopy*, Lecture Notes in Chemistry 97,
https://doi.org/10.1007/978-981-13-1607-4_7

The original version of this book was inadvertently published with errors. The corrections listed below are incorporated after publication

1. p. 158, Eq. (7.21)
   $Q^{(2)}_{0,sp} \rightarrow Q^{(2)}_{0,ps}$

2. p. 158, next line below Eq. (7.21)
   $-\dfrac{\partial}{\partial s} Q^{(2)}_{0,sp} \rightarrow -\dfrac{\partial}{\partial s} Q^{(2)}_{0,ps}$

3. p. 173, Eq. (7.69)
   $q_{sp}(\Omega) \rightarrow q_{ps}(\Omega)$
   $E_q(\omega_2) \rightarrow E_q(\omega_1)$

4. p. 173, Footnote 4 is added.
   unless otherwise noted. → unless otherwise noted.[4]

   [4] Note that the gradient of the generalized quadrupole moment $-\sum_q \partial_q q_{pq}(\omega)$ gives rise to the polarization, as we have mentioned in relation to Eq. (7.21). The current $\boldsymbol{J}$ is related to the multipole moments by

$$J_p = \left[\frac{d}{dt}\left(\boldsymbol{\mu} - \nabla\cdot\boldsymbol{q}^{\mathrm{E}}\right) + c\nabla\times\boldsymbol{\mu}^{\mathrm{M}} + \cdots\right]_p$$

A. Morita, *Theory of Sum Frequency Generation Spectroscopy*,
Lecture Notes in Chemistry 97, https://doi.org/10.1007/978-981-13-1607-4_12

$$= -i\omega\left(\mu_p - \sum_q \partial_q q^{\mathrm{E}}_{pq}\right) + c\sum_{q,r} \varepsilon_{pqr}\partial_q \mu^{\mathrm{M}}_r + \cdots$$

$$= -i\omega\left(\mu_p - \sum_q \partial_q q_{pq}(\omega)\right)$$

where we assumed the phase factor of $\exp(-i\omega t)$ (see Eq. (1.3)).

5. p. 176, Eq. (7.81)
   $q_{sp} \rightarrow q_{ps}$(4 terms)

6. p. 178, Eq. (7.87)
   $q_{sp} \rightarrow q_{ps}$ (4 terms)

7. p. 178, Eq. (7.90)
   $\beta'_{rpq} \rightarrow \beta'_{rqp}$ (2 terms)

8. p. 180, Eq. (7. 100)
   $q_{sp} \rightarrow q_{ps}$ (2 terms)

9. p. 181, Eq. (7.104)
   $\beta_{spq} \rightarrow \beta_{psq}$ (2 terms)

10. p. 181, Eq. (7.108)
    $\beta_{spq} \rightarrow \beta_{psq}$ (2 terms)

11. p. 182, Eq. (7.112)
    $\beta_{spq} \rightarrow \beta_{psq}$ (1 term)

12. p. 183, Footnote number changes. 4 → 5

13. p. 194, Footnote 6 is added.
    Here we explain its origin and physical meaning in an illustrative manner. → manner.[6]

    [6] χ^{IQB} could be interpreted as the contribution of electric quadrupole and magnetic dipole radiation. Here we present a heuristic explanation.

14. p. 195
    $\lim_{x\to -1+0} \frac{\mu}{1+x} \rightarrow \lim_{x\to -1+0} \frac{\mu}{1-x}$

15. p. 196, Footnote number changes. 5 → 7

The updated original version for this chapter can be found at
https://doi.org/10.1007/978-981-13-1607-4_7

Kolibri 360 GmbH
Ferdinand-Jühlke-Straße 7
99095 Erfurt, Deutschland
produktsicherheit@kolibri360.de